Innovation Africa

Innovation Africa

Enriching Farmers' Livelihoods

Edited by

Pascal C. Sanginga, Ann Waters-Bayer,
Susan Kaaria, Jemimah Njuki
and Chesha Wettasinha

publishing for a sustainable future

London and Sterling, VA

First published by Earthscan in the UK and USA in 2009

ISBN: 978-1-84407-671-0   hardback
      978-1-84407-672-7   paperback

Typeset by JS Typesetting, Porthcawl, Mid Glamorgan
Printed and bound in the UK by TJ International, Padstow
Cover design by Rob Watts at Rogue Four Design

For a full list of publications please contact:

Earthscan
Dunstan House
14a St Cross St
London EC1N 8XA, UK
Tel: +44 (0)20 7841 1930
Fax: +44 (0)20 7242 1474
Email: earthinfo@earthscan.co.uk
Web: **www.earthscan.co.uk**

22883 Quicksilver Drive, Sterling, VA 20166-2012, USA

Earthscan publishes in association with the International Institute for
Environment and Development

A catalogue record for this book is available from the British Library

Library of Congress Cataloging-in-Publication Data
Innovation Africa : enriching farmers' livelihoods / edited by Pascal C. Sanginga
... [et al.].
    p. cm.
  ISBN 978-1-84407-671-0 (hardback) — ISBN 978-1-84407-672-7 (pbk.)
1. Agricultural innovations—Africa. I. Sanginga, P. C.
  S494.5.I5I552 2008
  338.1'6096—dc22

                                        2008021477

Mixed Sources
Product group from well-managed
forests and other controlled sources
www.fsc.org Cert no. SGS-COC-2482
© 1996 Forest Stewardship Council

In memory of Ann Stroud

(6 May 1949 – 27 May 2007)

*Coordinator of the Africa Highlands Initiative (AHI)
from 1998 to 2006*

*Champion for mainstreaming innovative approaches to research and
development in natural resource management*

*Participant in the Innovation Africa Symposium
on 20–23 November 2006 in Kampala, Uganda*

Contents

PART I — INNOVATION CONCEPTS AND METHODS

List of Figures, Tables and Boxes

FIGURES

TABLES

BOXES

Foreword

Peter J. Matlon

The limited success of conventional research and development approaches in transforming African agriculture has led to a series of conceptual and methodological innovations throughout the last 40 years. Production systems approaches gave way to farming systems research, which in turn yielded to new participatory and farmer-first approaches, and then to broader livelihoods and knowledge systems approaches at household, community and meso levels. Each step in the uneven evolution expanded the unit of analysis and intervention, 'dynamized' the framework by acknowledging the non-linear and iterative nature of change processes, and introduced a larger scale and set of economic, socio-cultural, institutional and political factors to understanding and directing the drivers of technological change.

Perhaps the most current and complete expression of this evolution is the *innovation systems* approach. Among its most important contributions are:

- systematically incorporating functional linkages between stakeholders and organizations within the broader institutional and policy environment; and
- incorporating the profound but often subtle internal organizational changes that are necessary for effective linkage.

During the last several decades, the Rockefeller Foundation has worked to promote African agricultural development through capacity-building, research and pilot testing of interventions. The scale and breadth of this support was greatly increased recently through the establishment of the Alliance for a Green Revolution in Africa (AGRA) in partnership with the Bill and Melinda Gates Foundation. At nearly the same time, Rockefeller launched a new global initiative that aims to explore and promote new models of innovation – both technological and institutional – that advance the well-being of poor and marginalized people.

The confluence of these two foundation initiatives led Rockefeller to help sponsor a major international symposium on innovation systems organized in Kampala in November 2006. That meeting brought together some 140 research and development practitioners from Africa, Asia and Latin America to compare experiences in applying the innovation systems approach.

The chapters in this volume have been selected from among the papers presented. The collection traces the conceptual and methodological roots of innovation systems thinking; examines efforts to adapt and apply its components in different contexts; illuminates difficulties encountered in the field to promote multi-stakeholder communication and cooperation; explores the cultural and psychological barriers to institutionalizing innovation systems approaches; and considers how formal training programmes can be modified to reduce those barriers.

The chapters demonstrate well the progress being made in further developing and deploying innovation systems thinking. But they also reveal some of the major gaps and challenges that remain. From a review of the conversations that took place in Kampala and of the chapters in this volume, at least five significant challenges emerge that require further attention.

First, practitioners must establish greater synergy between 'modern' scientists and those who promote indigenous innovation. One can observe a counter-productive and sometimes ideologically based tension between the two camps. There is an urgent need for further evolution in the mind-sets and institutional cultures of both. Too often these are viewed as competing rather than complementary approaches, whereas the greatest gains are possible only when the two scientific approaches combine to create new opportunities and cross-fertilize the thinking of the other. If the innovations system agenda is to be advanced and mainstreamed into national policy and organizational good practice, both must learn to work together more effectively and with greater mutual respect.

Second, the perspectives and potential of the private sector need to be more systematically engaged. An innovation system must encompass the entire value chain – from basic, strategic, applied and adaptive research, through to product development, input delivery and output marketing – if real impacts are to be achieved at scale. Although a few papers discuss how to develop output markets, essentially none considers how to promote equitable and efficient input markets. For financial sustainability, public-sector entities and civil society need to be able to fully engage, complement and help shape private-sector activities and vice versa.

Third, issues of sustainability and scalability must be addressed with more rigour. Most of the country case studies present externally funded pilot-scale interventions that reflect varying degrees of success. And several mention the importance of taking these pilots to scale and sustaining them over time. But none systematically examines the specific barriers to scaling and durability, or reviews the accumulated experience to date to derive lessons for the future. Clearly, different national contexts are likely to require different methods. For example, in more state-oriented systems such as Ethiopia, scalability and sustainability will require different types of interventions compared with the more private-sector environment of, say, Nigeria. Regardless of the forms they take, evidence-based guidance on both issues is required urgently so that donor dependence can be reduced and greater long-term impact achieved.

Fourth, a more effective communications strategy to mainstream innovation systems approaches must be developed and implemented. Many of the papers treat communications simply as a means to disseminate new methods, to convince potential partners of their value and to train partners to deploy them. Far too little attention is given to how to use communications to stimulate dialogue and mutual discovery, to solicit feedback, and to listen and learn. Effective communications must emphasize two-way dialogue and non-confrontational engagement with farmers, collaborators and policy-makers. It is likely that some of the highest payoffs will result from close ongoing exchanges with policy-makers to understand their concerns, address them in more relevant ways and, through a dynamic dialogue, change mindsets and policy. This requires skilled advocacy and sensitive lobbying, and a clear understanding of policy-makers' incentives and the drivers of policy-making.

Finally, efforts to introduce innovation systems approaches should minimize complexity and bureaucracy. Several of the chapters in this volume describe efforts to mainstream innovation systems approaches using directive, top-down models, complete with national task forces and multi-institutional and multi-stakeholder steering committees. The problem here is that real innovation emerges by encouraging creativity, taking advantage of unexpected opportunities and exploiting serendipity. If efforts to introduce and implement an innovation systems approach are over-engineered and include multiple levels of bureaucracy, there is a great risk of stifling real discovery. Rather than building new structures, greater attention should be given to creating the enabling conditions and incentive structures that encourage information exchange, cooperation and policy changes that can unleash both bottom-up and lateral innovation.

Meeting each of these challenges is fully consistent with the philosophical foundations of innovations system thinking. Thus, there is reason to be optimistic that the approach will continue to evolve in these and other directions. Hopefully this volume, and the debates that it stimulates, will further catalyse this process, while at the same time helping to move the broader African agricultural community to embrace core elements of the innovation systems framework. When this occurs, all stakeholders will benefit, especially the poor African farmer and her household.

Acknowledgements

The Innovation Africa Symposium (IAS) held in November 2006 in Kampala, Uganda, was organized by a team of people from several institutions engaged in research for development: the editors together with Amanuel Assefa, Prolinnova-Ethiopia and AgriService Ethiopia (ASE); Dannie Romney, then with the International Livestock Research Institute (ILRI), Kenya; Kennedy Ikbowke, International Institute of Rural Reconstruction (IIRR), Uganda; Ponniah Anandayasekaram, then with the International Food Policy Research Institute (IFPRI) and ILRI, Ethiopia; Rahab Ngumba Njoroge, IIRR, Kenya; and Ronald Lutalo, Prolinnova-Uganda and Environmental Alert. Ronald Lutalo and Kennedy Ikbokwe also organized fascinating field visits for symposium participants to innovative farmers in Uganda. For managing the symposium logistics, including the poster session, we thank Annet Abenakyo, Jolly Basemera, Elly Kaganzi, Edidah Lubega, Resty Nagadya, Pamela Pali, Sylvia Perez, Nabintu Sanginga, and many others from the International Centre for Tropical Agriculture (CIAT)-Africa's administration and support team.

The symposium would not have been possible without the generous support of the Rockefeller Foundation. Special thanks go to Peter Matlon, then Managing Director in the foundation's Nairobi office, for his belief in Innovation Africa and for his thought-provoking words in opening and closing the symposium and in the foreword to this book. We are also grateful to several other organizations for contributing to the costs of the symposium: the Ford Foundation office for East Africa (via Milagre Nuvunga), the World Bank Indigenous Knowledge Centre for Africa (via Reinhard Woytek) and the co-hosting institutions CIAT, ILRI, IFPRI, IIRR-Africa and Prolinnova. We thank the International Research and Development Centre for funding a bulk order to allow distribution of the book to all symposium participants as well as to various research and development partners in Africa.

We are enormously grateful to the hundreds of participants in the Innovation Africa Symposium and their co-authors, who contributed freely their ideas and experiences. Their enthusiasm inspired the publication of this book. Their rich contributions were too many to be included in this book.

Over the past several months, the editors benefited from the valuable advice provided by members of the editorial support team – Ponniah

Anandajayasekum (ILRI, Ethiopia) Amanuel Assefa (ASE, Ethiopia), Guy Faure (Agricultural Research Centre for International Development (CIRAD), France), Niels Röling (Wageningen University and Research Centre, The Netherlands) and Laurens van Veldhuizen (Prolinnova International, ETC EcoCulture, The Netherlands) – who reviewed the contributions to the symposium and helped to select those to be published in this book. The IAS Secretariat at CIAT-Uganda continued to serve as a secretariat, alongside Desiree Dirkzwager, who provided secretarial support in the ETC EcoCulture office in The Netherlands. Many thanks to them all! As editors, we also enjoyed support from our colleagues and organizations – CIAT-Africa and ETC EcoCulture – allowing us to continue working on this book far beyond any paid time. In this way, they showed their commitment to enhancing innovation systems in smallholder farming.

We greatly appreciate the patience and cooperation shown by the contributors of chapters to this book as they went through several rounds of revision, which sometimes required drastic cutting and/or rewriting. Of course, none of this documentation would have been possible without the vision and immense efforts of numerous partners in governmental and non-governmental organizations and, above all, of the farmers and other land-users who engaged themselves in the innovation processes described in the chapters. This book is a tribute to all of them – the real heroes of innovation in Africa!

We are indebted to Tim Hardwick, the commissioning editor, and the editorial and production staff at Earthscan – particularly Hamish Ironside, Alison Kuznets and Olivia Woodward – for their interest in *Innovation Africa* and for their efficient communication and support in preparing this book for publication.

Last but not least, we express our gratitude to our family members for their forbearance when we were burning the midnight oil to meet deadlines for the symposium and for this publication.

List of Acronyms and Abbreviations

AAR	After Action Review
AESA	agro-ecosystems analysis
AGRA	Alliance for a Green Revolution in Africa
AHI	African Highlands Initiative
AIC	Association Interprofessionelle du Coton (Interprofessional Cotton Association, Benin)
AIS	agricultural innovation systems
AISSA	agricultural intensification in sub-Saharan agriculture
AKIS	agricultural knowledge and information systems
AKST	agricultural knowledge, science and technology
ANAFE	African Network for Agriculture, Agroforestry and Natural Resources Education
ARC	Agricultural Research Council
ARD	agricultural research and/for development
ARF	Agricultural Research Fund
ASARECA	Association for Strengthening Agricultural Research in Eastern and Central Africa
ASE	AgriService Ethiopia
ATDT	agricultural technology development and transfer
ATIRI	Agricultural Technology and Information Response Initiative
ATTA	Agricultural Research Council (ARC) Technology Transfer Academy
BASED	Broadening Agricultural Services and Extension Delivery (South Africa)
BLCA	brokered long-term contractual arrangement
BoLI	Bokora Livestock Initiative
BTLHA	Bokora Traditional Livestock Healers Association
CAADP	Comprehensive Africa Agricultural Development Programme
CABI	CAB International
CAHW	community animal health worker
CAP	community action plan
CAPRi	collective action and property rights
CAS	complex adaptive systems
CASE	Competitive Agricultural Systems and Enterprises approach
CDA	Community Development Agency (Kenya)
CEAD	Centre for Environment, Agriculture and Development, University of KwaZulu-Natal, South Africa

CFM	consolidated funding mechanism
CGIAR	Consultative Group on International Agricultural Research
CGS	competitive grant system
CIAT	Centro Internacional de Agricultura Tropical (International Centre for Tropical Agriculture)
CIDA	Canadian International Development Agency
CIKARD	Centre for Indigenous Knowledge for Agricultural and Rural Development
CIRAD	Centre de Coopération Internationale en Recherche Agronomique pour le Développement (French Agricultural Research Centre for International Development)
CIROP	Construction de l'Innovation et Rôle du Partenariat (Construction of Innovation and Role of Partnership)
CO_2	carbon dioxide
COARD	client-oriented agricultural research and dissemination
CORAF	West and Central African Council for Agricultural Research and Development
CoS	Convergence of Sciences programme
CS	capacity strengthening
CSO	civil society organization
CTA	Technical Centre for Agricultural Research and Cooperation
Cu	copper
CVM	Christian Veterinary Mission
DA	development agent
DACE	Department of Agriculture, Conservation and Environment (South Africa)
DAP	diammonium phosphate
DFID	UK Department for International Development
DGIS	Directoraat-Generaal Internationale Samenwerking (Directorate-General for International Cooperation)
EIAR	Ethiopian Institute of Agricultural Research
ERI	Enabling Rural Innovation
ESTA	Ethiopian Science and Technology Agency
EVK	ethnoveterinary knowledge
F4C	facilitation for change
FAIR	Farmer Access to Innovation Resources project
FAO	United Nations Food and Agriculture Organization
FARA	Forum for Agricultural Research in Africa
FFS	farmer field school
FGD	focus group discussion
FO	farmer organization
FOB	free on board
FPR	farmer participatory research
FSR	farming systems research
FUPRO	Fédération des Unions de Producteurs du Bénin (Federation of Producer Unions in Benin)

GFAR   Global Forum on Agricultural Research
GL-CRSP   Global Livestock Collaborative Research Support Program
GO   governmental organization
GTZ   Deutsche Gesellschaft für Technische Zusammenarbeit
              (German Agency for Technical Cooperation)
ha   hectare
IAASTD   International Assessment of Agricultural Science and Tech-
              nology for Development
IAC   InterAcademy Council (The Netherlands)
IAR4D   integrated agricultural research for development
IAS   Innovation Africa Symposium
ICARDA   International Centre for Agricultural Research in Dry Areas
ICRA   International Centre for development oriented Research in
              Agriculture (formerly International Course for development
              oriented Research in Agriculture)
ICRAF   World Agroforestry Centre (formerly International Centre for
              Research on Agroforestry)
ICRISAT   International Crops Research Institute for the Semi-Arid
              Tropics
ICT   information and communication technology
IDRC   International Development Research Centre
IDS   Institute of Development Studies
IED Afrique   Innovations, Environnement, Développement Afrique
              (Innovations, Environment, Development Africa)
IFAD   International Fund for Agricultural Development
IFPRI   International Food Policy Research Institute
IIRR   International Institute of Rural Reconstruction
IK   indigenous knowledge
ILEIA   Centre for Information on Low External Input and Sustainable
              Agriculture
ILRI   International Livestock Research Institute
INM   integrated nutrient management
INRAB   Institut National des Recherches Agricoles du Bénin (National
              Agricultural Research Institute of Benin)
INRM   integrated natural resource management
IPM   integrated pest management
IS   innovation systems
ISAR   Institut des Sciences Agronomiques du Rwanda (Institute of
              Agronomic Sciences of Rwanda)
ISD   Institute for Sustainable Development (Ethiopia)
ISNAR   International Service for National Agricultural Research
ISWC   Indigenous Soil and Water Conservation
K   potassium
KACHEP   Karamoja Christian Ethnoveterinary Programme
KARI   Kenya Agricultural Research Institute

KEVIN	Karamoja Ethnoveterinary Information Network
kg	kilogram
KIT	Koninklijk Instituut voor de Tropen (Royal Tropical Institute)
km	kilometre
KPG&MA	Kenya Potato Growers and Marketing Association
KTDA	Kenya Tea Development Authority
LA	learning alliance
LC	local council
LDA	Limpopo Province Department of Agriculture
LEIA	low-external-input agriculture
LISF	local innovation support fund
LP	livestock production
m	metre
M&E	monitoring and evaluation
MAK	Makerere University
MBCA	mutually beneficial collective action
Mg	magnesium
Mn	manganese
MoA	Ministry of Agriculture (Kenya)
MoARD	Ministry of Agriculture and Rural Development (Ethiopia)
MoU	memorandum of understanding
MRDP	Mpumalanga Rural Development Programme
MRIS	Mwea Rice Irrigation Scheme
n	total sample population size
N	nitrogen
NAADS	National Agricultural Advisory Services (Uganda)
NARDTT	National Agricultural Research for Development Task Team
NARF	National Agricultural Research Fund (Tanzania)
NARI	National Agricultural Research Institute
NARO	National Agricultural Research Organization (Uganda)
NARS	National Agricultural Research System
NDA	National Department of Agriculture (South Africa)
NEPAD	New Partnership for Africa's Development
NFE	non-formal education
NGO	non-governmental organization
NIB	National Irrigation Board (Kenya)
NPK	nitrogen, phosphorus and potassium
NRM	natural resource management
NSC	National Steering Committee (Ethiopia)
Nuffic	Nederlandse organisatie voor internationale samewerking in het hoger onderwijs (Netherlands Organization for International Cooperation in Higher Education)
ODI	Overseas Development Institute
OECD	Organisation for Economic Co-operation and Development
OPEC	Organization of the Petroleum Exporting Countries
OPM	Oxford Policy Management

PAR	participatory action research
PARI	Public Agricultural Research Institute
PARIMA	Pastoral Risk Management project
PDA	Provincial Department of Agriculture (South Africa)
PEA	participatory extension approach
PFI	Promoting Farmer Innovation
PHILA	Post-Harvest Innovation Learning Alliance
PID	participatory innovation development
PLAR	participatory learning and action research
PLT	peer-learning team
PM&E	participatory monitoring and evaluation
PPP	public–private partnership
PPR	project process review
PRA	participatory rural appraisal
PREPACE	Programme Régional de la Pomme de Terre et de la Patate Douce en Afrique Centrale et de l'Est (Regional Potato and Sweet Potato Improvement Network in Eastern and Central Africa)
PROFIEET	Promoting Farmer Experimentation and Innovation in Ethiopia
Prolinnova	Promoting Local Innovation in Ecologically Oriented Agriculture and Natural Resource Management
PTD	participatory technology development
PTF	village policy task force
PTLHA	Pian Traditional Livestock Healers Association
R&D	research and development
R4D	research for development
RAAKS	rapid appraisal of agricultural knowledge systems
RAWOO	Netherlands Development Assistance Research Council
RDE	rural development extensionist
RPF	resource-poor farmers
RULIV	Promotion of Rural Livelihoods programme
SA	South Africa
SAARI	Serere Agriculture and Animal Production Research Institute
SADC	Southern African Development Community
SFM	soil fertility management
SNA	social/system network analysis
SSSP	small-scale seed production
SWC	soil and water conservation
SWOT	strengths, weaknesses, opportunities and threats
T&V	training and visit
TaCRI	Tanzania Coffee Research Institute
TSP	triple superphosphate
UK	United Kingdom
UKZN	University of KwaZulu-Natal (South Africa)
UNDP	United Nations Development Programme

US	United States
USAID	US Agency for International Development
USh	Ugandan shilling
VEDCO	Volunteer Efforts for Development Concerns (NGO)
VIC	village information and communication centre
ZARDI	Zonal Agricultural Research and Development Institute
ZARF	Zonal Agricultural Research Fund
Zn	zinc
ZRDC	Zonal Research and Development Committee

CHAPTER 1

Innovation Africa: An Introduction

*Ann Waters-Bayer, Pascal C. Sanginga, Susan Kaaria,
Jemimah Njuki and Chesha Wettasinha*

SEEKING NEW WAYS OF ENRICHING FARMERS' LIVELIHOODS

Agricultural research, extension and education have the potential to make a great contribution to enhancing agricultural productivity in a sustainable way and, thus, contribute to reducing poverty in the developing world. However, their achievements in this respect have generally fallen short of expectations in Africa. Innovative groups within governmental and non-governmental organizations (NGOs) – both national and international – have been seeking alternatives to the linear transfer of technology model. On the one hand, growing economic and demographic pressures, coupled with the entry of new market forces and actors, have created a need for a more interactive approach to development. On the other hand, recent insights into socio-cultural realities and human behaviour have revealed the opportunities offered by recognizing the creative capacities of all actors in research and development, including (and especially) the farmers themselves. Understanding the existing innovation processes, recognizing the potential for catalysing such processes and learning how to support them will be key to the success of individuals and organizations involved in agricultural research and development.

It is increasingly recognized that innovation emerges out of the interplay of ideas from multiple sources. A growing number of projects and organizations are deliberately seeking to stimulate synergy between the various potential partners in agricultural innovation. Valuable experiences are being generated across the world on how diverse actors can be encouraged to work together, and how new ideas and products – whether from formal research or from other sources – can be transformed into innovations that benefit thousands of resource-poor farmers. Such processes prepare the path to sustainability: indeed, promoting continuing innovativeness is the only way to achieve sustainability in the face of rapidly changing conditions for agricultural production and human well-being. Enhancement of the innovativeness not just of individuals but rather of systems of interacting players is needed at all levels, from the grassroots to the globe. It is crucial to strengthen the capacities of farmers and their organizations to take the lead in collaborating with other players – whether they are local blacksmiths or international plant breeders – in order to develop new and better ways of using locally available resources. But such farmer-led processes of innovation can thrive only if changes are made at various levels within institutions, organizations and policy – and above all in building the capacity of formal researchers and development agents to play constructive roles within these agricultural innovation systems.

The international symposium Innovation Africa (IAS) held in Kampala, Uganda, brought together researchers, development practitioners and – albeit to a more limited extent – people from farmer organizations and the private sector to share their current thinking, experiences and lessons from initiatives to enhance innovation systems in agriculture and natural resource management, primarily in Africa. The meeting was, in itself, innovative in that it was a collaboration of rather unusual partners: three international agricultural research institutes – the International Centre for Tropical Agriculture (CIAT), the International Food Policy Research Institute (IFPRI) and the International Livestock Research Institute (ILRI); an international NGO that supports rural development (the International Institute of Rural Reconstruction, or IIRR); and a multi-stakeholder global partnership programme for research and development called Prolinnova (Promoting Local Innovation in Ecologically Oriented Agriculture and Natural Resource Management), which operates under the umbrella of the Global Forum on Agricultural Research (GFAR). The participants in the symposium came from research, development and academia, drawing on both theory and practice of innovation systems. This meeting of minds led to lively discussions and a better understanding of the implications of an innovation systems perspective for agricultural research and development – particularly the implications for capacity-building, organizational change and policy to promote innovation in ways that focus on helping the poor to enrich their livelihoods.

From over 100 contributed papers and posters, an international editorial group selected 23 to be included as chapters in this book. These were selected with a view to covering the main conceptual and methodological developments in agricultural innovation systems, and to showcase experiences, results and lessons from research and practice in different contexts in Africa. The contributions and discussions during the symposium revealed that concepts of innovation were fairly diverse, ranging from those who regarded innovation as the adoption of technologies introduced from research to those who regarded it as the outcome of social learning by many different actors.

In his keynote address, Röling took an autobiographical approach to following the 'story' of innovation – describing how the related concepts and methodologies have developed from the diffusion of innovations theory developed in the 1960s (Rogers, 1962) to the current concept of agricultural innovation systems (World Bank, 2006). This sees innovation arising out of a network of individuals and organizations that interact in both creating and applying agricultural knowledge. An innovation system also includes the institutions and policies that affect the behaviour and performance of the individuals and organizations involved. An innovation – the outcome of such a process of interaction within an innovation system – can be a new product, a new process or a new form of organization; but it can be regarded as an innovation only if it is actually used.

The chapters in this book reflect a shared understanding of innovation in this sense. They also share recognition of the need to create space and incentives for promoting collaboration between farmers, research and extension services and the private sector (input and output markets) in order to develop improved technologies and institutional arrangements that can alleviate poverty.

FIVE MAJOR THEMES

Within the broad framework of research and development in agriculture and natural resource management, the chapters of this book focus on five major themes, as set out below.

Innovation concepts and methods

Röling (see Chapter 2) sets the stage by giving a historical overview of how the concept of innovation has developed over the last half century, and what lessons can be learned from research on innovation and from the practical application of the concept in the field. The remaining chapters in Part I address the basic concepts, theories and principles of agricultural innovation, as well as methodological issues and challenges, with a view to alleviating poverty. The authors look at why research into innovation systems has flourished in the last few years and the alternative tools and methods that are being applied to translate the concepts into practice.

Strengthening social capital in innovation systems

The chapters in Part II examine partnerships and other forms of social capital in agricultural innovation systems. The authors describe approaches and challenges in building and managing multi-stakeholder partnerships for innovation; ways of integrating different disciplines and forms of knowledge; the role of farmer organizations and other local groupings of actors in these systems; and the synergies between local 'grassroots' and wider national or international innovation systems. They highlight the challenge of moving beyond 'participation' in the now conventional sense of the term – which refers to drawing farmers or other local resource users into research and development activities conceived by outsiders – to *partnerships* that recognize the contributions of the different actors and allow them an equal say in the collaboration.

Policy, institutional and market-led innovation

Part III deals with institutional change, policy-making and knowledge-sharing to support agricultural innovation systems. The authors of these chapters explore experiences in alternative ways and institutions to fund agricultural innovation processes, and ways of enhancing networking for mutual learning, as well as to influence thinking and policies about how agricultural research, extension and education should be done. They show how participatory approaches to analysing innovation systems can stimulate institutional learning and organizational change. Chapter 11, in particular, looks at the role of markets in catalysing innovation processes, and draws lessons from experiences in enhancing entrepreneurship and linking smallholder farmers to markets.

Enhancing local innovation processes

The chapters in Part IV reveal how local people initiate and manage innovation processes in agriculture and natural resource management. These local initiatives in informal experimentation to develop 'new things and ways that work' (Scheuermeier et al, 2004) are all too often overlooked or underestimated as engines of change in promoting broad-based agriculture-led economic growth and development. The authors examine what these local initiatives and strategies imply for the types of support that governments, civil society, the private sector and international agencies need to provide. They explore the link between farmer-led innovation and poverty alleviation, and draw lessons from experiences in enhancing community learning and change processes and in scaling up farmer-led participatory innovation processes.

Building innovation capacity

The chapters in Part V analyse different approaches to strengthening the innovation capacities of farmers and their organizations, other entrepreneurs, civil society organizations, universities, and government and private-sector organizations involved in agricultural research and development. They also address strategies and experiences in integrating innovation systems perspectives and approaches within institutions of higher learning and education at all levels.

The various chapters draw on diverse fields and disciplines of the social, agricultural and natural resource sciences, and present many examples of good practice in studying and enhancing the process of innovation for effective agricultural research, development and education. The concluding chapter, Chapter 25, highlights the main lessons learned from the symposium – above all, from the contributions presented in this book. It draws attention to gaps in knowledge and examines the prospects for research and practice of agricultural innovation systems in agriculture and natural resource management in Africa.

REFERENCES

Rogers, E. M. (1962) *Diffusion of Innovations*, Free Press, New York, NY
Scheuermeier, U., Katz, E. and Heiland, S. (2004) *Finding New Things and Ways That Work: A Manual for Introducing Participatory Innovation Development (PID)*, Swiss Centre for Agricultural Extension, Lindau, Switzerland
World Bank (2006) *Enhancing Agricultural Innovation: How to Go Beyond the Strengthening of Agricultural Research*, World Bank, Washington, DC

I
Innovation Concepts and Methods

CHAPTER 2

Conceptual and Methodological Developments in Innovation

Niels Röling

FOCUS ON INNOVATION

The Innovation Africa Symposium (IAS) focuses the spotlight on a subject that deserves all the attention it can get. Of course, innovation is a sexy concept that appeals to left and right, and young and old, including *Mzees* like myself. Innovation has promise. It sounds like a way forward. It is easy to get people behind it. But beware! The concept is used in different meanings. It can represent very different perspectives. It can lead to considerable confusion. It is a real 'battlefield of knowledge', as Norman Long once called it (Long and Long, 1992). Sometimes it is in need of innovation itself!

This conceptual overview is meant to put the subject on the map. I believe I am the right person to give it. Few people have fallen into more traps and were seduced by more meanings for innovation, innovation systems, system innovations and what not than I. So it seems a good idea to give this overview of conceptual and methodological developments as an intellectual autobiography. I use my own 'history of innovation' to take you through the minefield of meanings and perspectives. For each episode, I zoom in on implications for innovation in Africa.

1970: DOCTORAL STUDIES WITH EVERETT ROGERS

Diffusion of innovations

I obtained my PhD in the US during 1970 with Everett Rogers as my main supervisor. I was with his US Agency for International Development (USAID)-funded project The Diffusion of Innovations in Rural Societies, which operated in Brazil, India and Nigeria. I worked in Nigeria, where I spent four years, and later joined Rogers in Michigan, in the American Mid-West, where he developed one of the most influential theories of innovation.

Everett Rogers is called the father of the Diffusion of Innovations (Rogers, 2003). The paradigm which Rogers so successfully synthesized and promoted since his first overview in 1961 goes back to a study of the diffusion of hybrid maize in Iowa during the early 1940s (Ryan and Gross, 1943). Specific conditions in the Corn Belt led to the 'discovery' of diffusion as an autonomous process that multiplies the impact of research and extension. But before we all get excited about diffusion as the way to lift up African agriculture, we must be aware of the specific conditions that allow spontaneous diffusion of a novel idea or technology in a community of farmers:

- A large number of farms or firms all produce the same commodity for the same market.
- Each of them is too small to affect the price of the commodity. Hence, they all produce against the going price (price takers) and seek to improve their situation by producing more of the commodity, if possible against lower costs.
- Given the inelasticity of demand for most farm products, all these farmers trying to produce more, and more efficiently, exerts a constant downward pressure on product prices.
- All of the farms have access to credit, fertilizers, extension, farm journals and agri-business, and are members of farmer organizations to different degrees.

Introducing a new idea or 'innovation' (note how I use the term here – i.e. as a *noun* usually denoting a *technology*), such as hybrid maize, in such conditions (typically called a 'recommendation domain') can lead to a wave of 'innovation' (here used in the sense of a *process*) as individual farmers adopt. The wave of innovation is called the 'diffusion curve', which is usually depicted as an *S*: the diffusion process starts slowly, then gathers momentum and finally peters out when all farmers for whom it is relevant or feasible have adopted the innovation.

The 'discovery' of the diffusion of innovations led to a great deal of research. At one time, it was the most popular social science research subject and more than 2000 studies on diffusion had been completed by the time I last checked (quite long ago). One can imagine the excitement. Here was a spontaneous social process that multiplied the efforts of research and extension for free. Diffusion seemed key to social change and modernization. It explained the spread not only of agricultural technologies, but also of the hula hoop and contraceptives.

The agricultural treadmill

Little wonder that, when I did my PhD in Michigan, innovation could be talked of only in terms of diffusion. Apart from social scientists such as Rogers, economists also examined the phenomenon, the best known probably being Cochrane (1958) from Minnesota, who coined the phrase 'the agricultural treadmill'. To the existing theory, he added some important components related to farmers' incentives.

When a new technology begins to be adopted, it allows those using it to produce more, or more efficiently, against the going price, which is still initially determined by the old state of the art. This means that the few early adopters make a windfall profit. But as more farmers adopt (seeing the good results of the early adopters), the state of the art changes. Total production increases. Prices begin to drop. People who have not adopted the innovation see their incomes fall, even if they work as hard as before. The price squeeze finally forces them also to adopt. Hence, the diffusion process is *propelled by market forces*. This is called the 'treadmill'. Farmers who are too small, too old, too sick or too stupid to keep up eventually drop out. Their resources (such as land) are taken over by the 'stayers'. This process is called 'scale enlargement'. In my own country, The Netherlands, scale enlargement started as early as 1960 (Van den Ban, 1963). Since then, about 2 per cent of farmers give up annually. Those who survive in 2007 usually have large enterprises, a good education, an enormous working capital (tractors, buildings, livestock, etc.), and are highly organized and embedded in a network of supporting institutions, including input service cooperatives, farmer unions, truckers, processors, retailers (e.g. supermarkets), veterinarians and so on. Havelock (1973, 1986) was among the first to focus on this configuration of institutions as an essential counterpart to treadmill innovation.

The birth of a policy model: Transfer of technology

When it works, the diffusion process and especially the market-propelled treadmill have important consequences at the *macro* scale:

- Labour moves out of farming. In industrialized countries, only 3 per cent of the working population are still primary producers, even if 10 per cent are employed in agriculture-related activities. Each farmer can feed hundreds of people.
- Farmers cannot hold on to the benefits of technological innovation. Global food prices have continuously declined over the past 40 years (although, in 2007, as a result of climate change and a shift to biofuels, prices have started rising).
- A country becomes more competitive in the world market as its farming industry becomes more efficient.
- Farmers do not complain or rebel. The frontrunners who capture the windfall profits and benefit from scale enlargement are those who hold positions of power in farmer organizations.

Evenson et al (1979) established that investment in agricultural research and extension, the perceived drivers of innovation (given the conditions I have enumerated), has a high internal rate of return. This can be explained by the multiplier effect of diffusion and the macro benefits from the market-propelled treadmill.

An important observation: diffusion of innovations was a *research tradition* based on *empirical* studies that looked at what had happened in the *past*. But the macro benefits of the treadmill, as perceived by economists, transformed it into a *policy model* for what is desirable in the future. This model emphasizes technology transfer (technology supply push) and free markets as recipes for agricultural development – that is, the treadmill became the dominant guideline for how innovation should be promoted. This guideline is called 'the linear model' (e.g. Kline and Rosenberg, 1986) or the 'transfer of technology model' (Chambers and Jiggins, 1987): innovation is the end-of-pipe outcome of a linear process that runs from basic research, via applied and adaptive research, subject matter specialists, and extension and contact farmers, to widespread diffusion among 'follower farmers'. The training and visit (T&V) system tried to incorporate this model all over Africa. Throughout Africa, most policy-makers, ministry officials, research administrators, economists and researchers cannot imagine any other theory of innovation than the linear model and continue to adhere to it, even after years of failure in situations where it does not apply. Said one senior research administrator in West Africa: 'Our farmers get 1 tonne of maize, we get 7. The problem is transfer of technology.' To be sure: diffusion can be observed after it has taken place. But so far, it has not been possible to predict whether a technology will diffuse or not. The production of agricultural technologies by research, even if they 'work' in the experiment station, is absolutely no guarantee for diffusion.

Implications for Africa

Using the treadmill as a policy model in Africa has major shortcomings:

- The conditions in which the treadmill works usually do not apply. The political, social and ecological diversity in Africa allows no easy identification of recommendation domains. Most African farmers do not operate in well-developed commodity markets and do not have access to the information, inputs, capital, etc. required to capture the benefits from introduced technologies.
- Forcing people out of farming is not a good idea when alternative employment is not available. Even in the US and Europe, farm subsidies and tariffs were introduced to mitigate the effects of unfettered treadmill impact upon farm incomes.
- The treadmill forces farmers to externalize social and environmental costs (e.g. groundwater pollution, destruction of natural resources and biodiversity, human health impacts of pesticides), especially in conditions of weak legal institutions.

- The global treadmill means that African farmers have to compete with farmers in industrialized countries who have benefited from 60 years of efficiency gains and scale enlargement. Even if farmers in industrialized countries earn 25 times more than farmers in African countries, their labour productivity is 32 or more times higher, so they can out-compete African farmers anytime (Bairoch, 1997). Value added per agricultural worker in 2003 (constant 2000 US dollars) in developed market economies was US$23,081 with a growth from 1992 to 2003 of 4.4 per cent. For sub-Saharan Africa, the figures were 327 and 1.4 per cent, respectively (FAO, 2005). This imbalance is amplified by the subsidies that farmers in Organisation for Economic Co-operation and Development (OECD) countries receive. Some governments (e.g. Ghana) love the cheap imported food because it keeps their urban electorates happy; but that seems a short-term consideration that has negative long-term repercussions by destroying African food production capacity.

Conditions are rapidly changing. The US shift to biofuels to improve energy sovereignty, the increase in meat consumption and the drought in Australia and other extreme weather events brought about by climate change are driving up food prices. The consequences for Africa are not clear; but it seems unlikely that this rise in food prices will make African smallholders competitive in African urban markets.

The global treadmill prevents African farmers from contributing to global food security and African countries from gaining food sovereignty. On average, already 20 per cent or more of African food grain requirements are imported (IAC, 2004). These are serious issues, given a future marked by insecurity because of climate change, population growth, political instability, increasing fossil energy costs and, hence, increased non-sustainability of the industrial type of farming upon which world food security depends. The global treadmill, coupled to the generally accepted economic principle that 'relative advantage' should dictate who captures the market share, condemns African farmers to subsistence on a degrading and diminishing resource base.

Summary

In all, I came away from doing my PhD in the American Mid-West thoroughly imbued with diffusion of innovation and treadmill thinking, but also aware of the need to innovate this model of farm innovation. A good part of my professional life has been devoted to attempts to find alternatives to the treadmill and technology push. Daniel Benor, the father of T&V, used to say to me: 'Show me an alternative that works.' I have persisted in seeking alternatives because technology supply push so obviously fails in Africa. Meanwhile, the treadmill is increasingly running out of steam in industrialized countries as the externalities become politically unacceptable and as agriculture shifts from only producing commodities to also producing ecosystem services, such as drinking water, carbon dioxide (CO_2) sequestration, nutrient cycling, biodiversity and health.

My continued preoccupation with the linear model is justified in view of the continuing attempts to use technology transfer as the recipe for developing African agriculture. Recently, the Gates and Rockefeller Foundations have made US$150 million available over five years to fund the development and distribution of seeds that are 'suited to sub-Saharan Africa's parched climate, denuded soils and stubborn pests' through both research and distribution (*Economist*, 2006, p90).

THE SPECIAL RURAL DEVELOPMENT PROGRAMME IN TETU, KENYA

The marketing approach

In 1971, I joined the Institute of Development Studies (IDS) of the University of Nairobi to carry out postdoctoral research with a friend and colleague from the days at Michigan, Joe Ascroft from Malawi. He had been a market researcher in East Africa and was, for example, the originator of the Coca Cola Barometer, a tool to monitor sales. In Tetu Division, Central Province, Kenya, Ascroft was designing a project using the marketing approach to agricultural innovation, a project that turned diffusion research on its head.

The marketing approach is about the exchange of values between market partners, usually the exchange of goods and services for money. But non-profit marketing (e.g. Kotler and Andreasen, 2003) applies the approach to exchange of other values. Thus, an extension service can apply a marketing approach to designing services for different publics. Marketing research gives insight into the composition of likely markets: which categories of clients require which kinds of offering? Based on its objectives and what it can provide, an extension service can then decide about the nature of the clients it seeks to target and what offerings (services, products and messages) it needs to design for them, or perhaps even with them (in those days, we distinguished between *do to*, *do for* and *do with*).

Targeting the segment of 'forgotten farmers'

In our Tetu project (Ascroft et al, 1973; Röling et al, 1976; Röling, 1988), we targeted 'the forgotten farmers': we offered opportunities to those farmers who are likely to drop off the treadmill. In other words, we turned around the usual 'progressive farmer' approach, which heaps riches upon riches, and sought what Rogers used to call the 'laggards'. Our random sample survey allowed us to segment Tetu farmers on the basis of natural cut-off points on an index for 'innovativeness' (based on the number of innovations farmers had adopted). We then deliberately chose the segment that had adopted *least*. Together with agricultural specialists, we designed a hybrid maize package for 0.25 acres that seemed suited to the conditions of these 'laggards'. We made sure that the farmers had access to a suitable quantity of hybrid seed, 50kg of fertilizers, pesticides and a loan to purchase these inputs. We designed a 2.5-day training course at Wambugu Farmer Training Centre with an intake of 25 men and women farmers selected with the help of sub-location chiefs. In all, we trained a few hundred farmers. The training was the condition for farmers to receive the package. The whole field experiment was carefully monitored and evaluated.

The results were astounding. Virtually all of the 'laggards' who had done the course adopted the hybrid maize, many of their neighbours did as well, and repayment of the loan was over 90 per cent. It was very clear that even so-called laggards were innovative if they were offered the right conditions and opportunities. What is difficult is to create the conditions and access to resources to allow the 'laggards' (or, better, resource-poor farmers) to become innovative.

A farmer with 0.25 acres needs a large part of that land to produce food for the family. To be able to buy fertilizers, the farmer needs to produce a certain surplus for the market. The size of that surplus depends upon the price of fertilizer. If it goes up, the surplus needs to be higher and the risk of crop failure in a dry year or of needing the money from the surplus for a burial also becomes greater. Our Tetu study of the adoption of hybrid maize by resource-poor farmers made these mechanisms very clear. It raised serious questions about using costly high-input technologies for poverty reduction in highly diverse, risk-prone and variable conditions (Chambers and Jiggins, 1987). Just such insights led to the founding of the *ILEIA Newsletter*, where LEIA stands for low-external-input agriculture.

Implications for Africa

- Developing technologies and 'releasing' them to farmers, or 'delivering' them to 'end users' has not, on the whole, been very effective in Africa. In fact, the impact of agricultural research on smallholders has been remarkably limited (e.g. Gabre-Mahdin and Haggbladd, 2004). This is partly because formal research has paid so little attention to the (very) small windows of opportunity within which farmers can actually innovate. Much research output simply is not appropriate for farmers' conditions. Agricultural research has produced goods and services for which there is no market. Indeed, it does not engage in marketing research of its customers to see what they need and can use. An important reason why agricultural research in Africa has been able to continue like this for decades is that small-scale farmers are not organized and have no political clout.

- There is nothing wrong with small-scale African farmers. They are not traditional, stupid, uneducated, backward or whatever. They have only very few and small opportunities. The notion that African agriculture is stagnant because no great advances in agricultural productivity have taken place is wrong. African agriculture is incredibly dynamic. Smallholders are innovating and adapting all the time in order to cope with changing circumstances. African farmers have, on the whole, been able to produce food in keeping with the very rapid population growth over the past 50 years. They have done this with little use of external inputs or science-based knowledge, with little support from government (in fact, agriculture is a source of revenue for most African governments) and in the face of cheap food imports, climate change, conflict and disease. Hounkonnou (2001) has called these 'rural dynamics' the most hope-giving element in an otherwise dismal landscape.

- It is possible to design opportunities for small-scale farmers to innovate. The Kenya Tea Development Authority (KTDA) has for years provided farmers with a package of services, inputs and supervised credit that has allowed smallholders to benefit from the export opportunities for tea. The production societies for cotton in West Africa have also created income-earning opportunities for smallholders (Gabre-Mahdin and Haggbladd, 2004). A key point has been the supervised credit: the exporting organization can deduct the credit from the farmer's revenue. Since farmers cannot eat tea, cotton or cocoa, such mechanisms work. Schemes such as KTDA are among the most successful approaches to wide-scale innovation in rural Africa. Such brokered long-term contractual arrangements (BLCAs) seem to be the one mechanism that has so far been able to put money into the pockets of African smallholders because BLCAs create the conditions required for entering the market.

- Most governments have not been good at creating opportunities for small-scale farmers. Government-run supervised credit schemes soon become vehicles for revenue-raising, corruption and patrimonial networks. I myself learned that it is dangerous for a research project to create artificial conditions for farmers (as we did in the Tetu project) because most public agencies cannot replicate these conditions within their means and their bureaucratic systems. Their vulnerability to corruption and mismanagement has led international agencies to privatize BLCAs during structural adjustment ('abolish the parastatals'); but, in doing so, they have thrown out the baby with the bathwater.

BOARD MEMBER OF CENTRE FOR INDIGENOUS KNOWLEDGE FOR AGRICULTURAL AND RURAL DEVELOPMENT (CIKARD)

After my PhD and postdoctoral research, I became an academic, and my contact with the field became mediated through memberships of boards, professional training courses for international extension staff, supervision of MSc and PhD students, consultancy missions, scientific congresses and so on. One of these experiences was to serve as a board member (together with my wife Janice Jiggins) for the Centre for Indigenous Knowledge for Agricultural and Rural Development (CIKARD) in Iowa. It had been started by the late Mike Warren, an imaginative and concerned researcher who worked as an agronomist for the Peace Corps in Ghana and there met his wife Mary, a Nigerian gold trader. As a result of this mixed marriage (or maybe the marriage was the result – that is buried in history), Warren began to realize the extent to which colonialism had ignored indigenous wit, technology and knowledge. He became totally absorbed by this notion and started his CIKARD as a centre to study, conserve and pass on indigenous knowledge (Warren et al, 1991).

Indigenous knowledge

The linear model looks at research as the source of all innovation and is blind to the fact that farmers are researchers in their own right who, moreover, have to live by the results. They constantly try out things and, over generations, develop farming systems that satisfy their needs. They try to make optimal use of their environmental and market conditions. Even if farmers are not 'scientists' in the sense that we usually use this term, they are extremely good at discovering what works, although they cannot always explain why. And we are dealing with many generations, with collective intelligence and with millions of experimenters *in situ*. Indigenous knowledge is a very respectable and important source of innovation that had been ignored until people like Mike Warren brought it to our attention.

I remember visiting David Norman in northern Nigeria during the early 1970s. He had investigated why farmers who live near the experiment station at Samaru had, for as long as the British (and later the Nigerians) did their research there, refused to adopt mono-cropping and continued to practise mixed cropping. The study (Norman, 1974) showed that mixed cropping:

- creates a microclimate that is beneficial for crop growth;
- creates crop diversity, which reduces the spread of pests and diseases;
- is a clever risk-reduction strategy in a variable climate: if some crops die as a result of drought, one still has something to eat from the drought-resistant crops in the mix;

- optimizes the use of labour, which is the limiting factor in the farming system;
- optimizes the total monetary value of crop production.

In other words, the study should have looked at why the British and Nigerian scientists did not practise mixed cropping.

The design of farming systems that work in such intricate ways is something research cannot do. Research is good at developing component technologies, such as fertilizers and Bt-cotton. But farmers have designed systems within which these component technologies must provide a benefit. And all too often they do not because research has not bothered to analyse the systems into which the component technologies must fit.

Farming systems research and participatory technology development

Farming systems research (FSR) then became quite a movement (Collinson, 2000) and the International Farming Systems Association and its regional caucuses are still going strong. Farming systems research makes the following points:

- Farmers often know more than scientists when it comes to the characteristics and dynamics of the environment in which they farm, including risks of waterlogging, drought, pests, thieves, etc.
- Farmers know better than scientists the criteria by which innovations will be judged and the (possibly multiple) objectives that the innovations have to serve. Researchers usually assume that the objective is to become more productive or resource efficient. For farmers, many other criteria and objectives pertain. Since adoption is a voluntary act by farmers, it is their opinion that should prevail.
- Small-scale farmers (male and female) are intelligent beings. You can ask them about things and discuss things with them; you do not have to carry out costly and time-consuming extractive research to find out about these things yourself. This 'insight' led to the development of 'rapid appraisal' methods to provide outsiders with insights into the farming systems.

Soon the idea of an appraisal by outsiders gave way to the notion that resource-poor farmers could and should be *partners* in analysing their situation and designing their future. One of the best-known applications of this idea is participatory technology development (PTD) (e.g. Reijntjes et al, 1992), an approach in which scientists collaborate with smallholders in field experiments to develop technologies that are needed, wanted and appropriate to the conditions of the farmers concerned.

In this sense, insights have really changed compared to the linear model of technology supply push. What is remarkable is that most major national and international agricultural research centres have used these insights only to a limited extent and continue to believe in magic bullets and Green Revolution approaches for Africa. When the chips are down, core business for agricultural research remains breeding, biotechnology, smart farming, robotics, high-input agriculture and productivity per hectare. Non-profit marketing approaches, farming systems research and participatory research so far have had little impact on the dominant policy model of technology transfer. This persistence is an outcome of the lack of countervailing power of small-scale farmers over agricultural research and the lack of attention that African governments have paid to their farmers. Improving the impact of research is not so much a question of investing more in research, but of developing the farmers' ability to influence it.

Implications for Africa

- African farmers' innate tendency to innovate in their search for ways to improve their lives is a huge asset that we have barely learned to mobilize for agricultural development. By recognizing farmers as important sources of innovation, the Promoting Local Innovation in Ecologically Oriented Agriculture and Natural Resource Management (Prolinnova) programme is making an important contribution.
- African farmers are innovating not only in terms of component technologies, but also in terms of farming systems, something scientific research finds hard to do. An example is the development by the Adja in Benin of oil-palm fallow, a system of permanent land use in a very densely populated area that uses rotation of annual food crops with oil palm to restore soil fertility and suppress the grassy weed *Imperata cylindrica*. The rotation is economically feasible because of the sale of alcohol distilled from the palm wine (Brouwers, 1993).
- African farming systems are under constant pressure to innovate. One huge pressure is towards more permanent land use because of population growth. In its wake, this has brought serious problems, such as loss of soil fertility, emergence of pernicious weeds and reduced farm sizes. Technical solutions to these problems have hardly emerged. Formal research has been slow to pick up on such issues as weeds, which have become an important dimension of rural poverty (Vissoh, 2006). Farmers themselves have sought to solve the problems in other ways, such as the feminization of agriculture, where men go off to distant cities to find gainful employment and the women look after the farms. The lack of opportunity in rural areas has become destabilizing in that rural youths cannot replicate their cultural repertoire and have no future in their own communities (Richards, 2002).
- Experience across Africa shows that farmers are quick to grasp opportunities. The recent increase in the free-on-board (FOB) price of cocoa in Ghana from 40 to 70 per cent led to a doubling of

cocoa production within a few years without any technological breakthroughs. One can leave it to African farmers to make the best use of new opportunities provided by markets, employment generation, etc. Creating opportunity is not primarily a technical issue but an institutional one.

We move to a new notion of innovation, from technology supply push to opportunity development through institutional change. North (2005), who received a Nobel Prize for his contribution to the New Institutional Economics, distinguishes between policies for economic growth and policies for development. Growth policies assume that appropriate institutions are in place so that one can focus on letting market forces do their work. Market liberalization and feeding the treadmill with component technologies are typical growth policies. Development policies focus on creating the institutions that allow the market to function in the first place. Western neoliberal economists tend to apply growth policies also in situations where appropriate institutions still need to be developed. They forget that it took many years for their own modern market economies to develop such institutions as tenure laws, insurance, banks, quality control, farmer organizations and their influence on agricultural research and policy, cooperatives, etc. The brokered long-term contractual agreements for cocoa, cotton, tea, pineapples, French beans, etc. that allow African smallholders to participate in the global market are examples of – albeit often faulty – incipient institutional development that market fundamentalists have tried to privatize, often with negative consequences for African farmers.

FARMER FIELD SCHOOLS AND LANDCARE

Being an academic in a field such as innovation is an exercise in humility. It is very hard to actually design effective innovation of innovation. One can talk about it, develop criteria, recognize it when one sees it and help those who do it to understand what they are doing. But to actually create an alternative approach to the linear model has remained elusive. During the 1980s, I was fortunate through PhD students to come across two approaches that were based on totally different points of departure than the linear model: farmer field schools (FFSs) in Indonesia and Landcare in Australia.

Farmer field schools

The Green Revolution in irrigated rice in Indonesia was based on inputs of new varieties, subsidized fertilizers and pesticides. Careful control of market and irrigation conditions, credit packages and some strong-arm tactics in the beginning led to rapid diffusion of highly productive rice farming, typically moving from 1 tonne to 3 tonnes or 4 tonnes per hectare per crop (irrigation farmers can often grow two rice crops and a vegetable crop in a year). Huge areas and millions of farmers were blanketed with the same recommendations in terms of rice variety, and type and dosage of fertilizer and pesticides.

During the 1980s, a serious problem emerged. A small insect, the brown planthopper, which had never been a problem before, started to cause serious damage and threatened rice harvests over huge areas. This was a political issue of the first order. Suharto, the president at the time, had been able to take over from Sukarno because of rice shortages. It turned out that the pest problem had been induced by pesticides: the insect had become resistant to them, while its natural enemies had been destroyed. Hence, it could resurge very quickly after each spraying and more, and more potent, pesticides were required each time to control it (the 'pesticide treadmill'). Suharto reacted by banning 57 broad-spectrum pesticides, removing the 85 per cent subsidy on pesticides and asking for an extension campaign on integrated pest management (IPM). This emphasized control through natural enemies, growing a healthy crop and using pesticides on the basis of observation. The extension campaign with mass meetings, posters and T&V was a failure. IPM is too complex a message for this approach. Then United Nations Food and Agriculture Organization (FAO) staff and their Indonesian colleagues in the IPM programme designed the FFS as a radical alternative. I became involved through a PhD student who evaluated the approach (Van de Fliert, 1993; Röling and Van de Fliert, 1994). I had the good fortune to visit her in the field and to become a consultant to the FAO project as a result. My main contribution was to help them understand what they were doing already (Pontius et al, 2002).

The FFS has the following remarkable principles: grow a healthy crop, use natural processes, the farmer is an expert, and research linkage. About 25 farmers meet once a week during a growing season. They learn by dividing a field into two: in one plot, they grow rice the conventional way (i.e. with pesticides) and in the other, the IPM way. At each meeting, they split up into small groups and carefully observe a randomly selected number of rice plants in each field in terms of their growing conditions, health and the insects living on them. Each group discusses the results, makes a picture report and presents it to the plenary. This 'agro-ecosystem analysis' is the main activity during each meeting. The facilitator guides the process but does not intervene strongly. In addition, some special subjects are introduced (e.g. the life cycle of an insect) and experiments are performed (e.g. defoliation of rice to see how much of a rice plant can be eaten before it affects the yield). The farmers are completely free on the basis of their observations and discussions to decide how to manage their IPM plot. If they want to spray, they can spray. The farmer is the expert and carries out experiments.

Evaluations of IPM FFS projects (e.g. Van den Berg and Jiggins, 2007) tend to show that yields go up and costs go down, especially because of reduced pesticide use. Moreover, farmers learn to apply the same principle to other crops and situations. But most striking is that the FFS experience *empowers* farmers. They become opinionated and learn to organize and speak in public. Often, FFS alumni spontaneously organize themselves to provide FFS training to other farmers or to undertake development projects.

Higher-level policy-makers who visit an FFS in action are deeply impressed by the changes that it can bring about and often want rapid scaling up (Youdeowei, 2003). Alas, the effect of FFSs is very sensitive to the quality of facilitation. Rapid scaling up leads to poor quality because there is not enough time and money to train the facilitators well and to create the conditions for effective farmer education. Most governments find it hard to refrain from using FFSs as a vehicle for technology transfer.

Implications for Africa

- The IPM FFS has become a policy model that is often implemented by replication without adapting the FFS to specific conditions and needs. I have seen the rice FFS curriculum being applied lock, stock and barrel to bananas.
- Pesticide-induced pest problems are a typical second-generation Green Revolution phenomenon, which seldom occurs in Africa. African farmers have more problems with soil fertility, weeds, and pests and diseases that do not come about through pesticide use. Most FFSs in Africa therefore require careful experimentation and testing in farmer research groups to develop curricula that are appropriate in African conditions (e.g. Bruin and Meerman, 2001, describing a project in Zanzibar).
- Powerful bosses in ministries and research stations find it very hard to understand and accept the participatory processes upon which the effect of FFS is based and are all too ready to use FFS as an instrument to control farmers or to push their messages. Nederlof (2006) observed how a carefully developed FFS curriculum was turned into a technology push programme overnight by 'the boss'.
- It is difficult to scale up FFSs. Government agencies may be in the pay of pesticide companies which try to discourage IPM as much as possible. Typically, hierarchical government procedures do not easily fit the experiential learning approach and the type of facilitation that FFS requires.
- FFSs have little spin-off to farmers who have not taken part in an FFS. Complex practices such as IPM and integrated soil fertility management do not lend themselves to 'diffusion'. The FFS is a form of farmer education. In industrialized countries, farmer education and training have been central ingredients in making farmers effective partners in agricultural development. FFS is also a good basis for building farmer organizations and farmers' countervailing power. In fact, in systems such as the Gezira in Sudan, which are strictly run on

command and control, FFSs were discontinued because the alumni started complaining about things that they did not like (Khalid, 2002).

Landcare in Australia

The other innovative approach to innovation that deeply affected my thinking is Landcare (Campbell, 1994). This emerged in Australia when erosion, salination, desiccation and other environmental problems, as a result of applying European farming practices to a continent to which they were not suited, created awareness that new approaches were required. Land users were engaged in concerted action to ensure integrated management of watersheds, vulnerable soil types, saline groundwater tables, patches of native vegetation, etc. In other words, proverbially independent landowners in the 'outback' had to learn to manage their 'properties' in concert with others so that desirable ecosystem services at the landscape scale could be sustained. This required a great deal of collective action, organization, social learning and its facilitation. The Landcare movement pioneered methods and approaches that are now quite common across the globe (e.g. in the management of watersheds and communal forests). I myself instigated a major European project on Social Learning for the Integrated Management and Sustainable Use of Water at the Catchment Scale based on Landcare principles (see www.slim.open. ac.uk).

From these experiences, three key points emerged with respect to my understanding of innovation:

1   Innovation can be seen as the emergent property of interaction among stakeholders in a natural resource or ecosystem service (Bawden and Packam, 1993). Where the degradation of the resource or service is the collective outcome of each stakeholder's trying to satisfy his or her individual preferences, more sustainable management of that resource or service *necessarily* must emerge from collective processes – such as social learning, conflict and negotiation, agreement, reciprocal sacrifice of benefits and privileges, and leadership – that lead to concerted action. Technology and market incentives can play some role in getting there; but basically we are dealing with a totally different concept: innovation as the emergent property of interaction. This notion was already implicit in the FFSs; but in Landcare it became explicit and a subject of study. Soft systems thinking (Checkland and Scholes, 1990) became an important way to understand what was happening. I contributed by formulating the notion of a platform for resource use negotiation (Röling, 1994). My colleagues Paul Engel and Monique Salomon (1997) developed a methodology, Rapid Appraisal of Agricultural Knowledge Systems (RAAKS), for taking stakeholders through a systemic process of reflective action research, learning and decision-making that leads to the emergence of innovation from their

interaction. RAAKS is a tool to enhance the innovative performance of the actors in a theatre of innovation.

2   When innovation is the emergent property of interaction, promoting innovation becomes a matter of facilitating the interaction process. Of course, this was already important in such approaches as PTD; but where innovation is the result of collective voluntary behavioural change (instead of technical change to enhance productivity at the farm level), process facilitation becomes a key skill.

3   Landcare was largely run by groups of landowners facilitated by trained officers made available through state governments. But it remained a local phenomenon. The government of Australia did not give the policy and institutional support to create the conditions that allowed it to develop. Thus, it petered out when local people had done what they could do in their sphere of influence and ran into bottlenecks to consolidate, scale up or institutionalize the changes. From then on, I have been very aware that institutional support and favourable policies at higher levels are essential ingredients for success at the local level.

Implications for Africa

- Once we see that innovation can emerge from interaction among different stakeholders on platforms for deciding about concerted action towards some common objective, a great deal of African innovation becomes visible. Well known are typical African achievements such as 'harambee' in Kenya or town 'progressive unions' supported by 'sons abroad' in Nigeria. But this approach is becoming increasingly important for natural resource management as well. Dangbégnon (1998) describes successful action among arable farmers and pastoralists to agree on the use of common grazing lands. Hounkonnou (2001) reports a number of effective local self-development actions, such as a village succeeding in stopping robberies, and another going from strength to strength in developing its agriculture and establishing an effective local health service. In all of these cases, the basis for success was laid a long time ago when young people learned to trust each other, to organize and work together, and to withstand the 'big men'.

- The perspective on innovation through collective action opens up huge potential for change. Many of the worst aspects of poverty, such as the lack of services; rent-seeking and corruption; insecurity; lack of community amenities; and lack of access to education, drinking water, etc. could be remedied through effective collective action as so many successful projects across Africa have shown. The same applies to creating opportunities for smallholders to market their products and gain access to inputs and credit. Instead of trying to develop Africa's agriculture by introducing 'innovations' at the farm level, we now begin to see a possibility for innovation to create better opportunities for small farmers as a necessary condition for change at the farm level.

CONVERGENCE OF SCIENCES PROGRAMME

My most recent involvement with innovation in Africa has been the Convergence of Sciences (CoS) programme that worked from 2002 to 2006 in Benin and Ghana. CoS has greatly stretched the concept of innovation and brings together a number of the issues that I have presented above.

Initial focus on appropriate technology development

CoS addressed the problem of the low impact of science on the livelihoods of resource-poor farmers in West Africa. The Cocoa Research Institute in Ghana, for instance, has observed that cocoa farmers have adopted only 3 per cent of the technologies that it has developed (Ayenor, 2006). The response by scientists continues to be more of the same: investment in research, in training of scientists and in other ways of improving technology supply. Technologies by themselves are expected to generate opportunity: by increasing productivity at farm level, they are supposed to allow farmers to sell more and so increase their incomes. But the impact has been low. CoS tried to break through this impasse by developing a new 'pathway of science'.

To achieve this, CoS experimented in the field through the research projects of eight African PhD students, supervised by mixed social and natural science teams of both African and Dutch supervisors. A ninth PhD student drew the comparative lessons from the others' experiments.[1] The 'pathway of science' that CoS tested featured:

- a 'technography' for exploring the innovation landscape across the CoS themes and farming systems;
- diagnostic studies by each of the students to identify farmer communities with whom they could collaborate on concrete research issues that farmers considered relevant;
- joint field experiments with farmers; and
- platforms for discussion and learning composed of researchers, extension workers and others.

Most PhD students also created:

- forums of villagers not directly involved in order to monitor and learn from the work.

CoS showed that, through technography and diagnostic studies, one can identify niches of opportunity where agricultural science can make a useful contribution to PTD. In all CoS case studies, the research groups developed low-external-input technologies that built on a combination of farmers' knowledge and practice and scientific input. Many of the groups' technical experimental results have been published in international journals.

CoS started off with an emphasis on technological research. The technography and the diagnostic studies had made it possible to zero in on the small windows of opportunity that farmers have for beneficial technical change. But as CoS progressed, the need to stretch those opportunities was increasingly felt. For example, farmers who had been helped to increase their maize yields complained to the PhD student that they could not sell their surplus at a reasonable price.

Adjei-Nsiah (2006) and Saïdou et al (2007) provide an example of institutional factors that affect the windows of opportunity. After initially pursuing purely technical avenues to improving soil fertility management, they found that tenure contracts between landowners and immigrants created conditions in which the latter could do little but knowingly engage in unsustainable land use. The researchers then facilitated renegotiation of tenure contracts as the key intervention needed to improve soil fertility management. This, in turn, created room for technological change.[2]

Institutional innovation

In common parlance, the words 'organization' and 'institution' are used interchangeably. For instance, 'institution-building' typically refers to organizations such as universities, government departments, research organizations, etc. This easily leads to the idea that development requires the strengthening of the intervention power of *individual* organizations, agencies or institutes.

But institutions can also refer to the rules and agreements that reduce uncertainty in human interaction (North, 2005). Here, the focus is on *interaction* and on realizing the development gains that can be captured from the *interfaces* among important actors in rural development. This might include improving service delivery, reducing rent-seeking, strengthening mutual claim-making, building marketing chains, enhancing interdependence and empowering resource-poor farmers to have a voice in theatres of innovation. This interactive approach to institutional development fits with the innovation systems perspective that I will introduce below.

An illustrative example of institutions, as defined here, is the 'plimsoll line' (Jones, 2006). During the 19th century, many lives were lost at sea because the greed of unscrupulous ship owners drove them to overload their ships. Thanks to a committed philanthropist, Mr Plimsoll, a legally binding agreement was eventually reached about the safe maximum to which ships could be loaded. This maximum is marked on the hull of the ship and, to this day, is called the plimsoll line, although few people remember Mr Plimsoll for forging this life-saving institution as a normative agreement among stakeholders, backed by improved scientific understanding.

In developing countries, in accordance with the Washington Consensus, and with the supremacy of methodological individualism and rational choice economics, the focus is almost exclusively on competitive techno-logical change at farm level as the main driver of rural development. The institutional development in the public sector, such as veterinary services, extension agencies, cotton campaigns, agricultural education, marketing support schemes, etc., has been weakened or abandoned by governments under pressure to implement 'structural adjustment' (e.g. Stiglitz, 2006).

An example from CoS underscores the importance of institutional development in the sense discussed here. Dormon (2006) worked with cocoa farmers in southern Ghana on neem as an effective control measure against capsid bugs. Given the high prices for cocoa paid to farmers, they had keen interest in this technology. But it soon proved difficult to access and process neem seed, which is supplied from northern Ghana. As a result, Dormon and his farmers had to develop local capacity to purchase and process neem seed. This required organization and management beyond the level of the individual farmer. It required the involvement of a wider set of stakeholders who could develop income opportunities around this enterprise.

Figure 2.1 shows institutions and technology as two dimensions of innovation. It is the nature of their combination that requires attention. Institutions have been neglected in African development. Says Thompson (2006):

> Much of the failure of agriculture to achieve its potential is institutional and political. Support by the state has been unresponsive to the needs of the poor and inefficient in marketing producers' output, sometimes preventing the natural development of markets for producers. Public institutions need to be strengthened in their capacity to develop an appropriate blend of policies, regulatory frameworks and investments to re-launch the agricultural sector.

Innovation comprises not only hardware; it also includes software and 'orgware' (Leeuwis and Van den Ban, 2004). An important promise of the innovation systems approach is that it is able to address these institutional issues.

Innovation systems

I have been particularly attracted by innovation systems (IS) because Engel and I have spent a considerable part of our research careers on agricultural knowledge and information systems (AKIS), which we defined in a very similar way to IS (Röling and Engel, 1991). Both are multi-stakeholder processes that focus on creating synergetic configurations of stakeholders who make complementary contributions to concerted innovative action.

The current notion of IS came out of empirical studies of successful economic development in Asia, which noted that such development can best be explained as emerging from the interaction of key actors: 'An

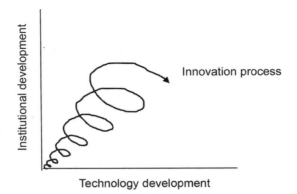

Figure 2.1 *Innovation as a function of institutional and technical change*

Source: adapted from Dorward et al (1998)

essential determinant of innovation was that suppliers of new knowledge were engaged with the users of that knowledge' (Barnett, 2006).

As happened with diffusion of innovations research, these empirical findings were transformed into a policy model. The World Bank (2006) recently published a paper on IS that has all the characteristics of a policy brief and therefore leaves open the possibility that IS is advocated in situations where it does not apply. The IS approach has not really been tested in African conditions. But doing so is very tempting, especially because the IS approach promises to provide a hands-on methodology for institutional development.

As a guiding concept, the IS approach focuses on *systems* (i.e. on relationships and linkages among elements within an arbitrary boundary). We are not speaking of a given system, such as the Ugandan IS, but of coalitions or configurations of actors that turn out to be important within a potential theatre of innovation, such as organic coffee. As in CoS, considerable time and effort will need to be spent on technographies and diagnostic studies of such theatres of innovation to establish who the actors are, to analyse their relationships and to identify the potential contribution of synergy among them. These studies need to be participatory: the actors concerned need to gel into effective 'communities of practice' with shared goals and, eventually, with the ability for concerted action towards rural poverty reduction. Concretely, such concerted action will focus on providing services, removing rent-seeking and corrupt practices, creating markets, and processing facilities and supply chains linking African farmers to emerging urban markets. Greater farmer influence and countervailing power will be an essential ingredient in making sure that development is appropriate and beneficial.

An IS approach needs to be applied across multiple scales. It is not enough to implement it at the decentralized level – say the district or *arondissement*. At the national and international level, institutional support and policies need to create framework conditions for success at the local level. The Washington Consensus, with its emphasis on methodological individualism, free markets, reduction of the public sector and promotion of the global treadmill, is under severe attack. This opens a perspective on international agreement to install the governance mechanisms that can constrain the destructive impact of the global treadmill on African agriculture. But, at the same time, African governments and elites must transform their own, often predatory, practices and create the support that allows African agriculture to develop and feed the growing urban population.

In the follow-up to CoS, we plan to invest much effort into learning from the experiments with the IS approach as a basis for creating a new understanding of innovation in Africa. This presentation will hopefully be continued at the next Innovation Africa Symposium.

CONCLUSIONS

My autobiographical approach to the challenge of providing an overview of conceptual and methodological developments in innovation has raised a number of key points:

- Policy thinking about innovation tends to be dominated by the linear model of technology supply push, partly because no other model promises such desirable macro-economic impacts. Yet this model has only very limited applicability in African conditions and has not led to significant development, let alone to poverty reduction. The transformation of *ex-post* empirical diffusion and treadmill research into an *ex-ante* policy model has been unprofessional. The continued support for the global treadmill unfairly pits small-scale African farmers against farmers in developed economies who have been helped by public resources to capture significant economies of scale. This leaves African farmers three options: to continue farming for subsistence on dwindling and degrading resources; to escape by emigration; or to participate in the opportunities that are sometimes provided through BLCAs. But current policies do not allow African commercial agriculture to develop on its own terms and in the protected conditions that OECD farming industries enjoyed when they were emerging.
- Many people who are practically involved in promoting agricultural innovation have little ability to think conceptually about innovation as such. They are as imprinted with the linear model as ducklings are with the sound of their mother's call. It is essential to throw open the concept of innovation to include endogenous development, social learning, concerted action, emergence from interaction and institutional change. Being at the end of my career, I regret not living to see the next generation of ideas that innovate the concept of innovation.

- Several alternatives to the linear model have been developed and tested in farmers' conditions. Many of these have demonstrated considerable potential to improve rural people's lives. Some others, such as the IS approach, still require substantial experimentation. One problem with such approaches is that their real achievements in terms of poverty reduction, empowerment, greater equity, more sustainable resource management, etc. do not show up in the national accounts that are used to monitor economic growth. Thus, the new approaches are seen as costing the earth while contributing nothing 'tangible'. Hence, they are considered 'fiscally unsustainable'. This has happened even with T&V, for a while the darling of the World Bank (Anderson et al, 2006), but also with FFS (Feder et al, 2004).
- A clear distinction must be made between technological change at farm level that leads to higher productivity *within* existing windows of opportunity, and institutional change at higher system levels that *stretches* these windows. Pushing technologies in conditions of limited opportunity is comparable to promoting a free market in a situation where essential market institutions, such as banks, do not function. In Africa, priority must be given to institutional change. It is not only farmers, but also national and international research organizations, local and national governments and especially international agencies that need to innovate. The IS approach promises to be an effective way to promote institutional development.
- Increasing the countervailing power of small-scale farmers is an essential condition for synergy among development actors. But it holds equally for national governments *vis-à-vis* international financial organizations. African leaders must thoroughly understand rural innovation processes if they are to stand up to misguided international policies and claim space for African agriculture to develop. Without jumping off the global treadmill, it is difficult to envisage African countries acquiring food sovereignty and African farmers contributing to global food security by making their resources more productive.

Currently, I am involved with hundreds of other authors, review editors and contributors in a fascinating process called the International Assessment of Agricultural Science and Technology for Development (IAASTD). This is driven by a multi-stakeholder bureau comprising national governments, the private sector and civil society organizations (CSOs). Hence, it is an arena of conflicting interests trying to reach agreement on policy options and investments that can make agricultural knowledge, science and technology (AKST) more effective in serving food security, poverty reduction and sustainable development. The IAASTD follows in the footsteps of the international assessments of biodiversity, climate change and ecosystems.

The IAASTD is a real 'battlefield of knowledge'. It makes clear that the controversies and contradictions which I have struggled with my whole life are as alive and kicking as 30 years ago. Some colleagues in the IAASTD, and not the least influential ones, still see poverty reduction and food security as a question of developing and pushing technologies. They resist including institutional change in AKST. Where some of us have emphasized changes in markets and trade as necessary conditions for poverty reduction, national governments have insisted on deleting all sentences containing such suggestions. This, in turn, is bitterly opposed by the CSOs. My explanation of the global treadmill as a mechanism that unfairly pits African smallholders against 'the most efficient farmers in the world' has been attacked from different sides. Neoconservative Americans call it 'dangerous spin', while organic agriculture supporters consider it ridiculous to call modern farming, with its wasteful and unsustainable use of fossil energy, 'efficient'.

And so 'la lutta continua'. Innovation remains a fascinating field of endeavour that never fails to raise the deepest emotions. I am glad I chose it for my career. I hope this chapter convinces others to do the same.

ACKNOWLEDGEMENT

The author is grateful to Ann Waters-Bayer for substantial editing of the earlier version of this chapter.

NOTES

1 The nine diagnostic studies appeared in 2004 in the *Netherlands Journal of Life Sciences*, vol 53, no 3/4, pp209–448.
2 The CoS experience with institutional development was published in 2007 in the *International Journal of Agricultural Sustainability*, vol 5, nos 2 and 3, pp89–264.

REFERENCES

Adjei-Nsiah, S. (2006) 'Cropping systems, land tenure and social diversity in Wenchi, Ghana: Implications for soil fertility management', PhD thesis, Wageningen University, Wageningen, The Netherlands
Anderson, J., Feder, G. and Ganguly, S. (2006) 'Analysing the demise of the Training & Visit system of extension', in A. Van den Ban and R. K. Samantha (eds) *Changing Roles of Agricultural Extension in Asian Nations*, B. R. Publishing, Delhi, India, pp149–174
Ascroft, J. R., Röling, N., Kariuki, J. and Wa Chege, F. (1973) 'Extension and the forgotten farmer', *Bulletin van de Afdelingen Sociale Wetenschappen*, vol 37, Wageningen University, The Netherlands
Ayenor, G. K. (2006) 'Capsid control for organic cocoa in Ghana: Results of participatory learning and action research', PhD thesis, Wageningen University, Wageningen, The Netherlands

Bairoch, P. (1997) 'New estimates on agricultural productivity and yields of developed countries, 1800–1990', in A. Bhaduri and R. Skarstein (eds) *Economic Development and Agricultural Productivity*, Edward Elgar, Cheltenham, UK, pp45–57

Barnett, A. (2006) 'Innovations: Lessons from the UK funded crop post-harvest research', in RAWOO (ed) *Knowledge Makes a Difference: Science and the Millennium Development Goals*, RAWOO Publication 30, The Hague, pp48–50

Bawden, R. J. and Packam, R. (1993) 'Systems praxis in the education of the agricultural systems practitioner', *Systems Practice*, vol 6, pp7–19

Brouwers, J. (1993) 'Rural people's knowledge and its response to declining soil fertility: The Adja case (Benin)', PhD thesis, Wageningen University, Wageningen, The Netherlands

Bruin, G. and Meerman, F. (2001) *New Ways of Developing Agricultural Technologies: The Zanzibar Experience with Participatory Integrated Pest Management*, Wageningen University and Research Centre/CTA, Wageningen, The Netherlands

Campbell, A. (1994) *Landcare: Communities Shaping the Land and the Future*, Allan and Unwin, St Leonards, Australia

Chambers, R. and Jiggins, J. (1987) 'Agricultural research for resource-poor farmers. Part I: Transfer-of-technology and Farming Systems Research; Part II: A parsimonious paradigm', *Agricultural Administration and Extension*, vol 27, pp35–52, 109–128

Checkland, P. and Scholes, J. (1990) *Soft Systems Methodology in Action*, John Wiley & Sons, Chichester, UK

Cochrane, W. W. (1958) *Farm Prices, Myth and Reality*, University of Minnesota Press, Minneapolis, MN (especially Chapter 5, pp85–107)

Collinson, M. (ed) (2000) *History of Farming Systems Research*, FAO, Rome/CABI, Wallingford, UK

Dangbégnon, C. (1998) 'Platforms for resource management: Case studies of success or failure in Benin and Burkina Faso', PhD thesis, Wageningen University, The Netherlands

Dormon, E. (2006) 'Actors' innovations and innovation processes in a changing policy and organisational environment: The case of cocoa production in Ghana', PhD thesis, Wageningen University, Wageningen, The Netherlands

Dorward, A., Kydd, J. and Poulton, C. (1998) *Smallholder Cash Crop Production under Market Liberalisation: A New Institutional Economics Perspective*, CABI, Wallingford, UK

Economist (2006) 'Food and government in Africa', *Economist*, 16–22 September, p90

Engel, P. and Salomon, M. (1997) *Facilitating Innovation for Development: A RAAKS Resource Box*, Royal Tropical Institute, Amsterdam, The Netherlands

Evenson, R. E., Waggoner, P. E. and Ruttan, V. W. (1979) 'Economic benefits from research: An example from agriculture', *Science*, vol 205, pp1101–1107

FAO (United Nations Food and Agriculture Organization) (2005) *The State of Food and Agriculture*, FAO, Rome, www.fao.org/docrep/008/a0050e/a0050e10.htm, accessed 23 February 2007

Feder, G., Murgai, R. and Quizon, J. (2004) 'Sending farmers back to school: the impact of Farmer Field Schools in Indonesia', *Review of Agricultural Economics*, vol 67, no 2, pp423–428

Gabre-Mahdin, E. and Haggbladd, S. (2004) 'Successes in African agriculture: Results of an expert survey', *World Development*, vol 32, no 5, pp745–766

Havelock, R. G. (1973) *Planning for Innovation through Dissemination and Utilization of Knowledge,* University of Michigan, Ann Arbor, MI

Havelock, R. G. (1986) 'Modelling the knowledge system', in G. M. Beal, W. Dissanayake and S. Konoshima (eds) *Knowledge Generation, Exchange and Utilization,* Westview Press, Boulder CO, pp77–105

Hounkonnou, D. (2001) 'Listening to the cradle: Local dynamics for African renaissance – Case studies from Benin and Ghana', PhD thesis, Wageningen University, Wageningen, The Netherlands

IAC (InterAcademy Council) (2004) *Realising the Promise and Potential of African Agriculture: Science and Technology Strategies for Improving Agricultural Productivity and Food Security in Africa,* InterAcademy Council, Amsterdam, The Netherlands

Jones, N. (2006) *The Plimsoll Sensation: The Great Campaign to Save Lives at Sea,* Little and Brown, New York, NY

Khalid, A. (2002) 'Assessing the long-term impact of IPM Farmer Field Schools on farmers' knowledge, attitudes and practices: A case study from Gezira Scheme, Sudan', International Learning Workshop on Farmer Field Schools: Emerging Issues and Challenges, 21–25 October, Yogyakarta, Indonesia

Kline, S. and Rosenberg, N. (1986) 'An overview of innovation', in R. Landau and N. Rosenberg (eds) *The Positive Sum Strategy: Harnessing Technology for Economic Growth,* National Academic Press, Washington, DC, pp275–306

Kotler, P. and Andreasen, A.R. (2003) *Strategic Marketing for Non-Profit Organizations,* Prentice Hall, New York, NY

Leeuwis, C. with Van den Ban, A. (2004) *Communication for Rural Innovation: Rethinking Agricultural Extension,* third edition, Blackwell Science, Oxford, UK

Long, N. and Long A. (eds) (1992) *Battlefields of Knowledge: The Interlocking of Theory and Practice in Research and Development,* Routledge, London

Nederlof, S. (2006) 'Research on agricultural research: Towards a pathway for client-oriented research in Africa', PhD thesis, Wageningen University, Wageningen, The Netherlands

Norman, D. (1974) 'Rationalising mixed cropping under indigenous conditions: The example of Northern Nigeria', *Journal for Development Studies,* vol 11, no 1, pp3–21

North, D. C. (2005) *Understanding the Process of Economic Change,* Princeton University Press, Princeton, NJ

Pontius, J., Dilts, R. and Bartlett, A. (2002) *From Farmer Field Schools to Community IPM: Ten Years of IPM Training in Asia,* FAO Regional Office for Asia and the Pacific, Bangkok, Thailand

Reijntjes, C., Haverkort, B. and Waters-Bayer, A. (1992) *Farming for the Future: An Introduction to Low-External-Input and Sustainable Agriculture,* Macmillan, London/ILEIA, Leusden, The Netherlands

Richards, P. (2002) *Fighting for the Rain Forest: War, Youths and Resources in Sierra Leone,* James Currey, Oxford, UK

Rogers, E. M. (2003) *Diffusion of Innovations,* fifth edition, Free Press, New York, NY

Röling, N. (1988) *Extension Science: Information Systems in Agricultural Development,* Cambridge University Press, Cambridge, UK

Röling, N. (1994) 'Platforms for decision making about eco-systems', in L. O. Fresco, L. Stroosnijder, J. Bouma and H. Van Keulen (eds) *Future of the Land: Mobilising and Integrating Knowledge for Land Use Options,* John Wiley & Sons, Chichester, UK, pp386–393

Röling, N. and Engel, P. (1991) 'The development of the concept of Agricultural Knowledge and Information Systems (AKIS): Implications for extension', in W. Rivera and D. Gustafson (eds) *Agricultural Extension: Worldwide Institutional Evolution and Forces for Change*, Elsevier, Amsterdam, The Netherlands, pp125–139

Röling, N. and Van de Fliert, E. (1994) 'Transforming extension for sustainable agriculture: The case of Integrated Pest Management in rice in Indonesia', *Agriculture and Human Values*, vol 11, nos 2 and 3, pp96–108

Röling, N., Ascroft, J. and Wa Chege, F. (1976) 'Diffusion of innovations and the issue of equity in rural development', *Communication Research*, vol 3, pp155–171

Ryan, B. and Gross, N. (1943) 'The diffusion of hybrid seed corn in two Iowa communities', *Rural Sociology*, vol 8, pp15–24

Saïdou, A., Tossou, R., Kossou, D., Sambieni, S., Richards, P. and Kuyper, T. (2007) 'Land tenure and sustainable soil fertility management in Central Benin: Towards the establishment of a cooperation space among stakeholders', *International Journal of Agricultural Sustainability*, vol 5, nos 2 and 3, pp195–213

Stiglitz, J. (2006) *Making Globalisation Work*, Norton & Co, New York and London

Thompson, J. (2006) 'The dynamics of changing rural worlds: Balancing income generation and household and community food security in an era of growing risk and uncertainty', Discussion Paper, Global Forum on Agricultural Research (GFAR) Third Triennial Conference Reorienting Agricultural Research to Achieve the MDGs, 8–11 November, New Delhi, India

Van de Fliert, E. (1993) 'Integrated pest management: Farmer field schools generate sustainable practices: A case study in central Java evaluating IPM training', PhD thesis, Wageningen University, The Netherlands

Van den Ban, A. W. (1963) *Boer en Landbouwvoorlichting*, Van Gorcum, Assen, The Netherlands

Van den Berg, H. and Jiggins, J. (2007) 'Investing in farmers: The impact of Farmer Field Schools in relation to integrated pest management', *World Development*, vol 35, no 4, pp663–686

Vissoh, P. (2006) 'Participatory development of weed management technology in Benin', PhD thesis, Wageningen University, Wageningen, The Netherlands

Warren, D. M., Slikkeveer, L. J. and Brokensha, D. (eds) (1991) *Indigenous Knowledge Systems: The Cultural Dimension of Development*, Kegan Paul, London

World Bank (2006) *Enhancing Agricultural Innovation: How to Go Beyond the Strengthening of Research Systems*, World Bank, Washington, DC

Youdeowei, A. (2003) 'Farmer Field Schools: science in action', *Pesticide News*, vol 61, PAN-UK, London, pp9–10

Comparison of Frameworks for Studying Grassroots Innovation: Agricultural Innovation Systems and Agricultural Knowledge and Information Systems

Amanuel Assefa, Ann Waters-Bayer, Robert Fincham and Maxwell Mudahara

GRASSROOTS INNOVATION IN SUB-SAHARAN AFRICA

The term 'grassroots agricultural innovation' refers to the interface of endogenous and exogenous innovation in the farming systems of smallholders and pastoralists, referred to here as 'farmers'. In sub-Saharan Africa, the economic base of these systems is usually subsistence farming. Immense natural resource challenges affect the lives of these people, and their actions, in turn, affect the environment. Since they struggle to survive and external forces of various interests intervene strongly, the dynamics of exogenous and endogenous innovation are complex. 'Exogenous agricultural innovation' refers to all innovation interventions to the local system that are initiated and controlled by outsiders and intended to improve local livelihoods and the environment – for example, the interventions of research, extension, the private sector, non-governmental organizations (NGOs), financial organizations, etc. to introduce new technologies, methods, services, products, processes and institutional arrangements. 'Endogenous agricultural innovation' refers to new initiatives and processes of local groups or individuals trying to address issues of poverty and the environment, and includes their technical, institutional, marketing or management innovation performances. In sub-Saharan Africa, a mix of both endogenous and exogenous components is typical for most grassroots innovation systems.

RECENTLY EMERGING PERSPECTIVES ON INNOVATION SYSTEMS

Agricultural knowledge and information systems

The AKIS perspective emerged in response to challenges in the theory of transfer, adoption and diffusion of innovations, which looked at why and how people adopt or do not adopt new agricultural practices (Leeuwis, 2004). This theory laid the basis for the National Agricultural Research System (NARS) concept and continues to dominate in the developing world. The legacy of the relative success of the Green Revolution in India and other Asian countries continues to influence many policy-makers, researchers and extension institutions in sub-Saharan Africa. The linear transfer of technology model allows accumulation of power at the centre. Politicians and policy-makers whose governments are characterized by centralized control of power and a command economy prefer this linear model as they feel they can control the system without having to be involved in the complex challenges of knowledge management. It is therefore not easy to bring about a paradigm shift in such developing countries.

The AKIS approach tries to overcome the limitations of the NARS concept and related policy and institutional arrangements. It emerged from numerous 'formative experiences' of applied social scientists trying to come to grips with the complexity of facilitating innovation, primarily in agriculture. Röling (1992) defined AKIS as 'the articulated set of actors, networks and/or organizations, expected or managed to work synergistically to support knowledge processes which improve the correspondence between knowledge and environment and/or the control provided through technology use in a given domain of human activity'. He developed the diagnostic framework to help discern the organizational forms that enable or constrain the generation, transformation and use of knowledge (Engel, 1997).

AKIS demands a radical policy shift from strengthening research or extension institutions, which is typical in the NARS perspective, to strengthening linkages and communication that should take place among the system actors. Unlike the NARS perspective, AKIS sees farmers not merely as recipients of technology from research via extension; rather, all system actors have a stake in the process of generating, disseminating and using knowledge. Learning about the stock of knowledge in these actors and creating a platform for their interaction to facilitate innovation are the main principles in AKIS. Many participatory approaches to (agricultural research and development (ARD) – for example, rapid appraisal of agricultural knowledge systems (RAAKS), farmer field schools (FFSs), participatory technology development (PTD) and participatory innovation development (PID) – developed out of this paradigm shift from the linear model to the AKIS model of multiple sources of innovation.

Critics of AKIS (e.g. Leeuwis, 2004) claim that it looks at knowledge generation and use without considering the influence of political and other forces and therefore cannot yield a complete and realistic analysis. According to Hall (2006), the AKIS concept still focuses on research supply but gives more attention to links between research, education and extension and to identifying farmers' demand for new technology.

Agricultural innovation systems

The agricultural innovation systems (AIS) perspective is increasingly used to explain how innovation takes place and how and by whom benefits are gained out of complex technological and institutional change processes. The theoretical framework of innovation systems was first used to explain processes in the 'developed' world that are governed by the rules of a free market economy and more or less democratic political systems. Industrial innovation is characterized by technology change in manufacturing, with emphasis on market opportunities and institutional change. In 1841, List first described 'a national system of political economy', a precursor of the innovation systems concept. It assumes that industrial production results from social and economic institutions such as education and infrastructure (Spielman, 2005). An innovation system can be defined as a network of organizations, enterprises and individuals focused on bringing new products, processes and forms of organization into economic use, together with the institutions and policies that affect their behaviour and performance. The innovation systems concept embraces not only the science suppliers but the totality of actors involved in innovation. It extends beyond knowledge creation to embrace factors affecting demand for and use of knowledge in novel and useful ways (Hall, 2006).

Lessons from applying the innovations systems concept in the industrialized world have been used to develop the AIS perspective. This has added value to the conventional linear perspective on ARD by providing a framework for analysing complex relationships and innovative processes that occur among multiple agents, social and economic institutions, and endogenously determined technological and institutional opportunities (Spielman, 2005). Given its industrial origin, current AIS studies emphasize the market and other institutional forces that affect innovation processes in agriculture. Commercially important agricultural commodities that have high value in national and global markets attracted the attention of many authors working in developing countries. The AIS analytical framework indeed helps in studying how innovation systems emerge, are coordinated and function, and how innovation performances are influenced by market and non-market forces in market-led economic systems.

Hall (2006) broadly summarized innovation and innovation processes as follows:

● Innovations are creations of social and economic significance that may be brand new but are more often combinations of existing elements.

- Innovation can comprise radical improvement, but usually consists of many small improvements in a continuous process of upgrading.
- These improvements may be of a technical, managerial, institutional (i.e. the way in which things are routinely done) or policy nature.
- Often, innovations involve a combination of technical, institutional and other changes.
- Innovation processes can be triggered in many ways (e.g. bottlenecks in production, changes in available technology, competitive conditions, international trade rules, domestic regulations and environmental health concerns).

KNOWLEDGE, INVENTION AND INNOVATION IN NARS, AKIS AND AIS PERSPECTIVES

Some authors distinguish between invention, which refers to the first occurrence of an idea for a new product or process, and innovation, which is the first attempt to put the invention into practice. Hall (2006) suggests that invention culminates in the creation of knowledge, while innovation encompasses factors affecting demand for and use of knowledge in novel and useful ways. The notion of novelty is fundamental to invention, but the notion of creating local change *new to the user* is fundamental to innovation. According to Anandajayasekeram et al (2005), inventions provide solutions to problems in the narrowly defined context of the designers, but may remain limited in application if they are not transformed into innovations by entering into the complex relations of people and institutions in wider socio-economic, cultural and political contexts.

The NARS perspective does not distinguish between knowledge, invention and innovation. It is based on a belief that scientific research is the sole supplier of knowledge. The types of knowledge created through the rigorous scientific process are inventions regarded as innovations. The process of generating new knowledge in scientist-controlled environments is called innovation, and innovations are the products of scientific research. They are not products of social processes that take place in interactions also outside of the formal research system.

In the AKIS perspective, innovation is the desired outcome not of researchers working in a controlled environment isolated from the bigger system, but rather of the knowledge system made up of multiple actors with complex and interrelated functions (Engel, 1997). It does not regard research as the sole supplier of knowledge, but rather as an important partner of other actors who make substantial contributions in generating knowledge. The purpose of an AKIS is to facilitate continuous innovation in agriculture-related practices. The individual innovation performances of farmers, particularly their creative practices in farming and natural resource management (NRM), bring new socio-economic values to the users and cannot be seen in isolation from the inputs of the other system actors. Röling (1996) describes innovation as a result of interaction among different actors making complementary contributions. Leeuwis (2004) shares this view and describes an innovation as a package of new social and technical arrangements and practices that implies a new form of coordination within a network of actors.

An important addition of AIS to the AKIS perspective is that AIS is concerned with a system made up of innovations that may take place at different knowledge fronts, such as the formal research system, the private sector, the technology-delivery agencies, farmers and other actors in the broader environment. This perspective reflects the influence of current world economic and social developments, characterized by transformations in knowledge-generation processes from elite control to a 'knowledge society', from using paper to store and share knowledge to using digital media and the web, and from research to searching and consultation as key tools to generate knowledge (Hall, 2006). In the AIS perspective, innovation management is given greater importance than knowledge management, which is an important aspect of AKIS.

In AIS, as in AKIS, the innovation process does not always start with formal research, and the knowledge coming from research does not necessarily create new practice or values. Rather, the AIS perspective underscores that it is only within the innovation system that knowledge and information from various sources interact to bring about new phenomena desired by the system actors. Leeuwis (2004) suggests that knowledge needs to be translated into skills and technologies and, thus, into sociotechnical innovation. Innovation is not just about research findings, but about the transformation of these into socially and economically valued products.

Methodological reflections on AKIS and AIS

During the last two decades, the development of organized procedures and tools to study relationships between system actors in generating, disseminating and using agricultural knowledge was the focus in AKIS. Different approaches were used to describe AKIS at national level or the specific interactions that take place at sector level. Conventional survey methods accompanied by interviews and focus group discussions could be used; but there are also well thought-through and carefully designed methodological approaches that go beyond describing the status of AKIS.

Rapid appraisal of agricultural knowledge systems (RAAKS) is a useful research methodology both to describe the status of AKIS and to facilitate agricultural innovation by focusing on the social organization of innovation (Engel, 1997). RAAKS has three distinct phases. Phase 1 focuses on defining the problem situation with the concerned actors in an interactive way and identifying the diverse system actors who have important stakes in the innovation process. Phase 2 is a detailed study of the linkages and communication between the system actors, knowledge

Table 3.1 *Defining features of NARS, AKIS and AIS perspectives related to agricultural innovation*

Defining feature	NARS	AKIS	AIS
Purpose	Planning capacity for agricultural research, technology development and technology transfer	Strengthening communication and knowledge delivery services to rural people	Strengthening capacities to innovate throughout the agricultural production and marketing system
Actors	National agricultural research organizations, agricultural universities, extension services and farmers	National agricultural research organizations, agricultural universities, extension services, farmers, NGOs and rural entrepreneurs	Potentially all actors in public and private sectors involved in creating, diffusing, adapting and using all types of knowledge relevant to agricultural production and marketing
Outcome	Technology invention and technology transfer	Technology adoption and innovation in agricultural production	Combinations of technical and institutional innovation throughout the production, marketing, policy research and enterprise domains
Organizing principle	Using science to create inventions	Accessing agricultural knowledge	New uses of knowledge for social and economic change
Mechanism for innovation	Transfer of technology	Interactive learning	Interactive learning
Degree of market integration	Nil	Low	High
Role of policy	Resource allocation and priority setting	Enabling framework	Integrated components and enabling framework
Nature of capacity strengthening	Infrastructure and human resource development	Strengthening communication between actors in rural areas	Strengthening interactions between actors; institutional change to support interaction, learning and innovation; creating an enabling environment

Source: adapted from Hall (2006)

networks, coordination and the broader environment in which innovation takes place. Phase 3 is a feedback session for the stakeholders to validate the findings and reach consensus on them, and to design joint action to enhance the innovation process in order to overcome constraints caused by weak or lacking linkages and communication, and limited sharing of knowledge and resources among the actors. The action plan also benefits from the interaction and debate of the system actors during the course of the study and in the stakeholder workshop. A known problem situation that constrains innovation processes serves as an entry point for a RAAKS exercise. The methodology is best used when diverse actors are involved in the action research and a well-trained facilitator carefully handles the interactive process.

PTD (Reijntjes et al, 1992), which is based on the principles and assumptions of the AKIS perspective, refers to joint experimentation and investigation by farmers and extensionists and, wherever possible, scientists to find ways of improving local livelihoods. It focuses on developing technological innovations through equitable interaction of these partners. PTD tries to overcome the limitations of the NARS perspective in which the research process – including agenda setting, data collection, analysis and reporting – is controlled and owned by the researchers and their peers in the formal sector.

As entry point, PTD takes agricultural problems identified often by farmers but also sometimes by researchers and other experts. The most challenging aspect of PTD is the process of blending farmers' and outsiders' knowledge so that they complement each other and bring new and locally relevant values. This requires an attitude of respect and honesty among the participating actors. If scientists cannot recognize the initiatives of resource-poor farmers, if they cannot appreciate the knowledge and reasoning behind the farmers' informal research and development efforts, if they cannot understand the social settings and motivations of the farmers, then they cannot engage effectively in ARD partnership with rural communities to alleviate poverty, increase food production and seek sustainable development (Waters-Bayer and Bayer, 2005). The PTD framework has six phases: getting to know each other, looking for things to try, designing experiments jointly, trying things out (implementing joint experiments), sharing results and sustaining the process (Reijntjes et al, 1992). Numerous other approaches such as FFS and farmer participatory research (FPR) share the same philosophy and principles, and can be grouped under the broad framework of PTD.

Participatory innovation development (PID) is a recently coined term to describe local innovation processes that take place with the support of outsiders (Waters-Bayer and Veldhuizen, 2004). It shares the same philosophy and principles but has wider dimensions than PTD, which is more limited to developing technology in the inner circle of farmers, researchers and extensionists. The unique feature of PID is that its entry point is not a problem but a local innovation that emerged out of the creativity of local people. The main purposes of PID are to enhance endogenous innovation processes, with outsiders' support on technical, methodological, institutional and policy dimensions, and to make use of emerging opportunities such as market, policy, networks, etc. Another distinguishing feature of PID is that it deals with not only technical but also institutional, cultural and other forms of social innovation. PID uses the broad framework of PTD and, in application, is being modified to accommodate the new values and dimensions of PID.

The AIS perspective focuses on the dynamics of innovation processes, departing from important aspects such as knowledge, institutional change, market forces and policy environment. Related methods are new and still being tested. AIS uses many social science research methodologies to analyse national innovation systems. The challenge is how to use the AIS perspective to facilitate innovation at sector level or in a given domain of human activity. Thus far, no specific methods and tools have been defined to do this. More often than not, studies are simply *ex-post* descriptions of the dynamics and complexities of some technological or institutional innovations – and, according to Spielman (2005), there the analysis ends. He suggests that a variety of methodological approaches could be used in AIS, including:

- analysis of the costs and benefits of knowledge production or dissemination, given the complexity of interactions among diverse agents;
- methodologies used in studying social learning processes among agrarian agents;
- benchmark or 'best practice' methods that have been used to study innovation systems in industrialized countries;
- game theory models that help to break down interactions into key decision points and payoffs (more applicable for knowledge-intensive sectors);
- agricultural technology management system analysis, which attempts to analyse interrelationships within and between organizations, as well as between organizations and external environments to improve organizational design and management functions.

Hall (2006) described the basic hypothesis of the methodological framework for AIS as follows: the capacity for continuous innovation is a function of linkages, working practices and policies that promote knowledge flow and learning among all actors within the sector. However, the aim is not only to identify the links or missing links in the system but to go beyond that and analyse the underplaying causes and impacts on the system.

LESSONS DRAWN FROM COMPARISON

Agricultural innovation systems

On the basis of country case studies by several authors, Hall (2006) analysed AIS according to five key issues (see Table 3.2). The reflections below are based only on the summary of findings presented in the main report. The issues presented in Table 3.2 are important for reflecting on the adequacy of the AIS perspective to study grassroots innovation in sub-Saharan Africa.

The issue of *innovation* was addressed in their study from the viewpoint of institutional linkages that do or do not support innovation processes – most notably, the functional linkages of the commercial sector and public research institutions. The study shows the importance of the private sector in innovation processes through either its own research or hiring high-level professionals. However, the types of technical and socio-institutional innovation that determine the level of complexity of the innovation system are not spelled out. The role of smallholders in the innovation process is addressed only in the Bangladesh shrimp industry case, where the ability of the poor to innovate is reportedly undermined. This suggests that the AIS framework – or the terms of reference for the study based on it – does not regard farmers' innovative capacity as an important pillar of the AIS that deserves closer examination. However, in grassroots innovation systems, the role of farmers as innovators is central.

The study looks at the private sector and related *market* issues and shows how market forces – not only the demand and supply relations, but also marketing infrastructure and other facilities – can make or break innovation performances. All case studies examined how market relations developed at local and international level, and the role played by governments. The AIS framework obviously gives adequate attention to market issues.

Human actions in innovation can have an impact on the *environment*, sometimes undesirable. It is vital to understand the relationship of human action and environmental impacts in order to be able to reverse or prevent these. In some case studies, innovation directly affected the environment (e.g. because of the lack of policy enforcement in the Bangladesh shrimp industry or the destruction of natural herbs when seeking medicinal plants). However, no information is given about environmental impacts in the other cases.

In grassroots innovation systems, where market influence is weak, environmental issues are particularly important. The further from industrialized settings, the stronger is the people's link with the environment as most of them live from natural resources. Moreover, social capital in the grassroots systems greatly influences natural resource management and how innovation in this respect affects sustainability.

The concepts of *empowerment* and *sustainability* are closely related: one may benefit the other, but can also have negative impacts on the other if

Table 3.2 *Major findings of the agricultural innovation systems study in five sectors*

Country	Sector focus	Issues addressed				
		Innovation	Environment	Market	Empowerment	Sustainability
Bangladesh	Commercial shrimp industry, dominated by export companies	Industry lacks confidence in research; weak interaction with it Ability of poor to innovate undermined Reactive approach to problem-solving	Environmental policies not enforced because of government negligence	Market relations well developed	Social equity needs to be addressed	Progress required in creating efficient sector-coordination body
Bangladesh	Home-based small-scale food processing sector that accounts for 80% of total industry	Poor access to information made it difficult to meet consumer preferences and quality standards Weak tradition of public research on food processing NGOs provide training in food processing	Not addressed	Many rural areas physically isolated from main market in Dhaka	Not addressed	Not addressed; recommendation focuses on capacity development

India	Exportable medicinal plants, grown in response to international demand	Large pharmaceutical companies developed associated research Small family-owned businesses lack research support Long-established mistrust among public research organizations, the private sector and NGOs Resistance to hybridizing scientific and traditional medicine	Sector lacked credit and technical skills to meet hygienic standards	Environmental degradation from destructive harvesting of plants	Sell traditional herbal medicine at local market Large pharmaceutical industries strongly linked with international market	Not addressed	Not addressed	Not addressed

Table 3.2 (continued)

Country	Sector focus	Issues addressed				
		Innovation	Environment	Market	Empowerment	Sustainability
Ghana	Pineapple export industry	Did own research or relied on foreign advisers	Not addressed	Private sector efficient in multiplying and distributing planting materials	Inclusion of small producers recommended	Not addressed; recommendation focuses on improving capacity and coordination
		Public research played minor role in developing pineapple export market		Export market continually growing despite growing competition and changing preferences of consumers		
		Assumption that scientists' work is commercially irrelevant				
		Universities disconnected from sector				

| Colombia | Cassava-processing agro-industry | Latin American consortium for cassava R&D and regional consortium of producer countries with strong links to national and international research organizations, including small-scale farmers

Apex organizations linked with processing and marketing innovations | Not addressed | In 1990 structural adjustment, brought policy-makers' attention to cassava

In mid 1990s, renewed government interest

Industry has strong links to domestic and international market | Not addressed | Not addressed; recommendation focuses on market research, technological innovation and more government investment |

Source: Hall (2006)

the relationship goes wrong. We refer here to empowerment related to grassroots innovation (i.e. farmers' opportunity and capacity to access, develop and use information, to make decisions on ARD priorities and to influence the decisions of others whose policies and practices affect their lives; and their resilience to respond to challenges exerted by human and non-human forces). Sustainability related to grassroots innovation refers to the extent that innovation is informed by local knowledge and is environmentally friendly, the cost and availability of technologies or services used to create new values, and the suitability of the innovation to the local socio-cultural context.

In Hall's (2006) study, the issues of environment, empowerment and sustainability are not central, although some comments were made as recommendations, such as inclusion of small-scale producers in the pineapple industry of Ghana and the need for social equity in the shrimp industry of Bangladesh. Hence, we ask: is the AIS conceptual framework adequate for studying the relationship between innovation, empowerment, environment and sustainability? The framework appears to be most attracted to the commercial sector, where market issues are more important than social capital and related issues.

Agricultural knowledge and information systems

Rivera et al (2005) assessed the status of AKIS in ten developing countries according to five parameters: policy environment; institutional structure for supporting innovation; conditions for expressing demand for innovation; partnership and networking; and finance systems for innovation. The results are shown in Table 3.3 by marking crosses (up to 3) to rank the magnitude of each parameter in the respective countries. The greater the number of crosses, the stronger the performance is.

This study pays particular attention to how public institutions of research, extension and education are aligned to develop public goods. In the AKIS cases, the role of public institutions in the innovation process is seen as very high and the private sector is not addressed as a source of innovation. In contrast, the role of public research in supporting the private sector in the AIS studies is seen as very low, perhaps because this perspective focuses on commercial rather than public goods. The AKIS studies give some attention to market issues, such as inputs (including credit) and infrastructural support. However, had the private sector been seen as a source of innovation, the linkage framework for AKIS would not have been triangular (research, extension and education) but rectangular, including the private sector. Although the methodological approaches in AKIS (PTD/PID) recognize and appreciate the potential of smallholders to innovate, the case studies did not address these issues. The fact that the studies were made at country level should not prevent them from commenting on farmers' role in innovation. The national picture should be able to reflect the dynamics of AKIS on the ground.

Table 3.3 *Status of agricultural knowledge and information systems in ten countries*

AKIS parameters	Cameroon	Chile	Cuba	Egypt	Lithuania	Malaysia	Morocco	Pakistan	Trinidad and Tobago	Uganda
1 Policy environment										
Existence of national AKIS policy	xxx	xxx	xxx	x	xxx	xxx	xxx	xxx	x	xxx
AKIS targets public goods	xxx	xxx	xxx		xxx	x	x			x
Attention to economic efficiency of agricultural sectors	xx	xxx	x		xx	xx	x			x
2 Institutional structure for supporting innovation										
Existence of AKIS units	xx	xxx	xxx	xxx	xx	xxx	xx	x		xxx
Central and branch supervision	xxx	x	x			xxx	xx	x	x	x
Initiatives to build institutional resources	xx		xxx		x	xxx	xx	x	xx	xxx
Sound strategy for programme decentralization	xx	xxx	xx		xx	x	xx	x	x	xxx
Monitoring, evaluation and impact assessment	xxx					xx	x	x		
Functional performance of AKIS entities	xx	xx	x	x	xx	xx				

Table 3.3 (continued)

AKIS parameters	Cameroon	Chile	Cuba	Egypt	Lithuania	Malaysia	Morocco	Pakistan	Trinidad and Tobago	Uganda
3 Conditions for expressing demand for innovation										
Demand-driven orientation	x	xxx	x		xx		xx			xxx
Agricultural market support		xx			xx	xxx			x	xx
Input (including credit) support	xx	x			x	x				
Physical infrastructure support	x	xx	xxx	xxx	xx	xx	xxx			xx
Joint planning/effective linkages	x	x	xx	x		x	x			x
Education for agricultural producers	x	X	xx	x	x	x	x	x		xx
Gender inclusion	xx		x	x		x	x			
4 Partnership and networking										
Structures for institutional cooperation	xx	xx	xx	x	x	xx	x	x		xx
Existence of strong public–private partnership	xx	xxx			x	xxx	x		x	xx
Programme participation by agricultural producers	x	xxx	xx		xxx	x	xx			xx

	1	2	3	4	5	6	7
Effective use of traditional communication	x				x	x	x
Effective use of computer internet technology		xx	xx	xx	xx		x
5 Financing systems for innovation							
Adequate funding for AKIS	xx	xxx	xx	x	xxx		x
Repartition of costs	x	xxx			xx	x	xx
Investment to develop stakeholder capacity	x	xx	xxx	xx	x	xx	xx

Source: adapted from Rivera et al (2005)

COMPLEMENTARITIES AND DIFFERENCES

Major complementarities

Both perspectives are grounded in the systems thinking of the constructivist paradigm. The AIS and AKIS frameworks are complementary; but each has its own strengths and limitations when used to study innovation in grassroots systems. Unlike the NARS perspective, in both AKIS and AIS, innovation is a social phenomenon that takes place in the complex interaction of diverse actors rather than in the isolated and controlled environment of researchers. Both of these perspectives recognize scientific knowledge coming from research organizations and other sources as an important, but not the only, input for innovation. Both, in principle, recognize farmers' innovative capacity; but the specific cases reviewed for this chapter suggest that neither perspective gives much attention to recognizing and developing this capacity. The problem may lie more in the methodological limitations of the studies than in the principles and assumptions of the theories. For example, the PID approach, which focuses on local people's innovative capacity, was essentially developed on the basis of the AKIS perspective but also shares many values and principles with AIS in the sense that it seeks diverse types of innovation rather than only technological ones.

Major differences

The main differences between the two perspectives are in emphasis and choice of areas of interest rather than in basic philosophies and principles. However, these differences have implications for policy formulation and institutional change and therefore cannot be ignored. Elucidating them will help to identify possible areas of methodological synergy to explain innovation systems that have a mix of the interest areas of AIS and AKIS.

Public institutions play a strong role in the innovation process from the AKIS but not from the AIS perspective. AKIS focuses on the linkages between the main public institutions relevant for agricultural innovation, while AIS addresses a wider set of actors, including the private sector and NGOs. In terms of technological innovation, the private sector takes the lead in the AIS case studies through building own capacities and buying in services. It depends very little upon public research institutions. Commercially important agricultural commodities with high value in national and global markets have obviously attracted the attention of the case study authors. Choices have to be made between developing commercial goods (in which AIS is most interested) and targeting public goods, which AKIS often does. The questions regarding policy choices may be:

- Should countries focus on development led by the private sector, with macro-economic and investment policies favouring this sector and a diminishing role of public institutions?
- Should countries have a strong public sector fostering production of necessary goods and services to people, possibly narrowing the chance for the private sector to develop?
- Should they have a combination of both?

The choices would depend upon the socio-economic level, governance system and political ideologies of the country. In the AKIS study, Rivera et al (2005) state that the private sector can play an increasingly important role in rural knowledge systems; but total privatization is not possible, even for commercial farming. The appropriate mix of public and private roles can best be determined through piloting and learning from experience. On the other hand, AIS thinkers focus on creating marketable value. Therefore, from the two sets of case studies, the main lesson drawn is: choice for AKIS or AIS frameworks to study national- or sector-level innovation systems depends upon the policy choices of the governments for development.

Unlike AIS, AKIS gives much emphasis to facilitating innovation and not just describing the process of how and why innovation occurs. This is an obvious advantage of AKIS over AIS as the approach helps to enhance innovation processes. RAAKS, PTD and PID all aim to facilitate innovation. Some authors have recognized the limitations of the AIS perspective in not being able to offer ways of doing this. Hall (2006), while drawing key insights from the case studies conducted on AIS, stated that:

> ...innovation can be based on different kinds of knowledge possessed by different actors: local, context-specific knowledge (which farmers and other users of technology typically possess) and generic knowledge (which scientists and other producers of technology possess). In an ideal innovation system, a two-way flow of information exists between these sources; but in reality this flow is often constrained because information is embodied in different actors who are not networked or coordinated. A central challenge in designing innovation systems is to overcome this asymmetry – in other words, to discover how to bring those possessing locally specific knowledge (farmers and entrepreneurs) closer to those possessing generic knowledge (researchers or actors with access to large-scale product development, market placement or financing technologies).

Because AKIS emerged to overcome the limitations of the NARS perspective on technology development, it is preoccupied with issues of how technologies are developed in a different way than in NARS. This history has kept AKIS focused on technological innovation, to which the major issues addressed by AKIS – knowledge and institutional change – are strongly related. In contrast, at least in theory, AIS encompasses technological, market-related, financial, institutional and other forms of social innovation. The PID approach, which grew out of the AKIS perspective, puts much the same emphasis as does AIS on going beyond only technological innovation; but a great deal of methodological and empirical work is still needed to put this into practice.

CONCLUSIONS

Because its history is rooted in the industrial innovation systems thinking, AIS is more attracted to commercial sectors, where market features are more important than aspects related to social capital. The cases selected for the AIS study were either commercially well developed and employed numerous people, sometimes including smallholder commercial farmers, or a rapidly growing sector that seized market opportunities at local and national level. The selection criteria for the sectors confirm that the target group addressed by AIS is not necessarily 'farmers' but all people involved in farming, processing, packing, trading/entrepreneurship, financial activities, brokering, etc. In contrast, farmers are at the centre of the AKIS perspective, and the institutions with which AKIS deals – the public research, extension and education organizations – are traditionally concerned with farmers' issues. The AIS case studies showed that public research organizations gave little or no support to most of the sectors considered. The emphasis of AKIS is not on commercial organizations but on small-scale farmers who are more linked with natural resources such as soil, water, vegetation and animals. In most cases, the farmers in developing countries are managing common property resources such as land, water and forest. Issues of sustainability and community empowerment are very important where NRM, poverty and environmental aspects are strongly connected. This leads us to conclude that the AKIS perspective is more relevant for grassroots innovation systems in sub-Saharan Africa.

It must, however, be noted that (in food-insecure parts of sub-Saharan Africa) the issue of market was often regarded as a second-generation problem, and both state and non-state actors concerned with food production paid little attention to it. Meanwhile, it has become evident that global changes in market regulations and government policies on market liberalization require changes in the conception and practices of innovation processes. Therefore, the addition of AIS on AKIS – especially its attention to market forces and more diverse actors and target groups – is very important. This suggests that the greater emphasis of AKIS on farmers, NRM and related sustainability and community-empowerment issues should be supplemented with important concepts in AIS about multiple sources of innovation and different types of innovation (technical, institutional, social, etc.). PID could, in principle, accommodate the basic concepts of both AKIS and AIS, but the approach would need further improvement to encompass the private sector and market analysis.

Both the AIS and AKIS perspectives still need to answer the question: should agricultural innovation performance be measured just by technical and institutional capabilities to yield economic benefits, or should they also be assessed from the standpoint of local and global environmental concerns, which also entail elements of sustainability and empowerment of the resource users? The environment in the grassroots innovation systems of sub-Saharan Africa is under great danger because of farmers' needs to achieve food security, sometimes at the expense of the environment. Empirical studies that show the relationship and interdependency of the innovation elements, including technical competencies, economic benefits and environmental friendliness (both natural and social), may help to answer this question.

REFERENCES

Anandajayasekeram, P., Dijkman, J. and Workneh, A. (eds) (2005) *Innovation Systems Approach and Its Implication to Agricultural Research and Development*, International Livestock Research Institute, Addis Ababa, Ethiopia

Drucker, P. F. (1985) *Innovation and Entrepreneurship: Practice and Principles*, Harper and Row, London

Engel, P. (1997) *The Social Organization of Innovation*, Royal Tropical Institute, Amsterdam, The Netherlands

Hall, A. (2006) *Enhancing Agricultural Innovation: How to Go Beyond the Strengthening of Research Systems*, World Bank, Washington, DC

Leeuwis, C. (2004) *Communication for Rural Innovation: Rethinking Agricultural Extension*, third edition, Blackwell, Oxford

List, F. (1841) *Das Nationale System der Politischen Ökonomie*, translated into English by Sampson S. Lloyd (1885) as *The National System of Political Economy*, available at http://socserv2.socsci.mcmaster.ca/~econ/ugcm/

Reijntjes, C., Haverkort, B. and Waters-Bayer, A. (1992) *Farming for the Future: An Introduction to Low-External-Input and Sustainable Agriculture*, Macmillan, London

Rivera, W., Omar, K. and Mwandemere, H. (2005) *An Analytical and Comparative Review of Country Studies on Agricultural Knowledge and Information Systems for Rural Development (AKIS/RD)*, FAO, Rome, Italy

Röling, N. (1992) 'The emergence of knowledge systems thinking: A changing perception of relationships among innovation, knowledge process and configuration', *International Journal of Knowledge Transfer and Utilization*, vol 5, no 1, pp42–64

Röling, N. (1996) 'Towards an interactive agricultural science', *Journal of Agricultural Education and Extension*, vol 2, no 4, pp35–48

Spielman, D. (2005) 'Innovation systems perspectives on developing-country agriculture: A critical review', ISNAR Discussion Paper 2, International Food Policy Research Institute, Washington, DC

Waters-Bayer, A. and Bayer, W. (2005) 'The social dimensions in agricultural research and development: How civil society fosters partnerships to promote local innovation by rural communities', International Conference on Agricultural

Research for Development: European Responses to Changing Global Needs, 27–29 April, Zurich, Switzerland

Waters-Bayer, A. and van Veldhuizen, L. (2004) 'Promoting local innovation: Enhancing IK dynamics and links with scientific knowledge', *IK Notes* 76, World Bank, Washington, DC (www.worldbank.org/afr/ik/default.htm)

CHAPTER 4

Applying the Innovation Systems Concept in the Field: Experiences of a Research Team in Uganda

Chris Opondo, Conny Almekinders, Jeremias Mowo, Rogers Kanzikwera, Pauline Birungi, Winnie Alum and Margaret Barwogeza

INTRODUCTION

The last two decades have witnessed renewed emphasis on the need for agricultural research and development to be more effective. Holistic and integrated approaches are advocated to meet the needs of farmer clients in their complex systems. Recently, the concept of innovation systems is gaining ground in agricultural research programmes. While the introduction of participatory approaches to technology development during the 1980s already meant a paradigm shift away from the conventional reductionist linear approaches (Chambers and Jiggins, 1987), the innovation system approach goes further in recognizing complexity and the need for more holistic perspectives to agricultural development (Hagmann, 1999; IAC, 2004; Dantas, 2005; Spielman, 2006; World Bank, 2006). In an innovation approach, technology development is an important ingredient in the process of bringing about agricultural development; but a range of other elements needs to be in place to make the improved technology meaningful to farmers.

The innovation systems approach has implications for the role of researchers. In the conventional approach, researchers are the providers of improved technology. In participatory approaches, much emphasis is placed on the researcher–farmer interaction and how they can jointly arrive at adapted technologies that are effective in the farmers' environment. In the innovation systems approach, the emphasis is on bringing all relevant players together in the process, not only farmers and researchers. The players, or stakeholders, when brought together presumably complement each other's knowledge and capacities, align their interests and commit to joint goals.

For researchers, the innovation systems approach implies a different way of doing research and a different role in relation to multiple new partners. It requires applying concepts and skills that researchers are normally not trained in. Although much importance is given to this approach, little is reported on how researchers who currently form the cadre of the institutions that have to transform and apply these new approaches can be trained for the new role that they are expected to play, what the new way of working means in day-to-day activities and what personal challenges are involved.

This chapter illustrates how intended transformations associated with adopting an innovation systems approach are put into practice by research teams in the context of rural Uganda. It aims to provide insights into enabling and impeding factors in bringing about such transformations. The experiences of researchers from the Bulindi Zonal Agricultural Research and Development Institute (ZARDI), part of the National Agricultural Research Organization in Uganda, are discussed. The chapter concludes that developing the competences for facilitating an innovation systems approach requires not only training, but also continued reflection, stimulation and learning in order to put these concepts and skills into practice, especially in an institutional environment that is still far from conducive.

RATIONALE FOR CHANGE

Shifting paradigms in agricultural research systems in Africa

The history of agricultural research in Africa represents a long-lasting search for a research paradigm that has the expected impact upon farmers' livelihoods (Byerlee, 1998; Eicher, 1989). Pursued by the perception that the impact of agricultural research has been unsatisfactory, agricultural research is pressed for tremendous reforms in this region (ASARECA, 1997; Omamo, 2003; Schreiber, 2003; IAC, 2004; FARA, 2005). The lack of clear articulation of impact from research investments has been raised as a concern by governments, donors and civil society alike and is leading to reduced budget allocation and to alternative funding mechanisms, such as the competitive grants schemes. In response to these concerns, a new way of working is being advocated that encourages researchers to design research and innovation processes that go beyond experiments and participation along linear models being promoted in NARS. The associated institutional change process calls for major reforms, such as:

- pluralism in agricultural research, reducing the dominance of public institutions and encouraging the formation of partnerships and innovation systems;
- decentralization of agricultural research – geographically and in decision-making;

- new funding mechanisms that encourage competition for research funds;
- realignment of research programmes to address and integrate research and development issues in order to ensure efficient value addition in commodity chains;
- establishment of stronger innovation systems that will enhance the scaling out/up of successful technologies (ASARECA, 2005).

The concept of innovation systems approach

The concept of 'innovation' refers to the search for, development, adaptation, imitation and adoption of technologies that are new to a specific context. The concept is based on the premise that an effective innovation system is one in which information flows between actors, allowing for new knowledge to be generated in the context of application (Gibbons et al, 1994; Biggs and Matsaert, 2004). An innovation system can therefore be seen as a network of actors that are directly involved in the creation, diffusion and use of new knowledge and technology, as well as the actors responsible for the coordination and support of these processes.

An approach that puts the 'innovation systems' concept central to the production of knowledge for agricultural development and change has gained ground in policy and academic circles over the last two decades (Dantas, 2005). It has been endorsed by an array of international and national bodies as well as non-governmental organizations (NGOs) and governments in both developed and developing countries. The approach presumes a major change in the way in which knowledge and technology are generated, viewed and utilized by different actors. It shifts attention away from research and the supply of scientific knowledge and technology to a process of change in which research for technology development is only one element.

The innovation systems approach in agricultural research and development takes an action and systems perspective (Hall et al, 2001; Flood, 2002; Biggs and Matsaert, 2004). It tries to overcome the weaknesses of the more reductionist and linear approaches of the conventional scientific paradigm. In response to the need to show impacts in non-conventional research within complex environments, NARS in Africa was compelled to take on the innovation systems concept.

Capacity-building for innovation systems approaches

The National Agricultural Research Organization (NARO) introduced the innovation systems concept as early as 2000 to make research relevant and impact oriented (Alacho, 2003; Clesensio, 2003). While reviewing the reform process in NARO, Hagmann and Blackie (2002) noted that in Uganda, as in other African countries, the process of generating knowledge and technology had become more 'scientific' and 'academic' and thus remained separated from the users. The division of research and extension had become increasingly strong and inhibited effective feedback loops in the system – with the same result as in most African research systems until today: low impact and supply- and discipline-driven agendas, rather than an interdisciplinary response to demand-led challenges that are cross-disciplinary and systemic by nature. Over time, the technology generation process had become bureaucratized, compartmentalized and control oriented.

A new initiative to reform NARO began in 2003. It was an initiative of NARO, Makerere University (MAK) and the International Centre for development oriented Research in Agriculture (ICRA), with support from the UK Department for International Development (DFID). The initiative called Learning Together endeavoured to build effective research systems within NARS that embrace the innovation systems concept, integrated agricultural research for development (IAR4D) and integrated natural resource management (INRM) approaches, thus addressing the multifaceted problems and needs of farmers and enabling stakeholder participation. The assumption was that the new approaches would help to avoid the 'projectization' of research through a multitude of interests from donors and other investors and from supply-driven disciplinary interests of researchers (Kibwika, 2006).

The overall objective of the initiative was to strengthen human and institutional capacity to undertake IAR4D as a new way of doing business. The more specific objectives of the initiative were to:

- enhance and mainstream the capacity of research teams to apply IAR4D approaches as a new way of working within NARO;
- strengthen and institutionalize the ability of MAK to provide capacity-enhancing opportunities in IAR4D for a range of stakeholders at various levels.

There are relatively few initiatives that have designed a process to prepare researchers and their managers to tackle the challenge of embracing new research and development approaches. NARO, ICRA and MAK designed a training trajectory with the support of an external consultant for process facilitation. A project implementation team and a group of workshop facilitators supported the training trajectory and provided mentoring to the trainees. Funding for this training was drawn by NARO through its various donor-supported projects and programmes, with ICRA providing staff time and materials free of charge.

A series of five residential learning workshops for researchers and managers of ZARDI teams was implemented from April 2004 to February 2005. In these workshops, the innovation systems concept was further elaborated and operationalized by the research teams. The new approach focuses on making NARS work as a whole, rather than strengthening certain components of it, as has been the case for many years. The synergistic interplay and the relationships between the components of the system are the crucial aspects to reach effectiveness and impact in the new NARS. Researchers in learning workshops realized that the boundaries and the 'members' of the new system are flexible, depending upon the product to be developed or the desired outputs/results. Pure research is just one component that cannot function unless all of the other components work and interact with each other.

The workshop participants acquired new skills and knowledge in aspects such as linking research to markets, implementing participatory monitoring and evaluation, and team and partnership building. The residential workshops were interspersed with application phases in the field, where the participating zonal teams applied the acquired skills and knowledge on their own or, in some cases, with coaching from a team of mentors. In the consecutive workshops, the participants reflected on their field application experiences and problems and received further training in preparing for the next field phase.

After each of the five learning workshops, members of the project implementation team and the workshop facilitators internally reflected on progress using three questions: what was going well, what was not going well and what modifications were needed.

PUTTING LESSONS INTO PRACTICE

Each ZARDI team that participated in the learning trajectory attempted to put their learning into practice. This chapter describes some reflections of the zonal team from Bulindi collected through focus group discussions and semi-structured interviews, with key informants working with the respective ZARDI.

'Before' situation

Back at their research station after the learning workshops, the three trainees from the research team of Bulindi ZARDI reflected on their current research practices under the conventional paradigm. They concluded that yield-increase goals dominated their research practice and that their research mode was linear (see Figure 4.1). It represented a relatively simplistic system of technology development that only involved researchers, extensionists and farmers. Most importantly, researchers were viewing themselves as the source of innovations, knowledge and approaches. Feedback from farmers to researchers was weak, if not absent.

The research team identified what they perceived as major limitations in this approach:

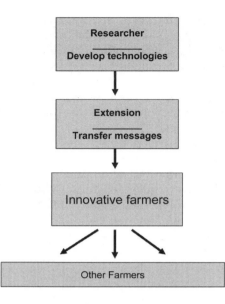

Figure 4.1 *Simplistic conventional research system with technology flowing from researchers to extensionists to farmers, with relatively weak feedback from farmers*

Weak communication and interactions with farmers
The communication between researchers and farmers was weak. As researchers, they tended to be 'locked up' in research institutes and had limited interaction with farmers or private enterprises. The only meetings with farmers were to explain what data to collect, to check on the experiments and to pick up the data sheets. One of the leading team members pointed out:

I have worked in a potato programme in which we generated good experimental data with support from farmers. The ideas for experimental treatments came from us or our colleagues, and we planned our work in order to get the required experimental data. There was little incentive or opportunity to apply our findings in a commercial context or to consider the needs of farmers and export markets. We also had rigid institutional distinctions between research and extension organizations that tended to reinforce this separation of research and application. We were expected not to tread on the extensionists' domain. The extension work was the intermediary of communicating and delivering information and technology to the farmers and then getting feedback to the researchers. But the extension staff we worked with were constrained in logistics; they did not have field cars, gasoline or travel allowances. As a consequence, the interactions with farmers and the feedback were weak. And input of the extension staff, the people who presumably were closer to the farmers than the researchers, was little if any.

Organization of research according to disciplines
The team emphasized the commodity orientation of the research work on the station undertaken by commodity research departments. The researchers supplied the technologies (or the information on these) to the extension officers who, in turn, transferred these to the farmers. This left the extensionists and farmers with the task of integrating these technologies within the complex realities of farm activities and livelihoods. The major concern of the researchers was to generate high-quality scientific data in their discipline to allow for scientific publications. Much of the research was therefore conducted on station. Those experiments conducted in the community were at plot level, only addressing the needs of that individual farmer. Interactions with research colleagues involved with other commodities were minimal. Researchers from the different sections and departments met occasionally during station review and planning meetings. Findings presented by one researcher at these meetings were not always relevant to the others. For instance, the results of the treatments of spacing potatoes or beans presented by one researcher had no particular relevance to another in whose area farmers were using mixed cropping patterns. Extensionists therefore found it difficult to translate these research results into messages that made any kind of sense to farmers.

Hierarchical structure and separated mandates
Decision-making in the organization is hierarchical and centrally directed, from the directors at the headquarters to the station managers and eventually to the researchers at the research station. The researchers gave prescriptive instructions to extensionists and, at times, to farmers on how to conduct experiments. The Bulindi team also noted the separation of roles and mandates between Public Agricultural Research Institutes (PARIs) for strategic research and the Zonal Agricultural Research and Development Institutes (ZARDIs) for downstream evaluation and diffusion. This created an 'artificial rift' in the operations of the two institutions and the outreach to farmers. For example, Kawanda and Namulonge stations are the lead PARIs that would 'develop' the technologies and 'bring them' to the ZARDIs, where the researcher assigned to that particular commodity would facilitate the evaluation of the technologies on station or contract farmers within the community to host experiments.

The team felt, in retrospect, that this, indeed, represented a system of supply-driven technology generation and testing. Any researchers joining NARS followed the normal line of command and aligned themselves with how others were 'going about things'. This reinforced existing mindsets, confirming the paradigm that structure determines people's attitude and practice, and where people continuously remake the structure. According to the team, a strong driver in this way of doing research is the reward mechanism within the organizations, which was (and still is) based on the technologies released and the number of scientific papers published.

Professional value
Researchers derived their professional value by being recognized as experts in their particular scientific disciplines. During discussions, the researchers agreed that in conducting research experiments on farmers' fields, they were the experts who either trained farmers or gave farmers instructions based on their scientific discipline. Some views expressed by the team on this method of research are found in Box 4.1.

Box 4.1 *Views on the conventional research system*

The team leader of the Bulindi team said that the researchers conducted their experiments based on their work plans, which they followed rigidly. No farmer meetings were conducted to solicit farmers' opinions on the constraints they faced and how to address these constraints. Bulindi and many other district institutes were used as demonstration centres for the technologies and information coming from the NARIs. In addition, the prevailing extension approach was transfer of technology, top-down, and a linear model of reaching farmers with the belief that research has the answers to their production problems.

Short-term blanket recommendations were given to farmers with hardly any room in the process for the voice and concerns of farmers. Social aspects such as partnership policies, collective marketing and management of technologies for equitable sharing of benefits were not catered for.

A Bulindi team member added that, in this reductionist approach, researchers controlled experiments and extracted information from farmers that was of interest for communication with peers in the scientific community The local context within which the farmers operated was largely ignored and farmers were considered to be people who knew nothing.

The researchers and managers had observed that many technologies released or recommended by them were not utilized by the farmers and felt that their work did not make a difference to the farmers. This was, however, not a major concern since they met the organizations' policies and conditions. Looking back, they felt there was nothing to be proud of as professionals. Their ambitions had been related to moving up the career ladder towards a better salary and a posting on a research station closer to Kampala.

'Now' situation: From commodity to innovation systems

The Bulindi team indicated that working along the lines of the innovation systems concept brings many non-traditional partners into the research process and many associated challenges. Although NARO policy was putting emphasis on integration and teamwork, they were not fully prepared for the challenges involved. The team leader mentioned that the researchers were left to organize themselves as well as the other stakeholders. The team recognized that they were very much on their own in trying to implement the new approaches, which went together with a changed mindset.

Researchers in the ZARDI Bulindi team now look at their research partners in a different way. During discussions, the team members indicated that, in trying to follow up on the assignments of the learning workshops, the abstract had become real. They have since experienced that it is not an easy job to get all of the partners on board for joint agenda setting. In practical terms, this meant that they had to meet all partners twice or more before the meeting – first to invite the partners and then to follow up on the invitation and pursue their attendance. In Bulindi, telephone and internet connectivity are unpredictable and the research budgets are tight and hardly allowed for transport to mobilize these various partners. The team also found that partners were not always interested in attending meetings on account of their busy agendas. This includes farmers, who in addition need to be transported to and from the meeting venue. Then, in the meetings, it is obvious that the diverse partners come with different perceptions, experiences and expectations. Theoretically, such meetings should generate institutional synergies while minimizing duplication of efforts. In practice, however, it is a very difficult exercise to maintain focus in the discussions and have meaningful interaction and outcome. The research managers argued that, with the innovation systems approach, social and technological institutions become larger and more complex, requiring substantial competence of the researchers to coordinate and facilitate the process. Despite the learning workshops, the researchers did not feel sufficiently competent in hosting such events, in being effective facilitators and in providing their disciplinary expertise. The lack of financial resources did not make the task any easier.

Box 4.2 provides some views of the team and farmers on the new ways of working within an innovation systems approach.

Changed research orientation

Within NARO, new themes have been developed as a mechanism to foster new ways of working that would bring changes: understanding people, livelihoods and the impact of technologies; strengthening innovation processes and partnerships; enhancing natural resource management; developing technologies that respond to markets; and linking producers, markets and policy. These research themes emphasize stakeholder involvement and consultations in research in order to develop relevant technologies. The transformation of the research work on beans is a concrete example of how the research orientation has changed.

Box 4.2 *Views of research team members and farmers
on the new ways of working*

The new ways of working include the application of new skills by the research team to broader topics than before. These include, for example, collective action for farmer-group marketing; managing production–market links; facilitating community and team meetings; participatory diagnosis; planning, monitoring and evaluation of technologies with farmers; and influencing policy-makers through feedback to programme director and senior management.

A technician stated:

> When farmers are happy, I am happy – more so when farmers tell me that they have improved their knowledge and farming practice as a result of my interactions with them on a given technological intervention. Farmers who used to host experiments for us now see a clear change in the way we interact with them. We go to them and ask their views on how the technologies are performing; we also encourage them to make modifications and adjustments on agronomic practices to fit their interest.

A leader of a farmer group explained:

> I had never seen or known what technologies NARO people have. Now I know, as I interact with the Bulindi team in our experimental plot or during our monthly group meetings. We also go to the station on field days or for annual planning meetings and see what Bulindi is doing for us.

Another technician stated:

> While collaborating with the public agriculture research institute's researchers in basic research activities, I used to take new crop varieties to farmers for experimentation, plant them and collect data. The farmers did not know what the experiments were all about as we used to contract their land and hire their labour. Now this has changed.

NARO Bulindi embarked on bean production and commercialization as its main research and development challenge under the NARO/MAK/ICRA learning initiative. This programme replaces the old agronomic and varietal improvement bean work and has two components:

1   working on bean-related innovations with conventional partners (extensionists and farmers) and non-traditional partners (ActionAid, community-based organizations, Makerere University, regional net-works, microcredit providers and traders in the area); and
2   analysis and design of an integrated bean commodity value chain to track the production–consumption process.

The team reflected on their experiences with the new bean research so far in order to draw some lessons for further improvement of their work.

One thing they noted was that the first stakeholder meeting that was organized to plan and implement the bean research actually initiated a kind of 'platform'. The stakeholders shared experiences that were useful to the researchers for defining new research proposals.

Second, the researchers have come to realize that the bean chain involves many actors, starting at the production site on the farm up to the consumers in and outside the zone. They have also observed that actors who operated in an *ad hoc* manner became organized once a clear market for beans was evident and guaranteed. The research team members have now experienced first-hand how putting local people's perspectives at the centre of research efforts creates ownership and effective collaboration with outsiders (Probst and Hagmann, 2003). A researcher in the team added:

> We still develop our normal annual work plans; but we do not stick to them as strictly as we used to. We change the planned activities according to new farmer priorities or as issues emerge. Currently, the zone is affected by banana bacterial wilt, which we did not anticipate. We are responding by giving information on how to best manage it. There was an outbreak of poultry disease this year, and we responded by meeting farmers and discussing better options for poultry management. These are topics we now work on, but were not planned for. They also are topics we are not prepared for. When the farmers ask us questions on topics of which we know little, we take the queries with us, we consult with experts on the station or sometimes in Kampala. We then bring this information to the farmers in subsequent meetings. Previously, we were only concerned with our own crop or disciplines and considered ourselves the experts. Now, we feel more open and do not play the expert anymore. Currently, we are dealing with issues related to new high-value crops such as vanilla, and market outlets for mushroom and *Aloe vera*. Some of these technologies have been introduced not by us but by NGOs, and are not within the mandate of ZARDI-Bulindi. Now, we go beyond our mandate to provide farmers with answers to retain our partnership with them and our desire to see impact and improved livelihoods.

The teams feel that the need to demonstrate impact is the key driving force behind the changed approach to their work. Other driving forces mentioned are lack of sufficient research funding and the demand from farmers. As for the latter, the researchers feel relieved that they now actually have the space to respond to farmers' needs.

CHALLENGES FOR RESEARCH TEAMS

Capacity

The new ways of working call for new skills, value systems and professionalism. The researchers and managers argued that skills for forging partnerships and for linking research outputs with policies and markets are needed since they were not trained in these activities. The learning workshops only addressed IAR4D topics and introduced the research teams to related concepts and skills. The discussions with the teams show that the workshops have had an impact upon the attitude and mindset of the researchers and managers. There is gradual appreciation of new research approaches. However, they also show that the skills to operate with this new attitude and mindset, and to put into practice the associated concepts and approaches, are still limited. Introduction to the coordination and facilitations skills in the learning workshops and the initial experiences in putting these into practice in Bulindi have made the teams aware of the level of skills required for satisfactory functioning. They are convinced that practice will further develop their skills.

Inter-institutional partnerships

The ZARDI teams concurred through small group discussion and a plenary debate in one of the learning workshops that forging and sustaining multi-institutional partnerships is crucial in operationalizing an innovation systems approach. The teams noted that weak partnerships exist because of incompatibility of vision and ways of conducting business (approaches to working with communities and time/activity schedules) among partners and time-consuming bureaucracy for approving procedures when ideas come from the bottom of the hierarchy.

Market orientation

All of the research stations in NARO are striving to ensure that the technologies they develop have a market value and can generate an income for farmers. However, the researchers, in general, and the biological scientists, in particular, do not have adequate skills to undertake market-oriented research by the nature of their professional training. Although they were introduced to market analysis and integration in the learning workshops, the researchers felt that one-time training on these topics is not adequate. They were also of the opinion that the current market-orientation policy is biased towards promoting export crops at the expense of subsistence production. This raised the question among researchers, farmers and extension staff about the future of food crops.

From these discussions, it becomes apparent that the researchers and their managers still feel that they have limited capacity in a number of new skill areas that are necessary in order to work successfully with the new concepts, such as team and partnership management; facilitation skills; general management skills; feedback culture; action research; process documentation; facilitation; and communication and negotiation. Understanding key concepts, such as innovation systems, IAR4D, INRM and market-led research, and integrating these within the design of research protocols come from building up their own experiences. These are not concepts and skills that they can come to grips with by only reading books or attending workshops intermittently, but by actually putting them into practice repeatedly. Implementation and experimentation do, however, require resources and professional space that can be provided only by committed research managers and policy-makers.

CONCLUSIONS

With this reflection on the experiences of the Bulindi team, the authors have attempted to share practical, on-the-ground implications of applying the innovation systems concept in the context of NARS in Africa. To the research team itself, it has become clear from these reflections that the learning workshops had an impact upon their own attitude and mindset. However, they have also realized that their skills to work in line with this changed attitude and mindset are still not yet well developed and that putting the new concepts into practice is very difficult.

The learning workshops addressed these topics and introduced the teams to the new concepts and skills. The coaching sessions in between the workshops were helpful in further developing these skills; but the teams felt that they still needed more practice. The experiences gained so far made them aware of what 'systemic' learning is: it does not go via books and short exposures, but through learning by doing. Now, after the series of learning workshops has finished, they are pretty much on their own, with multiple constraints to deal with.

What may be the most important gain from the learning workshops is taking on a more critical and reflective mode of thinking. This is the first and most essential ingredient for becoming a 'learning team'. This will help them to continue to learn and hopefully prevent them from falling back into the old routines, which is all too easy within an institutional structure that has not yet fully aligned itself to these new ways of working. For example, the reward structure is still based on 'scientific output'. An issue that the researchers are struggling with is to see the 'science' or research element in these multi-actor activities. Opportunities to publish or present their results from work with farmers are not abundant.

Apart from this desire to learn more to better contribute to research and development, other challenges are emerging or waiting to be addressed. How, for example, should indigenous knowledge and practices be integrated within market-led research? How can short-term advantages be created in order to keep farmers interested in longer-term processes that have less tangible results? Assuming that, in the future, they will not only be assessed as professionals on scientific output, how can one provide adequate and convincing evidence for the impact of their work? It is obvious that many of these challenges and questions go beyond the scope of research only and require modifications throughout society.

Many challenges also touch the researchers deeply, both in their belief systems and through the institutional–scientific cultures in which they have been raised. The change in rhetoric and the paradigms being advocated are obviously a first step; but the ongoing steps have to be made by the researchers themselves, under conditions which are not yet very supportive. Policy-makers and research managers need to be mindful of the great distance between the drawing board, where ideas for transformation are conceptualized, and the reality of the researchers in their stations and in the farmers' fields.

REFERENCES

Alacho, F. (2003) 'Meeting the needs of clients in research: A preliminary overview of NARO'S experiences', *Uganda Journal of Agriculture Sciences*, vol 8, no 10, pp11–16

ASARECA (1997) *Strategic Plan*, ASARECA Secretariat, Entebbe, Uganda

ASARECA (2005) *Strategic Plan 2005–2015*, ASARECA Secretariat, Entebbe, Uganda

Biggs, S. and Matsaert, H. (2004) 'Strengthening poverty reduction programmes using an actor oriented approach: examples from natural resources innovation systems', Agricultural Research and Extension Network (AGREN) Paper 134, Overseas Development Institute, London

Byerlee, D. (1998) 'The search for a new paradigm for the development of national agricultural research systems', *World Development*, vol 26, no 6, pp1049–1055

Chambers, R. and Jiggins, J. (1987) 'Agricultural research for resource-poor farmers; Part 1: Transfer of technology and farming systems research, Part 2: A parsimonious paradigm', *Agricultural Administration and Extension*, vol 27, pp35–52, 109–128

Clesensio, T. (2003) 'Delivering the promise: Impact of NARO and its technologies on agriculture research in Uganda', *Uganda Journal of Agriculture Sciences*, vol 8, no 10, pp1–10

Dantas, E. (2005) 'The "system of innovation" approach, and its relevance to developing countries', www.SciDev.net

Eicher, C.K. (1989) 'Sustainable institutions for African agricultural development', Working Paper 19, International Service for National Agricultural Research (ISNAR), The Hague, The Netherlands

FARA (Forum for Agricultural Research in Africa) (2005) *Strategic Plan*, Forum for Agricultural Research in Africa, Accra, Ghana

Flood, R. L. (2002) 'The relationship of "systems thinking" to action research', in P. Reason and H. Bradbury (eds) *Handbook of Action Research: Participative Inquiry and Practice,* Sage, London

Gibbons, M., Limoges, C., Nowotny, H., Schwartzman, S., Scott, P. and Trow, M. (1994) *The New Production of Knowledge: The Dynamics of Science and Research in Contemporary Societies*, Sage, London

Hagmann, J. (1999) *Learning Together for Change: Facilitating Innovation in Natural Resource Management through Learning Process Approaches in Rural Livelihoods in Zimbabwe*, Margraf, Weikersheim, Germany

Hagmann, J. and Blackie, M. (2002) *Report on Institutional Reform and Development Component*, National Agricultural Research Organization, Entebbe, Uganda

Hall, A. J., Sivamohan, M. V. K., Clark, N., Taylor, S. and Bockett, G. (2001) 'Why research partnerships really matter: Innovation theory, institutional arrangements and implications for the developing new technology for the poor', *World Development*, vol 29, no 5, pp783–797

IAC (InterAcademy Council) (2004) *Realizing the Promise and Potential of African Agriculture*, InterAcademy Council, Amsterdam, The Netherlands

Kibwika, P. (2006) 'Challenges of demand driven research and extension services in Uganda: Implications for competence', in P. Kibwika (ed) *Learning to Make Change: Developing Innovation Competence for Recreating the African University of the 21st Century*, Wageningen Academic Publishers, Wageningen, The Netherlands, pp85–105

Omamo, S. W. (2003) 'Policy research on African agriculture: Trends, gaps and challenges', Research Report 21, ISNAR, The Hague, The Netherlands

Probst, K. and Hagmann, J. (2003) 'Understanding participatory research in the context of natural resources management: Paradigms, approaches and typologies', AGREN Paper 130, Overseas Development Institute, London

Schreiber C. (2003) 'Sources of innovation in dairy production in Kenya', Briefing Paper 58, ISNAR, The Hague, The Netherlands

Spielman, D. J. 2006. 'A critique of innovation systems perspectives on agricultural research in developing countries', *Innovation Strategy Today*, vol 2, no 1, pp41–54

World Bank (2006) *Enhancing Agricultural Innovation: How to Go Beyond the Strengthening of Research Systems*, World Bank, Washington, DC

CHAPTER 5

Developing the Art and Science of Innovation Systems Enquiry: Alternative Tools and Methods, and Applications to Sub-Saharan African Agriculture

David J. Spielman, Javier Ekboir and Kristin Davis

INTRODUCTION

Agricultural education, research and extension can contribute substantially to enhancing agricultural production and reducing rural poverty in the developing world. However, evidence suggests that their contributions are falling short of expectations in sub-Saharan Africa. The entry of new actors, technologies and market forces, when combined with new economic and demographic pressures, suggests the need for more innovative and less linear approaches to promoting a technological transformation of smallholder agriculture. This chapter explores methodologies that can help to improve the study of agricultural innovation processes and their role in transforming agriculture in sub-Saharan Africa. Specifically, this chapter examines methods that address three key issues:

1 how agents interact in the production, exchange and use of knowledge and information within a system;
2 how agents respond individually and collectively to technological, institutional or organizational opportunities and constraints; and
3 how policy changes can enhance the welfare effects of these interactions and responses.

Methods for further exploration include social network analysis, innovation histories, cross-country comparisons and game theoretic modelling.

Recent attention given to these issues has focused on the wider 'innovation system', an increasingly popular concept in the study of how societies generate, exchange and use knowledge. An innovation system is broadly defined as the set of agents involved in an innovation process, their actions and interactions and the socio-economic institutions that condition their practices and behaviours (Freeman, 1987; Lundvall, 1992). The framework embeds technological change within a larger, more complex system of interactions among diverse actors, organizational cultures and practices, learning behaviours and cycles, and rules and norms.

Importantly, the innovation systems framework shifts the analytical emphasis from a conventional linear model of knowledge and technology transfer (from researcher to extension agent to farmer) to a more complex process-based systems approach. This shift is appropriate for the study of agriculture in sub-Saharan Africa given that the sector's growth and development are increasingly influenced by complex interactions among public, private and civil society actors, and by rapidly changing market and policy regimes that affect knowledge flows, technological opportunities and innovation processes.

Nevertheless, early applications of the innovation systems framework to developing country agriculture suggest that opportunities exist for more intensive and extensive analysis. There is scope for empirical studies to employ more diverse methodologies, both qualitative and quantitative, than are being used at present. In addition, there is room to improve the relevance of empirical studies to the analysis of poverty reduction and economic growth.

This chapter begins by detailing a conceptual framework based on theories of complex adaptive systems and innovation within these systems. It then looks at methodological challenges for innovation systems studies, followed by a discussion of alternative methodologies and conclusions.

CONCEPTUAL BACKGROUND

Complex systems

The basis for development is the ability of individuals, organizations and societies to improve on what they are currently doing (i.e. to improve their individual and collective capabilities). However, such improvements are contingent upon the environment within which innovation occurs. Individuals and their environments form complex systems characterized by a large number of actors, diverse interactions and relationships, as well as constantly changing influences emerging from technological, market, policy, cultural and other socio-economic factors.

Recognition of development as a complex process can have major consequences for the design and implementation of public policy; but such recognition is still relatively uncommon. Public policies still draw on conventional analyses that are based on modernist metaphors of hydraulics, machinery and factory production – metaphors that continue to exert a profound influence on the design of organizations, institutions and policies. As a result, scientists and policy-makers still emphasize control and predictability, and still design interventions that are built from the top down and are expected to be implemented through passive subordinated structures (Olson and Eoyang, 2001).

Complexity theories, on the other hand, emphasize the importance of self-organization, which results from the diversity of agents and decentralized nature of complex systems. Even though some agents have more influence than others, no agent or group of agents can totally control a system. Thus, policies in a complex system seek not to 'manage' the system but to operate on the probability of events, to increase the odds of desired outcomes and to reduce the chances of undesired results (Axelrod and Cohen, 1999).

In this chapter, we define a complex system as one whose properties cannot be analysed by studying its components separately. While there are several types of complex systems, the most relevant for the study of innovation processes are termed complex adaptive systems (CAS), or systems formed by many agents of different types in which each defines his or her strategy, reacts to the actions of other agents and to changes in the environment, and tries to modify the environment in ways that fit his or her goals (Kauffman, 2000). Behaviour patterns in this type of system often emerge from independent, spontaneous or unintended processes that render conventional mechanistic modes of analysis quite useless.

A CAS evolves through the combination of initial conditions, multiple interactions, trends and random variations in agents and their interactions. The strength of the trends and of the random effects changes along an evolutionary path. When the trends are strong, the CAS is more or less predictable; as the system evolves and new actors and interactions emerge, the CAS becomes less stable. Eventually, the random component may become more important than the trends and, at a certain bifurcation point, the system may become random and unpredictable (Nicolis and Prigogine, 1989).

Systems tend not necessarily towards chaos, but towards a situation that is inherently unstable and unpredictable. At any given moment, random variations occur with varying consequences and varying degrees of predictability. When the trends dominate, the probability that a random variation results in a minor event is high; when variations occur close to unstable configurations, the probability of catastrophic events is great.

Variation, selection and innovation

An important evolutionary force in CAS is the interaction between variation and selection, concepts borrowed from the evolutionary biology literature and characterized as follows. First, while new actors and strategies constantly emerge in a system, not all of them are adapted to the environment; selection enables 'survival of the fittest'. Second, changes in efficiency within these systems are discrete, interrupted by long periods of relative stability. Third, such changes do not stop in periods of stability but continue at least at the same rate as in the periods of adaptive innovation (Crutchfield and Schuster, 2003).

This leads to the notion of innovation. In this chapter we define an innovation as anything new successfully introduced into an economic or social process. In other words, an innovation is not just trying something new but successfully integrating a new idea or product within a process that includes technical, economic and social components.

This definition stresses three important features. First, innovation is the creative use of different types of knowledge in response to social or economic needs and opportunities (OECD, 1999). Second, a trial becomes an innovation only when it is adopted as part of a process; many agents try new things, but few of these trials yield practices or products that improve what is already in use. Third, innovations are accepted as such in specific social and economic environments (Bailey and Ford, 2003).

In the terminology of complexity theory, innovation results not just from variation (trying something new), but also from selection (finding things better than what is currently used) and incorporation within long, complex processes (Nickles, 2003).

Innovation can have an important socio-economic impact only when it is part of sustained processes involving many actors with different capabilities and resources. The reason is that if an innovation improves substantively – say, production – it must be accompanied by new managerial and marketing innovations to handle and sell the extra output.

Since individuals and organizations do not typically possess all the requisite capabilities and resources, they integrate into networks with other actors who can contribute the resources and expertise that they lack (Rycroft and Kash, 1999; Christensen and Raynor, 2003). Thus, a successful innovation process is determined by the extent to which networks gather sufficient variation in capabilities and resources from diverse agents.

Integration into these networks, however, is difficult because of known problems of implementing collective action: the difficulties of agreeing on rules, implementing common procedures, creating trust and monitoring opportunistic behaviour. Thus, networks form on the basis of relationships that evolve among agents and the institutional context within which they form.

Network structure and dynamics depend upon the innovation's complexity and maturity. In the case of simple or mature innovations, networks are loose. Because the economic and technical features of the innovation are relatively well known, members can relate to each other through formal contracts or markets. For new or complex innovations, actors have to interact often and informally to overcome unexpected problems and the technical and market uncertainty derived from the innovation (Rycroft and Kash, 1999; Christensen and Raynor, 2003).

Network effectiveness depends upon the collective capacity to facilitate exchanges of information and resources. In the terminology of network analysis, this capacity is known as the network's navigability and relies on the existence of 'central' actors (i.e. well-connected actors) interacting among themselves (Buchanan, 2002) and on the environment (i.e. laws or markets) in which the networks operate.

Network effectiveness also depends upon the ability of networks to search for and use existing information and, when it is not available, to generate it. This is, in turn, influenced by the network's ability to develop its organizational capabilities, or the individuals, technologies, shared norms and organizational routines needed to communicate information and coordinate resources (Zander and Kogut, 1995; Argote and Darr, 2000; Dosi et al, 2000; Bailey and Ford, 2003).

In the context of developing country agriculture, CAS can be used to describe a system of public extension agents, public researchers, market traders and farmers along with public policies on science, technology, agriculture, education and investment. One of the main hurdles that diminish small farmers' innovative capabilities is their inability to integrate within navigable networks comprised of such agents that provide access to technical and commercial information, markets and financing. Often, small farmers do not have adequate human and social resources to integrate within these networks or do not operate in an institutional environment where such networks easily form.

INNOVATION SYSTEMS AND AGRICULTURAL DEVELOPMENT

The contribution of systems-based approaches

Systems-based approaches such as those described above are not new in the agricultural development literature. The study of technological change in agriculture has always been concerned with systems, as illustrated by applications of the National Agricultural Research System (NARS) and the agricultural knowledge and information systems (AKIS) approaches.

However, the innovation systems literature, with its foundations in complexity theories, is a major epistemological departure from the traditional neoclassical studies of technological change that are often used in NARS- and AKIS-driven research. The NARS and AKIS approaches, for example, emphasize the role of public-sector research, extension and educational organizations in generating and disseminating new technologies. Thus, interventions based on these approaches traditionally focused on investing in public organizations to improve the supply of new technologies.

A shortcoming of this approach is that the main restriction to the use of technical information is not only its supply or availability, but also the limited ability of innovative agents to absorb it (Cohen and Levinthal, 1990). Even though technical information may be free and freely accessible, innovating agents have to invest heavily to develop the ability to use the information (Nonaka and Takeuchi, 1995).

Importantly, while both the NARS and AKIS frameworks made critical contributions to the study of technological change in agriculture, they are now challenged by the changing and increasingly globalized context in which sub-Saharan African agriculture is evolving (Science Council, 2005; World Bank, 2006). This includes such trends as the rapid growth of markets as the main drivers of technological change; new demographic and agro-ecological pressures; new economic regimes such as trade liberalization and regional trade integration; the growth of private investment in, and ownership of, knowledge, information and technology; and expansion of information and communication technology as a means of rapidly exchanging knowledge and information.

Hence, there is need for a more flexible framework to study innovation processes in developing country agriculture – a framework that highlights the complex relationships between old and new actors, the nature of organizational learning processes and the socio-economic institutions that influence these relationships and processes.

This brings us to the agricultural innovation systems (AIS) framework. This framework highlights how individual and collective absorptive capabilities translate information and knowledge into a useful social or economic activity in agriculture. The framework requires an understanding of how individual and collective capabilities are strengthened, and how these capabilities are applied to agriculture.

This suggests the need to focus far less on the supply of information (e.g. brick-and-mortar research organizations and universities) and more on systemic practices and behaviours that affect organizational learning and change. The approach essentially unpacks systemic structures into processes as a means of strengthening their development and evolution.

Applications of the agricultural innovation systems approach

The innovation systems approach is still nascent in the study of developing country agriculture. Biggs and Clay (1981) and Biggs (1989) offer an early foray into the approach by introducing several key concepts – institutional learning and change, and the relationship between innovation and the institutional milieu in which innovation occurs – that become central to later AIS studies.

Later studies by Hall and Clark (1995), Hall et al (1998, 2002, 2003), Johnson and Segura-Bonilla (2001), Arocena and Sutz (2002) and Clark (2002) introduce the innovation systems approach to the study of developing country agriculture and agricultural research systems. Regional and national applications of the innovation systems approach include Chema et al (2003), Peterson et al (2003), Hall and Yoganand (2004), Roseboom (2004) and Sumberg (2005) for sub-Saharan Africa; Vieira and Hartwich (2002) for Latin America; and Hall et al (1998) for India.

Several studies focus on the institutional arrangements in research and innovation – for example, Hall et al (2002) on public–private interactions in agricultural research in India; Porter and Phillips-Howard (1997) on contract farming in South Africa; or Hall et al (1998), Allegri (2002) and Kangasniemi (2002) on producer associations in South Asia and sub-Saharan Africa. Other studies focus on technological opportunities, such as Ekboir and Parellada (2002) on zero-tillage cultivation. A recent study by the World Bank (2006) contributes further to the development of the AIS approach with both conceptual and empirical evidence.

These studies are distinguished from many other works on agricultural research and development because they embed analyses of innovation within the wider context of organizational and institutional change processes. Furthermore, they offer some answers to certain research questions that the conventional literature is often unable to address. For example, Ekboir and Parellada (2002) offer a detailed look into the social and economic changes that encouraged the diffusion of zero-tillage cultivation in Argentina, a process that resulted from a complex series of events and interactions among farmers, farmer organizations, public researchers and private firms. Hall et al (2002) provide an in-depth study of the institutional and organizational learning processes that stimulated the diversification of agricultural research financing in India to include new actors (e.g. medium-sized firms and producer cooperatives) and new modalities (e.g. contract research and public–private partnerships). Clark et al (2003) unlock the mysteries of a successful donor-funded project in post-harvest packaging for small farmers in Himachal Pradesh, India, by studying the institutional learning and change processes that were incorporated within the project design.

Methodological limitations of the agricultural innovation systems approach

The AIS framework faces, however, several methodological limitations in its application to developing country agriculture (Spielman, 2006a, b). First, while the conventional innovation systems approach relies on a diversity of rigorous qualitative and quantitative methods in studies of industrialized countries, the methodological toolkit employed in the study of developing country agriculture remains fairly limited. Currently, the favoured methodology in the study of agricultural research in developing countries is the descriptive case study, typically drawn from an action research or stakeholder analysis exercise (Hall and Yoganand, 2004).

More often than not, studies are simply *ex-post* descriptions of the dynamics and complexities of some technological or institutional innovation. Powerful tools for systematic, replicable and consistent methods of analysis that could be used include in-depth social and economic histories; policy benchmarking, cross-country comparisons and best practices; statistical and econometric analysis; systems and network analysis; and empirical applications of game theory, to name but a few (Balzat and Hanusch, 2004). This methodological diversity and rigour could bring greater credibility and strength to the study of innovation systems in developing country agriculture.

Second, the AIS approach has not yet matured to a point where it can inform policy in developing country agriculture of specific interventions needed to enhance the potential for innovation and improve the distribution of gains from innovation (Spielman, 2006a, b). Although exceptions exist, the link between empirical analysis and policy recommendations remains either nascent or weak in the application of the innovation systems framework to developing country agriculture. With so many case studies conducted and so many lessons learned, researchers should be well positioned to advise governments on policy options and incentive structures that generate greater levels of innovation and improve the distribution of gains.

Third, few studies in the emerging AIS literature examine the poverty-related effects of innovation processes. This means asking whether an innovation increases efficiency in the production or utilization of knowledge directly relevant to those goods and services used by the poor in consumption or production; or whether an innovation improves the distribution of social surplus in a manner beneficial to the poor. Few studies make that leap from descriptive *ex-post* analysis of an innovation system to an *ex-ante* analysis of how an innovation system promotes institutional and technological changes that are explicitly pro-poor. There is still very little conceptual or methodological work within the wider literature on AIS to suggest a consistent focus on sustainability or equity.

NEW METHODOLOGICAL FRONTIERS

Given the complexity of innovation processes and systems, no single method can be used to analyse them. However, several methods could contribute significantly to the existing innovation systems toolkit. This section examines four possible methods: social network analysis; innovation histories; cross-country comparisons; and game theoretic modelling in the tradition of evolutionary economics.

These methodologies may be grouped into three categories: relational analysis, comparative analysis and policy-process analysis. When combined, these methodologies not only provide a valid, rigorous and replicable toolkit, but also possess the ability to influence decision-making on key issues in agriculture and rural development – enhancing productivity, increasing food security and nutrition, diversifying rural livelihoods and reducing poverty. While several of these methods are data intensive, others rely on combinations of qualitative and quantitative tools that make them viable even in light of the limited data availability or limited access that is common to many countries in sub-Saharan Africa.

SOCIAL NETWORK ANALYSIS

Social network analysis (SNA) allows researchers to study relationships among multiple actors by providing tools with which to visualize and analyse the relationships (Scott, 2000). SNA was developed by sociologists and further enhanced as an analytical technique in the fields of mathematics and statistics. By combining relational data with mathematical tools and concepts from systems theory, graph theory and complexity theory, SNA provides critical insight into the relationships between various people, groups or other entities. In the context of innovation, SNA offers a means not only to characterize, measure and map relationships between actors, but also to analyse the changes in those relationships and the knowledge flows contained within them (Davies, 2004).

The data used in SNA are unique because emphasis is placed on the relationships between actors rather than the attributes of the actors them-selves. Conventional data analysis typically focuses on actors, their char-acteristics and how they are similar or different. In SNA, the analysis focuses on the pattern of relations across actors. Therefore, the unit of analysis is the dyad: a pair of entities. Dyadic attributes of interest for innovation include social roles, interactions or flows of information between actors.

For the study of innovation, SNA provides tools that are unique and often absent in many of the cost-based tools. Given the focus on relational rather than attributional data, SNA provides holistic insight into the structure of a system and the interdependence between entities within that system: a 'molecular' rather than 'atomistic' view of the world (Borgatti, 1998).

For instance, Conley and Udry (2001) used SNA to show communication networks among villagers in Ghana. They found that geographic proximity did not determine how smallholders learned; rather, it was the social networks that smallholders were involved with. Muñoz et al (2004) mapped three networks in which Mexican commercial lemon producers participate: technical, commercial and social. These studies identified a few highly connected farmers that increased the network's navigability.

Innovation histories

Another strand of the literature emphasizes the 'innovation history' approach as a method of recording and reflecting on innovation processes as part of wider institutional learning and change (Douthwaite, 2002; Douthwaite and Ashby, 2005). The approach engages researchers in a stepwise process of identifying objectives and expectations of stakeholders, defining the innovation, constructing timelines and actor network maps, writing up the learning history and using the write-up as a catalyst for change. This approach tends to be internal to the organizations directly involved in the innovation process, but is a potentially useful means of documenting and disseminating analyses that can influence decision-makers in government and other sectors.

Comparisons across countries

Comparisons of AIS across countries using benchmarks, scorecards and indices suggest further methodological possibilities. The approach has proven itself as a valuable and effective tool for guiding innovation policy in many industrialized countries and regions. However, its application to developing countries is fairly novel.

Such comparisons provide a more subtle understanding of techno-logical change in sub-Saharan African agriculture and the key factors that help to explain the potential for continuous innovation in agricultural potential in individual countries. It offers a means by which to differentiate, rank and benchmark countries, while also providing tools with which to group countries, to demonstrate where interventions can be effective in several countries at once, and to illustrate the potential for spill-overs across countries. Most importantly, the exercise provides policy-makers, researchers and development practitioners with information and analysis that can guide investments and interventions into areas that contribute to the economic, social and environmental goals of the region.

However, there is much reservation about efforts to quantify innovation and develop comparisons across countries of what are essentially local context-specific processes that do not lend themselves to comparison (compare Balzat and Hanusch, 2004; Grupp and Mogee, 2004). Nonetheless, examples from the Organisation for Economic Co-operation and Development (OECD, 2005) and other international organizations suggest that such methods can be effective tools in understanding innovation across countries.

Game theory

Game theoretic modelling based on emerging work in evolutionary economics offers some insight into the value of the innovation systems framework. These models illustrate the spontaneous processes of social self-organization and the ways in which public policy and organizational structures can affect these processes. This perspective differs significantly from the neoclassical approaches to constitutional design and benevolent social planning: in a complex evolutionary approach, aggregate social outcomes are not the summation of individual maximizing behaviour; rather, they are the result of individual behaviour conditioned by the behaviour of others, the interactions among agents and by the institutional landscape that conditions these behaviour patterns.

For example, evolutionary models derived from biological population models as described by Maynard Smith (1982) are applied to describe the selection of socio-economic behaviours, both idiosyncratic and intentional, over time. The approach is described in detail by Nelson and Winter (1982) and pursued further by Andersen (2000, 2004), who models an innovation system with Schumpeterian characteristics to describe the strategic decision-making processes of diverse agents who cooperate, compete or otherwise interact over time.

CONCLUSIONS

Innovation is a complex process; for this reason, it cannot be characterized by simple models that relate quantities of inputs and outputs or simple monitoring systems that accumulate data on a limiting set of results-based indicators. As the dynamics of innovation processes are better understood, there is a need for more informative methods – both quantitative and qualitative – to analyse and foster innovation. This chapter has provided some insight into methodologies that can help to improve the study of agricultural innovation systems. These methods address the issues of:

- how agents interact in the production, exchange and use of knowledge and information within a system;
- how agents respond individually and collectively to technological, institutional or organizational opportunities and constraints; and
- how policy changes can enhance the welfare effects of these interactions and responses.

REFERENCES

Allegri, M. (2002) 'Partnership of producer and government financing to reform agricultural research in Uruguay', in D. Byerlee and R. Echeverría (eds) *Agricultural Research Policy in an Era of Privatization*, CABI, Wallingford, UK, pp105–121

Andersen, E. S. (2000) 'Schumpeterian games and innovation systems: Combining pioneers, adaptionists, imitators, complementors and mixers', Note for the IKE Seminar, 13 December (revised 1 April 2003)

Andersen, E. S. (2004) *Evolutionary Economics: Post-Schumpeterian Contributions*, Pinter, London

Argote, L. and Darr, E. (2000) 'Repositories of knowledge in franchise organizations', in G. Dosi, R. R. Nelson and S. G. Winter (eds) *The Nature and Dynamics of Organizational Capabilities*, Oxford University Press, New York, NY

Arocena, R. and Sutz, J. (2002) 'Innovations systems and developing countries', DRUID Working Paper 02-05, Danish Research Unit for Industrial Dynamics, Aalborg University, Aalborg, Denmark

Axelrod, R. and Cohen, M. D. (1999) *Harnessing Complexity. Organizational Implications of a Scientific Frontier*, The Free Press, New York, NY

Bailey, J. R. and Ford, C. M. (2003) 'Innovation and evolution: Managing tensions within and between the domains of theory and practice', in L. V. Shavinina (ed) *The International Handbook on Innovation*, Pergamon, London, pp248–257

Balzat, M. and Hanusch, H. (2004) 'Recent trends in the research on national innovation systems', *Journal of Evolutionary Economics*, vol 14, pp197–210

Biggs, S. D. (1989) 'A multiple source of innovation model of agricultural research and technology promotion', Agricultural Administration Network Paper 6, Overseas Development Institute, London

Biggs, S. D. and Clay, E. J. (1981) 'Sources of innovations in agricultural technology', *World Development*, vol 9, no 4, pp321–336

Borgatti, S. (1998) 'Social network analysis instructional website', www. analytictech.com/networks/, accessed on 1 March 2006

Buchanan, M. (2002) *Nexus: Small Worlds and the Groundbreaking Theory of Networks*, Norton, New York, NY

Chema, S., Gilbert, E. and Roseboom, J. (2003) 'A review of key issues and recent experiences in reforming agricultural research in Africa', Research Report 24, International Service for National Agricultural Research (ISNAR), The Hague, The Netherlands

Christensen, C. M. and Raynor, M. E. (2003) *The Innovator's Solution: Creating and Sustaining Successful Growth,* Harvard Business School Press, Cambridge, MA

Clark, N. (2002) 'Innovation systems, institutional change and the new knowledge market: Implications for third world agricultural development', *Economics of Innovation and New Technology*, vol 11, nos 4–5, pp353–368

Clark, N. G., Hall, A. J., Rasheed Sulaiman, V. and Guru, N. (2003) 'Research as capacity building: The case of an NGO-facilitated post-harvest innovation system for the Himalayan hills', *World Development*, vol 31, no 11, pp1845–1863

Cohen, W. M. and Levinthal, D. A. (1990) 'Absorptive capacity: A new perspective on learning and innovation', *Administrative Science Quarterly*, vol 35, pp128–152

Conley, T. and Udry, C. (2001) 'Social learning through networks: The adoption of new agricultural technologies in Ghana', *American Journal of Agricultural Economics*, vol 83, no 3, pp668–673

Crutchfield, J. P and Schuster, P. (eds) (2003) *Evolutionary Dynamics: Exploring the Interplay of Selection, Accident, Neutrality and Function,* Oxford University Press, New York, NY

Davies, R. (2004) 'Scale, complexity and the representation of theories of change: Part II', *Evaluation*, vol 11, no 2, pp133–149

Dosi, G., Nelson, R. R. and Winter, S. G. (2000) 'Introduction: The nature and dynamics of organizational capabilities', in G. Dosi, R. R. Nelson and S. G. Winter (eds) *The Nature and Dynamics of Organizational Capabilities*, Oxford University Press, New York, NY, pp 1–23

Douthwaite, B. (2002) *Enabling Innovation: A Practical Guide to Understanding and Fostering Technological Innovation*, Zed Books, London

Douthwaite, B. and Ashby, J. (2005) 'Innovation histories: A method for learning from experience', ILAC Brief 5, Institutional Learning and Change Initiative, Rome, Italy

Ekboir, J. M. and Parellada, G. (2002) 'Public–private interactions and technology policy in zero-tillage innovation processes – Argentina', in D. Byerlee and R. Echeverría (eds) *Agricultural Research Policy in an Era of Privatization: Experiences from the Developing World*, CABI, Wallingford, UK, pp137–154

Freeman, C. (1987) *Technology Policy and Economic Performance: Lessons from Japan,* Pinter, London

Grupp, H. and Mogee, M. E. (2004) 'Indicators for national science and technology policy: How robust are composite indicators?', *Research Policy*, vol 33, pp1373–1384

Hall, A. J. and Clark, N. G. (1995) 'Coping with change, complexity and diversity in agriculture: The case of Rhizobium inoculants in Thailand', *World Development*, vol 23, no 9, pp1601–1614

Hall, A. J. and Yoganand, B. (2004) 'New institutional arrangements in agricultural research and development in Africa: Concepts and case studies', in A. J. Hall, B. Yoganand, R. V. Sulaiman, R. S. Raina, C. S. Prasad, G. C. Naik and N. G. Clark (eds) *Innovations in Innovation: Reflections on Partnership, Institutions and Learning*, Crop Postharvest Research Programme/ICRISAT/National Centre for Agricultural Economics and Policy Research, New Delhi and Andhra Pradesh, India, pp105–131

Hall, A., Sivamohan, M. V. K., Clark, N. G., Taylor, S. and Bockett, G. (1998) 'Institutional developments in Indian agricultural research systems: Emerging patterns of public and private sector activities', *Science, Technology and Development*, vol 16, no 3, pp51–76

Hall, A., Sulaiman, R., Clark, N., Sivamohan, M. V. K. and Yoganand, B. (2002) 'Public–private sector interaction in the Indian agricultural research system: An innovation systems perspective on institutional reform', in D. Byerlee and R. Echeverría (eds) *Agricultural Research Policy in an Era of Privatization*, CABI, Wallingford, UK, pp155–176

Hall, A., Sulaiman, R., Clark, N. and Yoganand, B. (2003) 'From measuring impact to learning institutional lessons: An innovation systems perspective on improving the management of international agricultural research', *Agricultural Systems*, vol 78, pp213–241

Johnson, B. and Segura-Bonilla, O. (2001) 'Innovation systems and developing countries: Experiences from the SUDESCA project', DRUID Working Paper 01-12, Danish Research Unit for Industrial Dynamics, University of Aalborg, Aalborg, Denmark

Kangasniemi, J. (2002) 'Financing agricultural research by producers' organizations in Africa', in D. Byerlee and R. Echeverría (eds) *Agricultural Research Policy in an Era of Privatization*, CABI, Wallingford, UK, pp81–104

Kauffman, S. (2000) *Investigations*, Oxford University Press, New York, NY

Lundvall, B. A. (1992) 'Introduction', in B. A. Lundvall (ed) *National Systems of Innovation: Towards a Theory of Innovation and Interactive Learning*, Pinter, London

Maynard Smith, J. (1982) *Evolution and the Theory of Games*, Cambridge University Press, Cambridge, UK

Muñoz, M., Rendón, R., Aguilar, J., García, G. and Reyes Altamirano Cardenas, J. (2004) *Redes de Innovación: Un Acercamiento a su Identificación, Análisis y Gestión para el Desarrollo Rural*, Fundación Produce Michoacán, Michoacán, Mexico

Nelson, R. R. and Winter, S. G. (1982) *An Evolutionary Theory of Economic Change*, Harvard University Press, Cambridge, MA

Nickles, T. (2003) 'Evolutionary models of innovation and the Meno problem', in L. V. Shavinina (ed) *The International Handbook on Innovation*, Pergamon, London

Nicolis, G. and Prigogine, I. (1989) *Exploring Complexity*, W. H. Freeman and Co, New York, NY

Nonaka, I. and Takeuchi, H. (1995) *The Knowledge-Creating Company*, Oxford University Press, New York, NY

OECD (Organisation for Economic Co-operation and Development) (1999) *Managing National Innovation Systems*, OECD, Paris, France

OECD (2005) *OECD Science, Technology and Industry Scoreboard 2005*, OECD, Paris, France

Olson, E. E. and Eoyang, G. H. (2001) *Facilitating Organization Change: Lessons from Complexity Science*, Jossey Bass/Pfeiffer, San Francisco, CA

Peterson, W., Gijsbers, G. and Wilks, M. (2003) 'An organizational performance assessment system for agricultural research organizations: Concepts, methods and procedures', ISNAR Research Management Guidelines 7, ISNAR, The Hague, The Netherlands

Porter, G. and Philips-Howard, K. (1997) 'Comparing contracts: An evaluation of contract farming schemes in Africa', *World Development*, vol 25, no 2, pp227–238

Roseboom, H. (2004) *Adopting an Agricultural Innovation System Perspective: Implications for ASARECA's Strategy*, ASARECA, Entebbe, Uganda

Rycroft, R. W. and Kash, D. E. (1999) *The Complexity Challenge: Technological Innovation for the 21st Century*, Cassel, New York, NY

Science Council (2005) *Science for Agricultural Development: Changing Contexts, New Opportunities*, CGIAR Science Council, Rome, Italy

Scott, J. (2000) *Social Network Analysis. A Handbook*, second edition, Sage, London

Spielman, D. J. (2006a) 'A critique of innovation systems perspectives on agricultural research in developing countries', *Innovation Strategy Today*, vol 2, no 1, pp25–38

Spielman, D. J. (2006b) 'Systems of innovation: models, methods and future directions', *Innovation Strategy Today*, vol 2, no 1, pp39–50

Sumberg, J. (2005) 'Systems of innovation theory and the changing architecture of agricultural research in Africa', *Food Policy*, vol 30, no 1, pp21–41

Vieira, L. F. and Hartwich, F. (2002) *Approaching Public–Private Partnerships for Agroindustrial Research: A Methodological Framework*, ISNAR, Coronado, Costa Rica

World Bank (2006) *Enhancing Agricultural Innovation: How to Go Beyond the Strengthening of Research Systems*, World Bank, Washington, DC

Zander, U. and Kogut, B. (1995) 'Knowledge and the speed of transfer and imitation of organizational capabilities: An empirical test', *Organizational Science*, vol 6, pp76–92

II

Strengthening Social Capital in Agricultural Innovation Systems

Harnessing Local and Outsiders' Knowledge: Experiences of a Multi-Stakeholder Partnership to Promote Farmer Innovation in Ethiopia

Tesfahun Fenta and Amanuel Assefa

INTRODUCTION

Farmers in the Ethiopian highlands and pastoralists in the drier low-lands have long been challenged by food security problems. During recent decades, the situation has worsened, mainly because of human and naturally induced environmental problems (soil erosion and degradation) and population increase. In response, the Ethiopian government and international aid agencies have provided considerable support to these areas in the form of food aid, as well as development programmes aimed at reversing environmental damage and building sustainable livelihoods. Most of the interventions of government and aid agencies were externally driven, ignored the potential of local innovations and resources, and, in many cases, did not lead to sustainable development. Agricultural extension activities in Ethiopia, although often claiming to be participatory, remain delivery oriented rather than encouraging farmer innovation (Tesfaye, 2003).

Ethiopia is a country of ancient and diverse cultures and multiple ethnicities. Traditional land-use systems dominate. Although not yet well explored, Ethiopia is also the home of amazing systems of indigenous knowledge (IK) that helped the people to survive under adverse environmental conditions. The history of Ethiopian civilization provides evidence for the dynamics of IK. The domestication of cereal crops such as coffee, *teff* (*Eragrostis tef*) and *enset* (false banana, which provides a staple food for millions of people in the Southern Region) and the development of the bench-terrace system by the Konso people are among the important achievements of farmers using their IK. However, most researchers and development practitioners are not very interested in building on successful IK, nor do they recognize the dynamics of IK that emerge as local people confront new challenges. According to Tesfaye (2003), most development interventions give no attention to community IK on resource management, local institutions and coping mechanisms.

The government and aid agencies generally see farmers as recipients of technologies identified and packaged by outsiders, a term that refers here to all actors who try to influence the livelihood and value systems of local people with externally designed technologies, projects, institutional arrangements, etc., believing that their interventions will be effective. Not all interventions are useless; but what is damaging is the attitude of outsiders that what they introduce is the only right way to do things, while they consider the IK of the local people not worthy to be developed. Although some outsiders have begun to appreciate farmers' participation in technology development, many are still trapped in top-down, centre-outwards thinking and action, with researchers determining priorities, generating technologies and transferring them to farmers via extension (Teklu, 2001). The reality on the ground, however, is that millions of people in Ethiopia are farming under very diverse, complex and risk-prone conditions, and it would be impossible for formal research and extension to address every single agricultural problem throughout the country.

Realizing the potential in local people, some NGOs in Ethiopia took the initiative of establishing a national learning and advocacy platform to promote local innovation. This platform tries to stimulate institutions of research, extension and education to recognize the creativity and innovativeness of local people and to make adequate space for this in their policies and programmes. Significant changes are required in mind-set and in planning, budgeting, monitoring and evaluation in order to accommodate the innovation systems paradigm. NGOs have joined forces with like-minded people in governmental organizations (GOs) to achieve this goal.

HISTORICAL ACCOUNT OF INSTITUTIONAL DYNAMICS

Farmer Innovation in Africa (Reij and Waters-Bayer, 2001), a book based on the work of two Netherlands-funded programmes on Indigenous Soil and Water Conservation (ISWC) and Promoting Farmer Innovation (PFI), reported that ISWC Phase I covering 27 case studies in 15 African countries (including Ethiopia) found smallholders maintaining and expanding many indigenous practices of SWC, but few of the 'modern' SWC techniques promoted by development projects. ISWC Phase II explored the effectiveness of the indigenous and modern SWC practices through joint experimentation, involving farmers, scientists and development agents (DAs). Policy-makers were encouraged to acknowledge the work of innovative farmers linked with scientists in participatory technology development (PTD). In Ethiopia, ISWC–II operated in Tigray Region in the north and was coordinated by Mekelle University. The legacy of this programme gave rise to the formation of a national learning platform known as Promoting Farmer Innovation and Experimentation in Ethiopia

(PROFIEET), which was spearheaded by NGOs and is now composed of GOs and NGOs involved in agricultural research, extension and education.

The idea of initiating this platform arose after a two-day event to celebrate farmer innovation, organized in November 2001 in Axum by the Institute for Sustainable Development (ISD, a local NGO), Mekelle University (ISWC–II) and the Tigray Bureau of Agriculture. This event brought together local NGO staff and farmers and several government officials from all over Ethiopia. The main purpose was to publicly recognize local innovativeness by awarding prizes to outstanding farmer innovators and viewing their achievements in the field. However, the officials, NGO staff and farmers had little opportunity to interact critically and in a structured way on issues of local innovation and PTD, and on their potential added value to conventional research and extension.

The event in Axum ignited the inspiring idea of farmer innovation, which was a valuable take-home message. The participants from AgriService Ethiopia (ASE), in particular, a local NGO already practising PTD, went home with enthusiasm to organize a follow-on workshop to familiarize government staff from different regions of Ethiopia with concepts of promoting local innovation. Together with the organizers of the Axum event and with financial support from ISWC–II, ASE organized a two-day Awareness-Creation Seminar in Addis Ababa in January 2002. The experiences of ISWC–II on farmer innovation in Tigray Region, of ASE on PTD in Amhara Region and of the NGO FARM-Africa on farmer participatory research (FPR) in the Southern Region were presented to policy-makers and staff of the Ministry of Agriculture and Rural Development (MoARD) from 11 regions of Ethiopia.

This workshop, where the name PROFIEET was coined, was organized quickly without having thought critically about possible outcomes, particularly without having prepared for responding to needs that might emerge. The workshop stimulated the interest of many participants to identify farmer innovators and to work with them in joint experimentation. However, two key questions were raised:

1 Why were the federal and regional institutes of agricultural research not involved in the workshop, although they have a mandate for this type of work?
2 What can the organizers offer in terms of capacity development and backstopping support for those interested in promoting local innovation?

The workshop organizers continued informal discussion about the need to examine more closely the activities of various agencies in participatory research and development (R&D) and to define the support needed to scale up such activities. Meanwhile, international organizations that had been involved in ISWC and PFI had joined forces with others and had started a Global Partnership Programme called Prolinnova (Promoting Local Innovation in Ecologically Oriented Agriculture) under the umbrella of the Global Forum on Agricultural Research. They sought to identify or stimulate national initiatives such as PROFIEET and offered methodological advice and facilitated learning across countries. Through a key person in the Ethiopian Institute of Agricultural Research (EIAR) responsible for research–extension–farmer linkages, who had evaluated FARM-Africa's work in institutionalizing FPR in the Southern Region, they approached PROFIEET to explore interest in collaborating in Prolinnova.

The workshop organizers and the head of research–extension–farmer linkages in EIAR set up an interim steering committee to coordinate a state of the art study on participatory R&D in the country and to organize a national learning and planning workshop. The International Fund for Agricultural Development provided financial support through Prolinnova. After a locally recruited consultant had completed the study, a three-day workshop was held in August 2003 to discuss experiences and issues of participatory R&D in Ethiopia. Participants came from research organizations, NGOs, MoARD, regional agricultural bureaux, universities and the Ethiopian Science and Technology Agency (ESTA), and included experimenting farmers. Numerous case studies, including the farmers' experience in joint experimentation, were presented and discussed. A National Steering Committee (NSC) was then set up to guide the formation and functioning of a national learning and advocacy platform. This platform became the partner of Prolinnova in trying to integrate farmer-led R&D, building on local innovation and experimentation, within the work of relevant GOs and NGOs, with the ultimate aim of contributing to food security, sustainable rural livelihoods, poverty reduction and environmental protection.

The NSC – made up of people from GOs such as EIAR, MoARD, ESTA and Haramaya, Mekelle and Hawassa universities, and NGOs such as ASE, FARM-Africa, ISD, Pastoral Forum Ethiopia, SOS-Sahel and the Sustainable Land Use Forum – oversees the national multi-stakeholder platform, while ASE serves as secretariat and facilitator.

Over time, internal reflection by the platform led to the realization that it was giving too much attention to dialogue with federal policy-makers without having enough showcases of PID at grassroots level and without addressing regional-level policy-makers. Only the NGO members in the NSC were engaging in some PID activities in their working areas, while the GO members generally were not. In 2005, the NSC decided to decentralize, and members from the regions started setting up similar networks at regional (provincial) level. These were supposed to support farmer-led participatory innovation processes on the ground and to engage in dialogue with regional policy-makers.

CHALLENGES THAT PROFIEET ADDRESSES

Although Ethiopian farmers are sources of diverse knowledge and innovation, their potential had not been unlocked in the past because of several challenges:

- *Lack of appreciation by outsiders of farmers' knowledge and values* slows down development and democratization processes in Ethiopia. Conventional development thinking is based on the assumption that the poor and illiterate are ignorant and, if they are to be liberated from their poverty, formally educated people have to take care of all their development needs.
- *Insufficient opportunity for farmers to decide on R&D priorities.* Participation is an important dimension of knowledge management. The success of R&D initiatives depends largely upon the extent to which the initiators accommodate the knowledge and priorities of the intended beneficiaries. R&D agendas in Ethiopia are identified by formal researchers and approved by peer groups. The extension system promotes 'off-the-shelf' technology packages coming from research with little or no prior consultation with farmers (Tesfaye, 2003).
- *Lack of financial support to encourage local innovation* has constrained development of the local economy and sustainable natural resource management (NRM) practices. Providing financial support to enhance local innovation processes is one way of expressing outsiders' appreciation of and trust in farmers' knowledge and potentials. Traditionally, Ethiopian farmers do not claim financial support from the government or aid agencies to improve their innovations. This is mainly because of the long-standing paternalistic tradition of these agencies, which made many farmers undervalue their own knowledge and innovations and depend upon outsiders' technology. Resources for agricultural R&D are entirely controlled by the formal institutions.
- *Limited skills and experience in facilitating participatory learning processes.* The main purpose of recognizing and providing support to local innovation is to enhance farmers' capacities to continue to innovate so that they can overcome site-specific problems that cannot be precisely addressed by the formal R&D services. However, this does not mean that farmers have adequate answers for all the problems that they face. Farmers can apply their knowledge more effectively if they are supported with complementary knowledge of outsiders and are facilitated in integrating different types of knowledge. Scientists and policy-makers can help farmers to improve their situation more quickly. However, in R&D agencies in Ethiopia, skills and experience in facilitating participatory learning are limited.

In order to meet these challenges, PROFIEET has been trying, since the August 2003 workshop, to:

- create a receptive environment at local and national level to appreciate and stimulate local innovation;
- systematically identify, document and promote local innovations in the highland, pastoral, coffee-growing and *enset*-growing areas;
- make funds available to innovative farmers to support farmer-led participatory innovation processes;
- encourage scientists to help farmers develop their innovations and expose them to new and relevant knowledge and technologies in a spirit of mutual respect and learning;
- help policy-makers to appreciate the knowledge and innovativeness of local people and stimulate policy-makers to provide continued support to farmers' efforts.

Accomplishing these tasks demands a clear theoretical and methodological framework as well as guidance for PROFIEET partners to help them engage in theory-informed practice.

FRAMEWORK FOR THINKING AND DOING

The concepts

The conceptual framework for Prolinnova is derived from those of the ISWC and PFI programmes. PFI was developed by the United Nations Development Programme (UNDP) Office to Combat Desertification and Drought and funded by the Netherlands Directorate General for International Cooperation (DGIS) to identify and support farmer in-novation in Kenya, Uganda and Tanzania. The partners developed a working definition of 'farmer innovators': farmers who have developed or are testing new (in local terms) ways of land husbandry that combine production with conservation (Critchley, 1999). ISWC, likewise supported by DGIS, operated in seven countries in Africa, and each country programme developed its own working definition. For example, ISWC-Ethiopia defined an innovator as someone who develops or tries out new ideas without the support of formal extension services. 'New' was defined as something that had been started within the lifetime of the farmer, not something that she or he inherited from parents or grandparents. In contrast, ISWC-Tunisia decided to include inherited technologies in its inventory of local innovations. In general, however, ISWC–II partners regarded farmer innovation as 'something new to a particular locality, but not necessarily new to the world' (Reij and Waters-Bayer, 2001).

PROFIEET has its roots in the ISWC-Ethiopia work. Its working definition for innovative farmers is: farmers who have tried out new, value-adding agricultural or NRM practices using their own knowledge but also by appropriating outsiders' knowledge. While recognizing IK as an important asset in development, PROFIEET partners see that innovative farmers are not people who are using IK as it was during their ancestors' time, but rather people who act on IK and/or outsiders' knowledge by conducting informal experiments and making the knowledge more usable or better fitted to their current realities. Therefore, the main focus of PROFIEET is not on IK as a static asset, but on the *dynamism* of IK that brings new values to the users. Innovative farmers are not the 'model farmers' trained by DAs in specific techniques. Innovativeness is the capacity of individuals or groups to look into given situations from different angles and create new and positive values. Farmers trained by DAs may also be innovative if they add value to the incoming knowledge or technology by making it fit better into the local setting or by blending it with pre-existing practices or technology and thus making something new out of it. Innovative farmers add value to existing practices through experimentation in the quest for changes that are beneficial in economic, social and environmental terms.

Local innovation is also a way of life for poor Ethiopian farmers who are challenged by constant changes in policy, markets and the environment. Innovation is an inherent characteristic of people striving to make a living in a difficult situation: they must innovate in order to survive. Farmers continuously experiment, adapt and innovate (Chambers et al, 1989). This is especially the case with resource-poor farmers. Ayelech Fikre, an innovative woman in the central highlands of Ethiopia, when asked what motivated her to innovate in SWC, said: 'The problem [soil erosion and difficulties of getting more land] taught me to do all these activities. Otherwise I could not survive' (Alemayehu, 2001).

The significance of local innovation ranges from being useful only to the individual farmer, sometimes even limited to specific circumstances (e.g. plot of land and type of animal) to being widely applicable and useful to many farmers. Therefore, innovation facilitators need to give attention to two levels:

1 stimulating the creativity of farmers, which has been suppressed for years because of undesirable socio-psychological influences of outsiders;
2 helping to scale up local innovations that are economically, socially and ecologically sound, with due recognition of intellectual property rights.

Another important dimension of the concept of 'local innovation' is that it embraces not only technological innovation, but also new ways of managing livelihoods, in general, which include new ways of communication, organization, accessing resources, etc. Technological innovation often leads to new institutional arrangements (i.e. to changes in the 'rules of the game': routine practices, community agreements, legal frameworks, value systems, etc.).

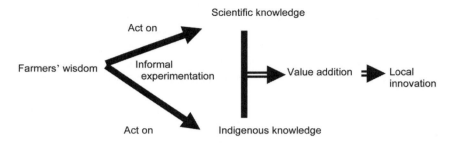

Figure 6.1 *Conceptual framework for local innovation*

PROFIEET uses a schematic presentation to explain the conceptual framework for local innovation (see Figure 6.1).

The approach

PID in agriculture and NRM are the overarching methodological approaches that PROFIEET uses to put this theory into practice. PID is a farmer-led and expert-supported process of innovation development that usually takes a local innovation as starting point. It commonly involves collaboration of farmers, DAs and researchers, in which farmers who are already trying out new things take the lead. The process includes not only research but also application of the results. The primary goal of PID is not to scale out farmers' innovations emerging from PID in a transfer of technology mode. Rather, it is to scale out the spirit of innovativeness so that all farmers are encouraged to try new things to see what could work in their own situation. PID has its roots in PTD and shares most of the same methods and tools. However, some new values in PID include the following:

- Outsiders seek to work with those farmers who have tried or are trying out something new and thus are already in the midst of an innovation process. Innovative farmers are not just 'participating' in the process; they are recognized as lead researchers.
- PID is not only about technical innovation but also about new ways of thinking and doing things (social, institutional, cultural, economic, etc.) that may help to improve farmers' lives.

The PID activities thus far have only started to develop the new dimensions of social, institutional and economic innovation. The following summarized description of different phases in PID is therefore skewed to technological innovation.

Getting started
This involves building relationships and discussing partnership for PID with grassroots institutions, including farmer groups. Stakeholders

involved in innovation in the area are identified, and their interest and experience in engaging in PID are assessed.

Discovering things to try out
The agricultural innovation subsystem is studied, focusing on local innovations, and the findings are discussed among farmers and other stakeholders in order to stimulate their interest and commitment to work jointly on some of the local innovations. Farmers and DAs identify the types of local innovation that impress local farmers most. They decide which innovations should be shared with other farmers and which ones should be explored further (*local innovation-based joint experimentation*). Farmers are also supported in analysing their problems in agriculture and NRM and identifying issues that they want to research with outside support (*problem-based joint experimentation*). Farmers are given opportunities to visit and learn about technologies developed by formal research and by farmers elsewhere, and to choose those technologies they want to examine more closely and test in their own reality (*trying out new ideas of others*).

Designing experiments
Farmers prioritize local innovations that they want to develop further and/or new technologies to try out in their own settings. 'Experts' (researchers and extensionists) are invited to farmers' meetings to help design the research. Farmers present how they intend to try something out, the assessment criteria they plan to use and the type of information they want to collect. The experts suggest how to make the experimental methods simpler and more valid, and how they can help in analysing data. The plan for the experiments indicates when, where, how and by whom the innovation development process will be monitored. Sometimes experts may be involved in laboratory analyses or literature studies to provide information to support the process.

Joint experimentation
Farmers lead participatory research based on local innovations, locally identified problems or externally developed techniques that interest them. They report progress and results to the farmer group or a community learning forum which meets periodically, using evaluation criteria agreed jointly at the outset. The experimenting farmers keep records of their work or, if they are illiterate, the PID facilitators help them to do the recording. External experts provide support to the individual or group innovators during the research.

Sharing results
The facilitators help farmers to document the PID process and results and find ways of sharing their experiences with other farmers and field practitioners locally and elsewhere in the country. The local innovations that the farmer group or community learning forum thinks could benefit

others are shared by organizing farmer field days, farmer-to-farmer visits, farmers' workshops, innovation markets, etc. DAs and researchers support the innovative farmers to present the results of their work in higher-level workshops and conferences.

Sustaining the process
This is about ensuring that PID becomes a culture of the community and the supporting agencies. Both farmers and outsiders need to appreciate the process and results and see the importance of sustaining the process so that PID can continue to improve farmers' livelihoods. Some activities that may help in this regard are:

- training DAs in PID and helping them to facilitate PID with farmers;
- organizing visits by influential policy-makers to expose them to local innovation and PID;
- encouraging farmers to think of ways of financing their PID work to achieve self-reliance;
- organizing events to give recognition to innovative farmers and to DAs and senior GO and NGO staff who support the process.

MAJOR ACHIEVEMENTS AND WORK IN PROGRESS

In 2003, PROFIEET organized a national workshop to discuss relevant experiences in Ethiopia and to establish the national learning and advocacy platform. The inventory of experiences of various institutions in the country that are engaged in participatory R&D provided opportunities to map out existing and potential linkages between the stakeholders for future collaboration. The most interesting parts of the workshop were the presentations by farmers on their work and the subsequent reflections by separate groups of farmers, extensionists, researchers and policy-makers on what actions need to be taken to improve linkages and mutual learning. The workshop outputs fed into the design of a country proposal on scaling out and institutionalizing PID in major stakeholder organizations in Ethiopia. In 2004, this proposal received funding from The Netherlands government through the international Prolinnova programme and, in 2007, also from ActionAid-Ethiopia as a result of direct negotiation by PROFIEET.

As part of the international Prolinnova programme, PROFIEET organized the first international workshop of Prolinnova partners in Ethiopia in early 2004. This brought together people with similar experiences from around the world and offered a good opportunity to pass messages to Ethiopian policy-makers. The interactions of participants in the meeting and the interviews of some key facilitators of PROFIEET on national television and radio conveyed the need to reorganize R&D in Ethiopia in order to give more attention to local innovation and initiatives. Initially, these ideas were not welcomed by some key officials in agricultural research, who thought that PROFIEET had no mandate to comment publicly on how research is done. This led to EIAR's resigning as chair of the NSC. However, after repeated meetings and lobbying with concerned people in EIAR, it came back on board with renewed spirit to collaborate in PROFIEET. Today, this federal agricultural research organization puts great emphasis on an innovation systems approach as a strategic direction and is reorganizing itself for impact-oriented service to farmers.

Consolidating PROFIEET as a multi-stakeholder platform has not been easy because institutions with different and sometimes conflicting perceptions and philosophies about R&D must be accommodated. Much time was spent discussing institutional arrangements, drafting guidelines, preparing proposals and seeking additional donor support as the funds provided through Prolinnova could serve only as 'seed money'.

In late 2005, PROFIEET adopted a strategy to set up regional forums to coordinate work on the ground. It identified four broad farming systems: the typical highlands (mixed cereal-based crop–livestock farming); the coffee-growing areas in the southwest; the *enset* and other root-crop growing areas in the south; and the pastoral areas in the lowlands. To facilitate formation of the regional platforms and to initiate the PID work, PROFIEET held three workshops: in Axum for the highlands, in Jimma for the coffee-growing areas and in DireDawa for the pastoral areas (the fourth workshop in Awassa for the *enset*-growing area was only held in 2007). Each workshop was preceded by a day-long seminar to share concepts of local innovation and PID with relevant GOs and NGOs. The regional partners selected ten innovative farmers and invited them to make presentations in the workshop. This had two parts: training on PID concepts and planning PID activities. In each workshop, the participants selected three innovations for joint experimentation involving farmers, scientists and DAs. These were, in the highlands:

1   rotary water-lifter from hand-dug well;
2   reducing waterlogging by digging underground canals; and
3   improving 'modern' beehives.

In coffee-growing areas, the three innovations involved:

1   farmer-made hydroelectric power;
2   manually operated dry coffee de-husker;
3   coffee plant rejuvenation techniques.

In pastoral areas, the innovations comprised:

1   mixing camels', goats' and cows' milk to avoid curdling;
2   transfer of papaya pollen by hand;
3   repulsion of retained placenta in cows.

The PROFIEET partners are documenting the PID processes and results, and starting to expand the process to other issues identified by farmers. In a largely illiterate rural society, audiovisual documentation is vital for promoting local innovation and sharing the basic philosophy of the platform. To this end, most of the local innovations identified by the regional platforms have been documented by professional filmmakers.

Currently, the activities facilitated by PROFIEET fall under four major categories:

1   the DGIS-supported activities of identifying innovative farmers in various agro-ecosystems and initiating PID;
2   the Farmer Access to Innovation Resources sub-project supported by the French government, which involves action research into mechanisms of providing farmer groups the means of funding the research and innovation development activities that they have prioritized;
3   ActionAid-funded pilot activities in six districts to introduce PID to DAs working at farmer training centres;
4   the diverse work related to PID being carried out by member institutions in their own domains (these activities are not centrally planned and coordinated by the PROFIEET Secretariat, but form part of the overall efforts to accomplish PROFIEET's mission, and include providing support to farmer field schools, facilitating FPR, organizing training on PID-related methodologies, and using various forums to advocate change in institutions of research, extension and education towards supporting farmer-led R&D processes).

MAJOR LESSONS LEARNED

In most cases, local innovations in agriculture and NRM are not visible unless one takes time to discuss with farmers in a learning spirit. Identifying local innovations is therefore a difficult task until the involved agencies develop experience in this. Most important is to have the attitude that farmers are creative in adapting to changing circumstances. Often, local innovators do not realize that they are doing something that can have significant benefits for others because they innovate not for academic or even commercial purposes, but simply to overcome the challenges that they face. It is up to the outsiders to observe such innovations and to make them more widely known for further development and improvement. Sometimes, this seems to focus so much on the technologies which farmers have developed that less attention is given to the innovation process. However, experience shows that recognizing local innovation encourages other farmers to reveal their own innovations that they had not previously thought would interest outsiders. This exercise is a springboard to stimulate the process of PID and to initiate policy dialogue about this.

PID can be best implemented if researchers are involved right from the start. However, in most parts of Ethiopia, few researchers are available to work directly with farmers in PID. It is therefore imperative to encourage DAs and NGOs to play the supportive role. In essence, PID is not about conducting research, but rather about locally initiated development supported by external knowledge, often provided by extension staff. However, it is important to change the attitudes of DAs and to build their confidence that they can play another role than just transferring technologies coming from research.

A multi-stakeholder process to promote local innovation is slow in terms of seeing results on the ground. However, it helps in institutionalizing the approach. The multi-stakeholder structure is an innovation in itself. Institutional changes and creative ideas that may have significant policy implications are generated in the interaction between the stakeholders. It is important to appreciate the contributions of the actors who build on the collective agenda and keep the initiative going. Engaging different stakeholder organizations in joint activities heightens the feeling of collective responsibility and ownership of results so that the platform does not become the 'show' of a few organizations who control the resources.

Starting with a national platform versus starting with regional (provincial) platforms and then forming the national body at a later stage are different strategies with their own strengths and limitations. Beginning with a national body makes it more difficult to connect to the grassroots dynamics. Many NSC members come from organizations in Addis Ababa. If some come from the regions, it is expensive to bring them to the capital for meetings. However, a strong national platform has better opportunities for advocacy, fundraising and international communication. Starting the network at the regional level makes it easer to involve organizations working directly with farmers so that results can be demonstrated more quickly. However, in Ethiopia's federal system of government, in which the regional states are autonomous, the regional platforms cannot communicate so well with national policy-makers. PROFIEET includes both types of structure and seeks to strengthen the regional platforms to be self-reliant but in good communication with the national platform.

Government organizations and NGOs differ in their working modalities and organizational cultures. GOs follow bureaucratic financial and administrative procedures and are therefore less flexible. NGOs observe procedures, but generally in a more flexible way, so it is easier to make things move faster. A coalition of these two different systems is not easy to manage. Experience has shown that the NSC needs considerable time to discuss procedural issues at the expense of discussing concepts and making concrete plans.

Conflicts are inevitable in a multi-stakeholder platform such as PROFIEET. One function of the platform should be to manage conflicts. There must be transparent rules and guidelines in place from the outset. A culture of expressing frustrations needs to be encouraged, and the platform must feel responsibility to respond quickly to the frustrations expressed. However, even if all precautions are taken, there will still be conflicts of interest. It is therefore necessary to learn to live with differences as long as they do not block progress in achieving the platform's goals.

CONCLUSIONS

The conventional transfer of technology model, which has long dominated and is still practised by most public institutions in developing countries, has failed to respond to the needs of farmers living in very diverse and risk-prone areas. The newly emerging innovation systems approach offers an alternative that can better respond to the multiple needs of such farmers. This approach embraces the contributions of many different actors, rather than being confined to formal research and extension institutions. Where a market economy prevails, private-sector actors influence the innovation process significantly. Where subsistence farming prevails, the innovation systems approach must ensure that the knowledge and priorities of smallholder farmers are given enough space. This is the main reason why PROFIEET – operating in a country where more than 85 per cent of the people try to live from farming less than 1ha of land per household – makes local innovation the central agenda of the partnership.

Some international initiatives to build multi-stakeholder partnerships try to mould these in a predetermined way. This may led to disappointing results as the national agendas and socio-cultural situations often differ from what the international partners assume. The support that PROFIEET enjoys from the international Prolinnova team is, however, primarily in terms of capacity-building, fund acquisition and international advocacy. PROFIEET is clearly a national affair managed by Ethiopians to meet collectively set goals. The institutional arrangements and nature of activities of the various Prolinnova country programmes are very diverse; each programme is unique. This policy of the international network not only stimulates innovative approaches in each country, but also provides an example to the people coordinating the national initiatives to act likewise in decentralizing the approach within the country.

In Ethiopia, it is not common to see a platform comprising both state and non-state institutions. PROFIEET is one of few such examples in the country. Despite the challenges of dealing with differences in institutional culture, the multi-stakeholder platform has become an important forum for joint and mutual learning. A more receptive spirit is being developed between the GOs and NGOs involved.

PROFIEET has launched a new learning front on Farmer Access to Innovation Resources (FAIR) to pilot local innovation support funds (LISFs). An LISF is a sum of money that is available for farmers to use to conduct their own research and develop their innovations, bringing in external support as they deem necessary. Two pilot LISFs have been set up, one in the south (Amaro) and one in the north (Axum) of the country. These pilots will provide lessons on how farmers could access resources for innovation and how such resources could be recharged from various sustainable sources. This pilot will add to the international experiences on how farmers can best be supported to decide on R&D matters that affect their lives and their environment.

REFERENCES

Alemayehu, M. (2001) 'Ayelech Fikre, an outstanding woman farmer in Amhara Region, Ethiopia', in C. Reij and A. Waters-Bayer (eds) *Farmer Innovation in Africa*, Earthscan, London, pp28–32

Chambers, R., Pacey, A. and Thrupp, L. A. (eds) (1989) *Farmer First: Farmer Innovation and Agricultural Research*, Intermediate Technology Publications, London

Critchley, W. (1999) *Promoting Farmer Innovation: Harnessing Local Environmental Knowledge in East Africa*, RELMA, Nairobi, Kenya

Reij, C. and Waters-Bayer, A. (2001) *Farmer Innovation in Africa: A Source of Inspiration for Agricultural Development*, Earthscan, London

Teklu, T. (2001) 'Towards farmer participatory research', in Gemechu Kenani et al (eds) *Proceedings of Client-Oriented Research Evaluation Workshop*, Ethiopian Agricultural Research Organization, Addis Ababa, Ethiopia, ppix–xii

Tesfaye, B. (2003) *Understanding Farmers: Explaining Soil and Water Conservation in Konso, Wollaita and Wollo, Ethiopia*, Wageningen Agricultural University, Wageningen, The Netherlands

An Innovation System in the Rangelands: Using Collective Action to Diversify Livelihoods among Settled Pastoralists in Ethiopia

D. Layne Coppock, Solomon Desta, Seyoum Tezera and Getachew Gebru

INTRODUCTION

In 2000 we discovered dynamic pastoral women's groups in remote northern Kenya. They comprised formerly poverty-stricken women who joined together during recent decades to improve their lives. They used innovative forms of collective action to accumulate money, diversify livelihoods, fill gaps in public service delivery and mitigate negative impacts from drought. In contrast, less than 50km to the north in Ethiopia, pastoral women continued to live in a very traditional way. These women were often very poor and depended largely upon pastoral production despite increasing risks to their livelihoods. Differences between the Kenyan and Ethiopian women were remarkable, given the short distance of separation and because most were from the same ethnic group: the Boran. In 2001, a team of researchers and development agents brought the Kenyan and Ethiopian women together to share experiences. During a cross-border tour, 15 Ethiopian women visited five Kenyan women's groups. Much has changed in southern Ethiopia as a result of this tour. At least 59 collective action groups have formed with over 2000 members (76 per cent female). The Ethiopian pastoralists have become empowered, their incomes have increased and their livelihoods diversified. This six-year process was supported by an innovation system involving 46 partners from various sectors sharing an emphasis on action-oriented outputs and authentic community participation. Despite notable development achievements, the sustainability of the innovation system is not a given. Sustainability will require vigilance, leadership, maintenance of inter-institutional relationships and stakeholder incentives.

FRAMING THE PROBLEM

Traditional pastoralism on the Borana Plateau in the Oromia Region of southern Ethiopia had long been sustainable (Coppock, 1994). In years past, the Boran were able to subsist on milk and meat from livestock herded across vast rangelands. Periodic deficits in livestock output could be addressed through opportunistic cultivation or trading animal products for non-pastoral foods. Prior to the 1980s, human and livestock populations were in balance with seasonal forage and water resources. Recently, however, steady growth in the human population has led to a decline in the per capita availability of natural resources and, hence, made the traditional system unsustainable (Desta and Coppock, 2004).

What can be done to improve this situation? Agricultural technology offers little hope here. Spurring emigration is not promising either, as Ethiopian pastoralists have few employment opportunities outside of the rangelands. Another option is livelihood diversification to generate or expand non-traditional sources of income and assets.

It is in this context that we began to search for a viable livelihood-diversification process. Early on, we discovered the utility of collective action for livelihood diversification among settled pastoral women in northern Kenya. We then imported this approach to Ethiopia. Facilitating the diffusion of collective action among Ethiopians became the core of our project. In this process, a large collaborative network was created to help make change happen. This network is essentially an innovation system.

Here we give some background about the Pastoral Risk Management (PARIMA) project and an overview of literature concerning collective action and innovation systems. We then highlight some project achievements and relate these to the use of an innovation system. We describe challenges that we faced in helping to create and maintain an innovation system, as well as the constraints to sustaining this system into the future.

Pastoral Risk Management project background

The Pastoral Risk Management (PARIMA) project has existed since 1997. The project operates in a 124,000 square kilometre region in northern Kenya and southern Ethiopia bisected by the international border (see Figure 7.1). This region is home to many pastoralists and agro-pastoralists who endure poverty, insecurity, lack of public services and drought. The overall goal of PARIMA has been to identify ways of improving pastoral livelihoods through risk management.

In early 2000, a routine trip by PARIMA staff in remote northern Kenya revealed the presence of dynamic women's groups in small towns and settlements. It was reported that, before joining in groups, many of these women had barely survived by selling charcoal and firewood. They formed groups to improve their lives, and the preliminary evidence was

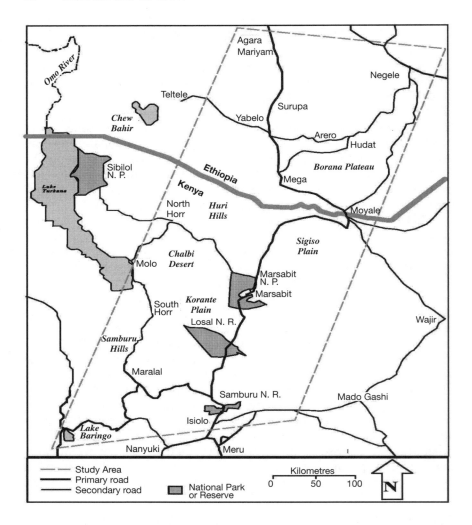

Figure 7.1 *Study region of the Pastoral Risk Management project in northern Kenya and southern Ethiopia*

Source: Layne Coppock

impressive. The women we observed were well dressed, physically robust and could point to a variety of group achievements, including installation of culinary water systems, improved housing, local education centres and the establishment of small businesses.

Some members of the PARIMA team had worked in southern Ethiopia since the 1980s. We were aware that pastoralists there continued to live as they had for centuries despite increasing resource-based pressure

(Coppock, 1994). Women in southern Ethiopia continued to dress traditionally and often appeared gaunt, in marked contrast to the Kenyan women. The Ethiopian women had no tradition of collective action. Cooperative behaviour among Ethiopian women was largely limited to mutual assistance in livestock management (Coppock, 1994). The differences between these sub-populations were remarkable, given that most of the women were Boran and separated by only 50km of rangeland. The international border appeared to be a formidable barrier for such women to interact.

Collective action

Collective action is defined by Meinzen-Dick and DiGregorio (2004) as 'a voluntary action taken by a group to achieve common interests'. Ramírez and Berdegué (2003) noted that 'collective action is a strategy designed to achieve particular objectives that correspond to public goods' and that there are many forms of collective action. Development of a collective action culture is often gradual, and group objectives may include increasing incomes and well-being of members, modifying social relations, influencing policies, developing human and social capital, and fostering social networking (Ramírez and Berdegué, 2003). Grootaert (2001) noted that group formation is valuable to build social capital and enhance income generation among the poor. Johnson and Berdegué (2004) argued that there is increasing need for collective action to build competitive small business capacity among rural producers. Panda (2007) noted that factors such as institutional arrangements, incentives and leadership are important in the viability of collective action schemes.

Women are a common constituency of collective action groups in developing societies. Indian water management projects have included thousands of women over long periods of time (Panda, 2007). Women's engagement in formalized cooperative behaviour can be promoted by strong grassroots institutions, and collective action has allowed women to overcome social barriers. Women have benefited from collective action in terms of increased income and improved family livelihoods (Panda, 2007). Place et al (2004) describe self-help groups emerging in Kenyan farming systems. These groups are often dominated by women and undertake many activities to benefit group members and their communities. Ramírez and Berdegué (2003) noted that collective action does not ensure the reduction of gender inequality in rural areas.

Sedentary cultivators and urban dwellers appear to be the most likely types of people to undertake collective action. There is little evidence of collective action in the dry rangelands, where pastoralists have valued social independence and household mobility (Coppock, 1994). Yet settlements in pastoral areas have recently grown because of social and economic factors (Fratkin, 1992). As more pastoralists become sedentary, this may provide new opportunities for collective action.

Innovation systems

Innovation systems are multi-stakeholder problem-solving partnerships involving research, extension, development agents and rural communities (Sanginga, 2006). One core value of innovation systems is that expert knowledge and the felt needs of rural people are placed at centre stage (Hall, 2006). Innovation systems have emerged to address urgent needs to reduce poverty and improve agricultural productivity. Innovation systems use collaboration to generate and apply knowledge more rapidly and to build capacity among partners (Hall et al, 2006; Sanginga, 2006; see also Chapter 2 in this volume). Complex problem-solving is interdisciplinary. Traditional disciplinary research and development approaches have generally failed to solve complex problems (Hall et al, 2006; see also Chapter 5 in this volume).

There is no one process to create an innovation system (Hall, 2006). Research is needed to clarify how institutions and stakeholders collaborate and what aspects could be made more efficient (Hall, 2006). According to Sanginga (2006), documentation of lessons learned from using innovation systems is rare. He postulates that success of an innovation system is related to:

- the degree to which partners share a vision and have complementary problem-solving abilities;
- dynamic leadership;
- incentives for stakeholders to collaborate;
- prospects for scaling up and institutionalizing partnerships;
- the scope for investing in human and social capital; and
- the extent of resource sharing among stakeholders.

Challenges for innovation systems include:

- high staff turnover among stakeholder organizations;
- clashes among stakeholders;
- coping with high expectations;
- high transaction costs of maintaining an innovation system; and
- challenges for public–private sector partnerships.

PROJECT OBJECTIVES AND METHODS

Research on Kenyan pastoral women's groups

Our main purpose in studying the pastoral women's groups of northern Kenya was to confirm our initial observations that such groups were highly successful (Desta and Coppock, 2002). We interviewed 16 groups in Marsabit and Moyale districts. We wanted to find out how and when the groups were formed, how they are governed and what activities they have pursued.

Connecting Ethiopians with the Kenyans

The objective of bringing the Ethiopian and Kenyan women together was simply to see what would happen. We took 15 Ethiopian women leaders to Kenya for one week during 2001. The travellers visited Kenyan women's groups at five sites in Moyale and Marsabit districts (Desta and Coppock, 2002). After the tour, we followed up with the Ethiopian women to record their impressions and subsequent initiatives.

Forming sustainable collective action groups in Ethiopia

The process of creating collective action groups in southern Ethiopia began at the same time as our discovery of the women's groups in northern Kenya. We were initially unaware of how one component could influence the other. Ultimately, they fitted extremely well because, as will be shown, the Kenyans provided the hard evidence the Ethiopians needed to illustrate benefits of collective action.

During 2000, we initiated participatory research to better diagnose pastoral problems in southern Ethiopia and identify possible interventions. We applied participatory rural appraisal (PRA) (Lelo et al, 2000), where researchers, development agents and community members were engaged for a week of introspective analysis at various locations. The outcome of a PRA is a community action plan (CAP), which sets a path for change based on solving problems with a reliance on local resources. The PRAs in the different communities yielded similar results – namely, that while the communities indicated their biggest problem was scarcity of food and water, the solutions focused on a need for education and finding realistic ways to increase and diversify incomes (Desta et al, 2004). With the PARIMA project serving as facilitator, funds from the US Agency for International Development (USAID) Mission to Ethiopia were used to support the CAPs. A panel of development professionals was assembled to review and prioritize CAPs for funding consideration. Local development partners – governmental organizations (GOs) and non-governmental organizations (NGOs) – assisted communities in implementing the CAPs. The CAPs initially focused on non-formal education (NFE) of youth and adults along with training in micro-finance. Communities donated labour to construct earthen-walled classrooms where a flexible instructional mode was developed to target key literacy problems (Tezera et al, 2003). The micro-finance model involved two levels of organization (Desta et al, 2004): primary groups normally had five to seven members, and five primary groups were merged into secondary groups with up to 49 members. Members started saving regularly according to group bylaws and developed a savings culture. The first micro-loans were distributed in 2002, about a year after the savings programmes began. The interest rate was 10.5 per cent. Later, other capacity-building courses were implemented that dealt with small business management, group leadership, livestock marketing and procedures for preparing animal products for profitable sale. Courses were designed and administered by several GO and NGO partners and supported by USAID mission funds.

After implementing the CAPs, the groups were monitored in an action-research mode (Greenwood and Levin, 1998). Group progress has been tracked on a quarterly basis for the past six years. If a group problem occurs, then corrective measures are implemented. Repeat visits have also allowed for data collection on group financial performance. The Kenyan women have served as important mentors overall. By late 2007, the 59 collective action groups in Ethiopia had graduated into legally recognized multi-purpose cooperatives according to federal and regional development policies. This required an increasing collaboration with government.

It was also important to lay a foundation for economic sustainability of the groups based on livestock marketing since livestock production is the key output for pastoral regions. Since 2003, PARIMA and its partners organized various meetings, exchange tours and workshops to link pastoralists with livestock exporters and policy-makers. This occurred against a backdrop of growing demand for export of small ruminants to the Gulf states, rapid development of the private export industry in the Ethiopian highlands and availability of funds to capitalize the livestock trade (Desta et al, 2006). These interactions allowed pastoralists to learn about the product requirements for export markets and how they could participate in a marketing chain. Policy-makers and leaders of export firms also learned about the production potential of the rangelands. Livestock purchase agreements were forged among buyers and sellers, and several exporting firms began to operate on the Borana Plateau. One outcome was the creation of a northbound supply chain involving the movement of hundreds of thousands of small ruminants from northern Kenya and southern Ethiopia to abattoirs in the Ethiopian highlands (Desta et al, 2006). Dressed carcasses are then chilled and exported via air transport. Many of these animals have been traded by collective action groups in southern Ethiopia.

Creating an innovation system

The PARIMA project began seeking partners early in 2000 to assist with project development. This was simply because the task at hand was too large for PARIMA to handle alone. The project is small. For example, the Ethiopia-based team has only three professional staff, four support staff, two vehicles, two small offices and a modest operating budget. The team also viewed itself as a temporary entity. These factors encouraged

PARIMA to embrace the role of institutional collaborator and facilitator to build an effective innovation system.

In 2001, the PARIMA team in Ethiopia lacked many abilities or connections to enable it to undertake a broad-based effort to diversify pastoral livelihoods. These deficiencies required remediation via institutional partnerships. For example, some of the major deficiencies included:

- an inadequate background in participatory research and outreach methods;
- the inability to design and implement capacity-building courses for illiterate people;
- a lack of cross-border connections that would allow ready access to the Kenyan women's groups;
- a lack of influence amongst policy-makers to allow unrestrained cross-border movement of citizens in support of development activities;
- the inability to fund or implement CAPs; and
- a lack of strong linkages to the public and private sectors that would be instrumental in creating new livestock marketing chains.

In terms of institutional partners, Egerton University in Kenya was important in instructing the PARIMA team in the use of participatory methods. The Kenyan NGO Community Initiatives Facilitation and Assistance (CIFA) helped PARIMA to make contacts with the women's groups in northern Kenya. The Southern Tier Initiative of the USAID Mission to Ethiopia provided funds to support the PRAs and CAPs. The Ethiopian NGO Action for Development (AFD) joined as one of several implementing partners to support the CAPs, a list that grew to several agencies from the Oromia regional government as well as international NGOs such as Save the Children-USA. The Furra Institute of Development Studies in southern Ethiopia initially helped to implement short courses. Federal and regional policy-makers provided input to project activities, and this encouraged their buy-in. Efforts to connect to livestock exporters allowed us to add private-sector firms such as LUNA and ELFORA Agro-Industries to the innovation system.

PROJECT RESULTS

Kenyan pastoral women's groups

The following highlights concerning attributes of Kenyan pastoral women's groups are derived from PARIMA unpublished data. The 16 groups interviewed had existed for an average of ten years. The number of charter members averaged 24 – all women. About 85 per cent of charter members were illiterate. Major objectives of group formation included the reduction of poverty by increasing incomes via micro-enterprise development and livelihood diversification. About half of the groups were formed after people got the idea from a development agent, while the others formed spontaneously. The groups are self-governed and have elected officers, constitutions and bylaws. Responsibilities of members include attending meetings, contributing labour, making monthly payments to group accounts and supporting important functions. Groups can accumulate large sums of money. The groups mitigate drought effects on their members by providing water and food for the needy, promoting restocking with small ruminants and extending low-interest loans. Group characteristics that promote sustainability reportedly include unity of purpose, transparency and accountability of the leadership, and making wise business decisions that lead to well-diversified micro-enterprises. Future plans of the groups are diverse and ambitious. They aspire to create improved housing, meeting halls, shops, schools, training centres, health centres and water facilities.

Connecting Ethiopians with Kenyans

The initial reaction of the Ethiopian women during their tour in northern Kenya was a combination of amazement, frustration and hope (Desta and Coppock, 2002). One means of capturing the impact of the first cross-border tour is via interviews conducted in 2001. The stories were similar and two segments are reported here.

As a result of the Kenya tour, a 40-year-old married woman with five children wanted to send her youngest (six-year-old) daughter to school. She said:

> The tour to Kenya was the first time for me to leave home to travel out of the country, and it was quite an amazing experience ... what fascinated me most was the unity, hard-working spirit and courage of the Kenyan women ... they have changed their lives enormously and developed an unflinching spirit of helping each other.

She explained that the tour had also changed her personal life in many ways. She had replaced her dilapidated house with a new one. She mentioned that, until recently, she had used the same cattle hide to wash clothes on, to serve as a camel pack for loading and to sleep on at night. She said:

> I have changed it now. I bought a plastic basin to wash my clothes and my body. The hide is exclusively used now as a camel pack, and I made by hand a simple and comfortable mattress from locally available material. I have a plan to purchase a foam mattress soon. A few women from my village have already bought a mattress. We are changing a lot.

She thought she would soon build a tin-roofed house. She noted: 'At first, I was ashamed of myself during the tour to Kenya, but then I began counting myself as a human being ... we were encouraged by many people and organizations. Before the tour, I realized that we were living in isolation ... empty and desolate.' She had observed during her trip that women in Kenya had benefited from the rotating savings and credit schemes; she said that she wanted to establish such informal rotating savings and credit systems in her community. She said her group could even surpass what the Kenyan women had achieved.

A 38-year-old woman noted that, as a result of the tour, she regretted not having sent her children to school. Following her return, she convened a number of meetings with her neighbours and tirelessly shared all that she had learned. Her group members strengthened their savings activity and became involved in social support and helping each other in the case of members' sickness, marriage or birth, as they had seen in Kenya. Her personal life had improved because she had bought a foam mattress and some household utensils and clothes for her family and had also improved the hygienic condition of her home. She increased her livestock assets and planned to change her hut into a tin-roofed house. Remarkable changes had also occurred in her savings culture. She said:

> In the past, once we sold livestock, we finished up all the money within a day for no reason; we spent the money carelessly and senselessly as a child or spread it out like feed for a chicken. But now we have learned the importance of saving and we have even grasped how to use our household resources efficiently, let alone not to misuse our money.

She concluded: 'All ambitions, dreams and ideas are the result of the tour.'

Collective action groups in Ethiopia

Between 2001 and 2005, we oversaw the formation of 59 secondary savings and credit groups. These groups have been aggregated into ten pilot projects that include an education and capacity-building programme. The statistics for the pilot projects are impressive (PARIMA, unpublished data). In total, the ten pilot projects had attained a total membership of 2085 by September 2007 – all women. A total of over 800,000 Ethiopian birr had been saved, equivalent to US$93,000. Funds had been distributed in about 4500 micro-loans averaging 1062 Ethiopian birr (US$123) each. Loans had been used for a variety of purposes; but promoting livestock trade dominated. Animals in poor condition were often being bought at low prices and then fattened up for sale. Other activities included loans for starting small businesses such as butchers' shops, bakeries, teashops and commercial vegetable production. Almost all loans had been repaid, including interest; total accrued interest was over 200,000 Ethiopian birr. Overall, this process has followed a five-step model, as depicted in Figure 7.2.

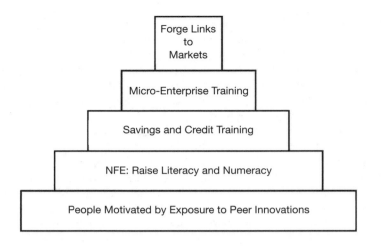

Figure 7.2 *Stepwise process of capacity-building for pastoral collective action groups in southern Ethiopia as created by the Pastoral Risk Management project*

Source: Layne Coppock

In terms of livestock marketing, we observed that groups were willing and able to trade animals and enter a supply chain. Eleven groups sold over 25,600 head of goats and sheep to two export firms during 2004 to 2005, just part of a much larger flow in the region (Desta et al, 2006). The groups have been moderately profitable and income-generation opportunities have been created, although market involvement has in some cases been risky. Overall, our observations suggest that, given high demand, careful investment in capacity-building and reduction in marketing risks, Boran pastoralists can move aggressively to market small ruminants and increase their incomes.

The action research revealed a few threats to group survival over the past six years (PARIMA, unpublished data). In general, the major threat to the sustainability of groups is not drought, but politics as well as the challenges of managing people and finances in a group context. There have also been malicious efforts by some traders to undermine livestock-marketing involvement by inexperienced groups (Desta et al, 2006). In each case where a group experienced problems, the ability of PARIMA and its partners to intervene carefully, manage conflict and restore confidence was vital in sustaining collective action. This illustrates the value of long-term mentoring and monitoring. Not one group has failed.

Structure and function of the innovation system

Overall, an innovation system has evolved in southern Ethiopia that includes four types of members who are involved in one or more of the following project functions: advising, implementation, training and funding. These functions have been performed by a total of 46 formal and informal institutions, including federal and regional governmental organizations, international and local NGOs, pastoral communities and private-sector participants. A general summary of the membership is shown in Table 7.1. There have been 15 international or regional members

Table 7.1 *Innovation system around the Pastoral Risk Management project in southern Ethiopia, 2001–2007*

International- or regional-level members	Local- or district-level members				
	Yabelo District	Moyale District	Liben District	Dugda Dawa District	Dire District
ALRMP	AFD	BZA-M	COOPI	DA-DD	DA-D
AU-IBAR	BZA-Y	DA-M	DA-L	EPGs-DD	EPGs-D
BTL	DA-Y	EPGs-M	ELFORA	OCPB-DD	OCPB-D
CIFA	EO-Y	LUNA	EPGs-L	OPaDB-DD	OPaDB-D
FIDS	EPGs-Y	OCPB-M	GTZ		
Immigration	LUNA	OPaDB-M	GZA		
KARI	OCPB-Y		OCPB-L		
KPWGs	OPaDB-Y		OPaDB-L		
LMA			SAVE/USA		
OARI					
OCPC					
OPaDC					
PARIMA					
Policy					
STI					

Notes: International- or regional-level members: ALRMP = Arid Lands Resource Management Project (Kenya); AU-IBAR = African Union Inter-African Bureau for Animal Resources; BTL = Borana traditional leadership (Aba Gada); CIFA = Community Initiatives Facilitation and Assistance (Kenya and Ethiopia); FIDS = Furra Institute of Development Studies; Immigration = federal immigration officials (Kenya and Ethiopia, at Moyale); KARI = Kenya Agricultural Research Institute (Marsabit); KPWGs = Kenyan pastoral women's groups; LMA = Livestock Marketing Authority (Ethiopia); OARI = Oromia Agricultural Research Institute (Yabelo); OCPC = Oromia Region Cooperative Promotion Commission; OPaDC = Oromia Region Pastoral Area Development Commission; PARIMA = Pastoral Risk Management team of GL-CRSP; Policy = federal and regional policy-makers in Ethiopia; STI = Southern Tier Initiative of USAID Mission to Ethiopia.
Local- or district-level members: AFD = Action for Development; BZA = Borana Zonal Administration; COOPI = Italian development NGO; DA = district administration; ELFORA = private exporting firm; EO = Education Office; EPGs = Ethiopian pastoral groups; GTZ = Deutsche Gesellschaft für Technische Zusammenarbeit (German Agency for Technical Cooperation); GZA = Guji Zonal Administration; LUNA = private exporting firm; OCPB = Oromia Cooperative Promotion Bureau; OPaDB = Oromia Pastoral Area Development Bureau; SAVE/USA = Save the Children-USA (international NGO).
Districts: Y = Yabelo; M = Moyale; L = Liben; DD = Dugda Dawa; D = Dire.

Source: Layne Coppock

and 31 local- or district-level members overall from 2001 to 2007. Some of the partners have morphed over time into other forms (e.g. the 59 pastoral community groups have graduated into legally recognized producer cooperatives). Members have also varied in terms of the duration of their involvement with the project. Some have been short-term members, providing only formal advisory or training inputs as needed. Some have been medium-term, providing local support for implementing CAPs. Private-sector participants such as LUNA or ELFORA have been involved over the last four years; but the intensity of their efforts is dictated by market forces, which vary annually. Other members of the innovation system have emerged as long-term stakeholders; they prominently include the pastoral communities themselves as well as the Oromia regional government agencies.

Importantly, the large area covered by the rangelands as well as the low population densities and limited infrastructure have always encouraged spatial segregation of development processes. For example, the district town centres of Yabelo, Negelle and Moyale in Oromia Region are separated by an average distance of over 100km. International NGOs such as CARE have operated out of Yabelo since the mid 1980s, while Save the Children-USA has operated out of Negelle over a similar period of time. More recently, the livestock export firms LUNA and ELFORA have spatially segregated their animal collection efforts, with LUNA focusing more on Yabelo and Moyale, and ELFORA on Negelle. The 59 pastoral groups originally created by the PARIMA project and its partners are located in or near each of the three district centres, as well as at several other sites. Each pastoral group was part of a CAP that required local development partners to assist with implementation. As a consequence, the PARIMA innovation system is not one regional monolith; rather, it has been replicated in several places with different local partners (see Table 7.1).

DISCUSSION AND CONCLUSIONS

The collective action achieved in northern Kenya and southern Ethiopia has been impressive. Key elements include education, micro-finance, micro-enterprise and income diversification – and all relate to the creation of new forms of human and social capital. Similar observations of the achievements of collective action have been made elsewhere (Grootaert, 2001; Place et al, 2004; Panda, 2007). Our greatest challenge at present is how to sustain and scale up collective action where appropriate, especially with a goal of improving food security and risk management in relation to drought and population growth. The sustainability of the collective action process in southern Ethiopia is affected by local, regional, national and international factors. Sustainable access to livestock markets is an especially important and dynamic challenge.

While our project has led to some important impacts in southern Ethiopia, this would not have occurred without our partners in the innovation system. It is difficult to separate out the relative importance of advising, implementing, training and funding. If one of these four categories had been missing, however, no impact would have occurred. It is important to note that the advising realm may have played the key role overall because – in our typology – advising includes leadership and mentoring. Leadership is a crucial aspect of an innovation system (Sanginga, 2006). Leadership has been observed to occur in several forms during our project. One form has been via the Kenyan women's groups: without the Kenyan women as role models, it is unlikely that there would have been any endogenous ambition for change in Ethiopia. Another form of leadership was found in the Ethiopian women who went on the northern Kenya tour and then took charge back home after seeing what the Kenyans had accomplished. Both of these examples are a testament to having the project beneficiaries – namely, rural farmers and herders – comprise the centrepiece of an innovation network (Hall, 2006). A third form of leadership was provided by the staff of the PARIMA project and its key partners. These people provided the vision and energy to build and maintain relationships and move people and their organizations towards fulfilling larger goals. In this process, the PARIMA staff made conscious efforts to operate transparently, putting the needs of project beneficiaries ahead of their own, enduring very high transaction costs and yet also showing generosity in distributing accolades for project achievements. These attributes are among universally accepted attributes of organizational trust-building and leadership (Kouzes and Posner, 2002).

Finally, PARIMA staff sought the funding needed to pull the pieces together. Funding provided incentives for many partners to collaborate. Prior to the arrival of PARIMA on the scene, there were virtually no traditions of embracing community-led initiatives or forging inter-institutional collaboration on the Borana Plateau. Rather, an atmosphere of competition and mistrust seemed to prevail. Lack of partnership-building among pastoral communities and development agents had been exacerbated by frequent change in federal and regional government agencies, rapid turnover in the staffing of other local organizations, ever-present political and ethnic tensions in the field, and frequent crises such as drought that moved donor agendas towards relief rather than development activities. These factors all promoted instability and uncertainty for pastoral development programming.

Although the innovation system in southern Ethiopia has led to significant achievements, the future of the system is open to debate. In one sense, the mature collective action groups (e.g. cooperatives) and their agency mentors should be able to facilitate replication by other groups. Pressing development needs include continued investment in capacity-building, market development, and the establishment of supportive legal and policy frameworks. Innovation systems will be highly dynamic. Partners having high value two years ago may not have high value in the future. New partners may also emerge. Incentives for partners to collaborate may change over time.

So, is this innovation system sustainable? We sense that the working culture for collaborative pastoral research and development in the region is gradually changing for the better. Government is devolving funds and decision-making to district levels and the potential for inter-organizational communication is increasing through enhanced use of email and the internet. Once PARIMA departs from the scene, however, we suspect that the innovation system in support of collective action may not be sustained. This is simply because many of the partner institutions lack the incentives, leadership, organizational stability and/or resources to maintain collaborative relationships. This situation may be addressed when donors or development ministries begin to link project performance – and positive impact upon pastoral people – with their continued support and when improved project performance, in turn, is clearly linked to benefits derived from innovation systems.

ACKNOWLEDGEMENTS

We thank all partners in Kenya and Ethiopia. We appreciate comments from anonymous reviewers that helped us to improve this chapter. This project has been supported by US universities, host country institutions and the Office of Agriculture and Food Security, Global Bureau of USAID. Work was conducted under Grant No PCE-G-00-98-00036-00 as awarded to the University of California, Davis, in support of the GL-CRSP. Funding support also came from the USAID Mission to Ethiopia. Opinions expressed are those of the co-authors and may not reflect official views of USAID or our partner organizations.

REFERENCES

Coppock, D. L. (1994) *The Borana Plateau of Southern Ethiopia: Synthesis of Pastoral Research, Development and Change, 1980–91*, International Livestock Centre for Africa, Addis Ababa, Ethiopia

Desta, S. and Coppock, D. L. (2002) 'Linking Ethiopian and Kenyan pastoralists and strengthening cross-border collaboration', *Ruminations: Newsletter of the Global Livestock Collaborative Research Support Program* (University of California, Davis), winter, pp4–7

Desta, S. and Coppock, D. L. (2004) 'Pastoralism under pressure: Tracking system change in southern Ethiopia', *Human Ecology*, vol 32, pp465–486

Desta, S., Coppock, D. L., Tezera, S. and Lelo, F. (2004) 'Pastoral risk management in southern Ethiopia: Observations from pilot projects based on participatory community assessments', Research Brief 04-07-PARIMA, Global Livestock Collaborative Research Support Program, University of California, Davis, CA

Desta, S., Getachew G., Tezera, S. and Coppock, D. L. (2006) 'Linking pastoralists and exporters in a livestock marketing chain: Recent experiences from Ethiopia', in J. McPeak and P. Little (eds) *Pastoral Livestock Marketing in Eastern Africa: Research and Policy Challenges*, Intermediate Technology Publications, Rugby, UK, pp109–128

Fratkin, E. (1992) 'Drought and development in Marsabit District, Kenya', *Disasters*, vol 16, pp119–130

Greenwood, D. and Levin, M. (1998) *Introduction to Action Research: Social Research for Social Change*, Sage, Thousand Oaks, CA

Grootaert, C. (2001) 'Does social capital help the poor? A synthesis of findings from the local level institutions studies in Bolivia, Burkina Faso and Indonesia', Local Level Institutions Working Paper 10, World Bank, Washington, DC

Hall, A. (2006) 'Public–private sector partnerships in an agricultural system of innovation: Concepts and challenges', Working Paper Series 2006-002, United Nations University, Maastricht, The Netherlands

Hall, A., Mytelka, L. and Oyeyinka, B. (2006) 'Concepts and guidelines for diagnostic assessment of agricultural innovation capacity', Working Paper Series 2006-017, United Nations University, Maastricht, The Netherlands

Johnson, N. and Berdegué, J. (2004) 'Collective action in agribusiness', Brief 13, 2020 Vision for Food, Agriculture and the Environment, CAPRi (CGIAR System-Wide Program on Collective Action and Property Rights), International Food Policy Research Institute (IFPRI), Washington, DC

Kouzes, J. and Posner, B. (2002) *The Leadership Challenge*, Jossey-Bass, San Francisco, CA

Lelo, F., Ayieko, J., Muhia, R., Muthoka, S., Muiruri, H., Makenzi, P., Njeremani, D. and Omollo, J. (2000) *Egerton PRA Field Handbook for Participatory Rural Appraisal Practitioners*, third edition, Egerton University, Njoro, Kenya

Meinzen-Dick, R. and DiGregorio, M. (2004) 'Collective action and property rights for sustainable development: overview', Brief 1, 2020 Vision for Food, Agriculture and the Environment, CAPRi, IFPRI, Washington, DC

Panda, S. M. (2007) 'Women's collective action and sustainable water management: Case of SEWA's water campaign in Gujarat, India', Working Paper 61, CAPRi, IFPRI, Washington, DC

Place, F., Kristjanson, P., Makauki, A., Kariuki, G. and Wangila, J. (2004) 'Assessing the factors underlying differences in achievements of farmer groups: Methodological issues and empirical findings from the highlands of Central Kenya', *Agricultural Systems*, vol 82, pp257–272

Ramírez, E. and Berdegué, J. (2003) *Collective Action to Improve Rural Living Conditions*, Fondo Mink'a de Chorlavi, Santiago, Chile

Sanginga, P. (2006) 'Enhancing partnerships for enabling rural innovation in Africa: Challenges and prospects for institutionalising innovation partnerships', Innovation Africa Symposium, 20–23 November, Kampala, Uganda

Tezera S., Desta, S. and Coppock, D. L. (2003) 'Improving pastoral livelihood security through education: experiences of the PARIMA project in southern Ethiopia', *Ruminations: Newsletter of the Global Livestock Collaborative Research Support Program*, University of California, Davis, winter, pp3–5

Social Networks and Status in Adopting Agricultural Technologies and Practices among Small-Scale Farmers in Uganda

Robert Mazur and Sheila Onzere

INTRODUCTION

In the context of changing environmental and economic realities, agricultural innovation constitutes a cornerstone in efforts to develop agriculture and improve the livelihoods of small-scale farmers in Uganda (Sanginga et al, 2004). The reconfiguration of agricultural research and extension in Uganda means that positive outcomes are now particularly dependent upon strengthening the roles that farmers play in innovation systems (Wennink and Heemskerk, 2006). At the farmer level, social networks and their changes have emerged as crucial elements in defining the nature of those roles and in delineating the conditions for success or failure of innovations. For farmers, social networks facilitate and incubate innovations by providing a space where knowledge sharing, experimentation and risk mitigation can be embedded.

Many studies have shown how social networks are important in the successful adoption and adaptation of agricultural technologies and practices. However, there is a gap in understanding how the adoption of these agricultural technologies and practices affects structural elements of social networks in non-instrumental ways (German et al, 2006). Research on these issues is important for several reasons. First, it can help to develop an understanding of the social and farming system niches in which certain technologies fit best. Second, research that goes beyond traditional categories can help in the quest to identify bottlenecks constraining particular types of individuals and social groups. Third, such research can identify major leverage points. Finally, there is a need to assess the positive and negative impacts of technologies on resource access and livelihoods.

This chapter addresses the gap mentioned above by examining how a set of innovations adopted from an NGO affected individual and group status for small-scale farmers in Luwero and Kamuli districts in Uganda. It is based on research conducted in June and July 2005 as part of a project designed to understand the social mechanisms and forms of social capital that support adoption and innovation. First, we develop a typology of the farmers interviewed, their livelihood strategies and the constraints they faced in the adoption of agricultural technologies and practices. Then, we look at how farmers used their social networks to support activities related to adoption. Finally, we examine how these activities initiated transformation of social status among the farmers.

METHODS AND DATA COLLECTION

Operationalization of key concepts

Innovations are seen as extending beyond new technologies to include new skills and ways of organizing. They are conceptualized not as isolated phenomena but as necessarily supported and embedded in context-specific social relations (Lindkvist, 1998). In this study, innovations are defined as the adoption, adaptation and use of new agricultural materials and practices by farmers in order to improve their livelihoods.

The agricultural technologies and practices were introduced to Luwero and Kamuli farmers by Volunteer Efforts for Development Concerns (VEDCO), an indigenous NGO that promotes food security and sustainable agriculture through rural development assistance to small-scale producers in Uganda. Farmers were asked to identify and discuss the technologies and practices adopted from VEDCO. Four main areas were identified:

1   farming and animal-rearing practices, including mulching, pruning, planting in straight lines, and confined poultry- and pig-keeping;
2   improved traditional crop varieties such as banana, orange-flesh sweet potato and rice;
3   export crops such as okra, sunflower and vanilla; and
4   market linkages for export crops.

Social networks are the web of relationships among farmers spanning familial bonds and voluntary associations (Fairhead and Leach, 2005). Social networks have discernable boundaries and a normative order (Scott, 1986). In both Luwero and Kamuli districts, social network boundaries were articulated using both formal and informal criteria, including farmer-group membership, friendship, kinship and household membership. In both districts, farmers reported relying most on family, farmer-group members and extension workers for material and social support in their farming activities. The boundaries of a social network can be interactional, spatial or temporal (Scott, 1986). As distinct from spatial and temporal boundaries, interactional boundaries are formed when people interact on particular activities or objectives. Social networks with interactional boundaries related to agricultural activities were relatively new in Kamuli District, while they were more established in Luwero District.

Farmers interviewed

A qualitative approach to data collection was utilized to understand the social processes involved. Through observation, conversation and interviews, respondents are able to describe their situation in the way they see it; from this, grounded theory can be derived (Glaser and Straus, 1967). An interview guide was used to stimulate conversations that were directed by farmers as they related experiences that mattered to them and offered their perspectives.

In-depth interviews were conducted with 26 farmers from Kamuli and Luwero districts. Four categories of farmers were identified by local farmers and VEDCO using wealth ranking in both districts. Table 8.1 indicates the number of farmers interviewed in each category:

Table 8.1 *Farmers interviewed in Luwero and Kamuli districts*

Type of farmer interviewed	Luwero District		Kamuli District	
	Women	*Men*	*Women*	*Men*
Food secure/agricultural trade	3	2	–	1
Food secure	1	2	2	–
Moderately food secure	4	3	1	1
Food insecure	2	2	2	–

- food-secure/agricultural-trade farmers who had enough food for the household and a surplus to sell regularly, either on the domestic market or to the European market through a produce export company (IceMark);
- food-secure farmers who produced enough food for household consumption but had no regular surplus to sell, except occasionally on the domestic spot market;
- moderately food-secure farmers who had enough food for the household, but the situation was precarious; a 'shock' to their livelihood would quickly relapse them into food insecurity;
- food-insecure farmers who did not have enough food to satisfy household needs.

The initial stage of the study was conducted in Kamuli District, in east-central Uganda, where VEDCO had been assisting farmers for just six months. These interviews facilitated an understanding of how social networks were used by farmers in the initial stages of adoption. The study then moved to Luwero District, where VEDCO has been active since 1986. This chapter therefore focuses primarily on the analysis of interviews in Luwero District, while including insights from Kamuli. Two main areas were covered in the interviews. A contextualizing set of questions stimulated farmers to discuss the technologies and practices they had adopted and the experiences and factors that they took into consideration in doing so. Another set of questions encouraged exploration of the impacts of adopting technologies and practices on the social status of farmers. The farmers ranged in age from 24 (male) to 63 (female). The majority had at least primary school education, with only two farmers never having attended school. While several male farmers had attended secondary school, only one had completed it. Only one female farmer had attended secondary school.

RESULTS

Livelihood strategies

Agriculture constituted the main livelihood source for all respondents. All of the farmers were involved in mixed-crop farming and practised small-scale livestock rearing. The most common crops grown in the two districts were bananas, yams, cassava, potato, beans, okra, vanilla, maize and upland rice. Livestock commonly reared included poultry (layers and broilers), pigs and dairy cattle. For three-quarters of the farmers, production was primarily oriented towards household subsistence. One-quarter of the farmers were food insecure. Just over one-half of the farmers interviewed were food secure or moderately food secure. Whenever there was a good harvest, they sold the surplus on the domestic spot market. Farmers waited for traders who routinely scouted the area to approach them and negotiate a price for the desired produce. In this case, transportation costs to the market were borne by the trader. Alternatively, farmers transported their produce to the market and hoped to find a buyer.

About one-quarter of farmers were involved in agricultural trade. These included five farmers from Luwero District (two female and three male) and one male farmer in Kamuli District. The female farmers were involved in rearing animals for the domestic market: one kept pigs and the other kept chickens. Two male farmers in Luwero grew okra that was sold to the European market. Prior to this, both farmers had sold produce on the domestic market. The other male farmer in Luwero sold cassava to a local boarding school. In Kamuli District, only one farmer interviewed (a 32-year-old man) belonged to a cooperative and regularly sold his maize on the domestic and regional markets. No one in Kamuli was involved in export trade.

Moderately food-secure, food-secure and agricultural-trade farmers also engaged in direct marketing to neighbours and people in the area, sometimes with elements of value addition. Several women farmers, for instance, reported selling traditional beer made from bananas to neighbours. Other livelihood resources included remittances from house-hold members – mainly adult children – who had a wage-earning job in nearby towns or the capital city, Kampala. Although not favoured, providing casual labour to other farmers was also mentioned as a source of income.

Vulnerability context

In both districts, environmental changes, labour, financial capital, transportation, markets and information were mentioned as major constraints to adopting and sustaining agricultural technologies and practices. Environmental changes included increased pests and diseases, soil degradation and irregular precipitation patterns. Isaac, a moderately food-secure farmer from Luwero, talked about the challenge of environmental change:

> There are challenges of weather that have really disrupted us. If I plant a crop that necessitates very little rain and then there is too much of it, I mean, I lose. And when there is too much of one thing, rain or sunshine, I end up not meeting my expectations because of the weather changes. (Interview, 8 July 2005)

Jane, also a moderately food-secure farmer in Luwero, when asked the same question, replied:

> There is a problem of weather changes where I plant, let's say, beans and then they are hit by a dry spell. Then when the rains come, there is too much, I cannot harvest anything. I need moderate rainfall and then sunshine at optimal levels. Also the pests and diseases are now multiplying and affecting crops. If I don't spray the beans, then I can hardly harvest anything. (Interview, 11 July 2005)

At the time of this study, coffee and the traditional variety of banana had been badly affected by wilt diseases, depleting traditional sources of income. Seeking to mitigate the effects of this shock, many farmers reported that they were attracted to VEDCO workshops because the NGO was offering disease-resistant banana varieties. New crop varieties, agricultural practices and animal-rearing practices, however, required more labour, money and time inputs. Resource limitations prevented moderately food-secure and food-insecure farmers from adopting technologies and practices. John, a moderately food-secure farmer, explained how limited finances and labour restricted even farmers with abundant land to subsistence:

> The major challenges are capital [financial] and labour. I can manage to cultivate 4 acres; but then I have no money to employ somebody to help me manage those 4 acres. So I end up doing just half an acre, which may not even be enough for household consumption. So I am left with nothing to sell off in the long run and I remain in that vicious cycle of poverty. I have no starting point as a farmer. (Interview, 8 July 2005)

Those farmers involved in agricultural trade were able to use part of their regular income from the sale of produce or animals to hire labour during planting, weeding and harvesting times. The ability to hire extra labour during these household labour peaks was a crucial element in determining whether farmers could maintain the innovations they adopted. Labour constraints were of special concern to female farmers. While men could access the labour of all household members, female farmers relied mostly on labour provided by children in the household. For this reason, many female farmers reported waiting until school holidays when children could help to start a new project. A food-secure female farmer from Luwero, Becka, explained the different contributions of her household members to her farming activities:

> My family has greatly contributed, especially the children during holidays. If VEDCO assigns me tasks that I cannot do when they are in school, I wait until they are on holiday. They provide labour, but they also like doing work with their mother. I mean, I am not forcing them to do what they don't want. And then my husband, he is not involved in my activities. He does not ask me: 'Why are you doing this, where are you going?' I am alone in operating my business with the children. He does not encroach on my output, even after the sales. (Interview, 14 July 2005)

Lack of transportation, markets and information were also reported as major challenges to adoption. Erasto, an agricultural-trade farmer, talked about problems of domestic marketing because of transportation, information and price difficulties. At the time of the interviews, he was selling okra on the export market.

> I used to produce and market groundnuts. Buyers would not look for them in the village. I [first] had to go out to a place called Kasana and then find someone to buy the groundnuts and then negotiate the prices and then take the [groundnuts] there ... The problem I faced was low prices. I mean, sometimes I was in need [of money] or I had a problem, so someone [the buyer] could charge me, and then change the prices. So I had a problem of prices and, generally, the market wasn't there. The other thing was transport. Trying to transport the produce to the buyer was a big problem. Because the incomes were very low, I could not maintain a bicycle. (Interview, 7 July 2005)

Farmers in both areas also viewed adoption of new crop varieties, livestock and practices as a way of addressing these challenges. Many farmers indicated going to VEDCO's sensitization workshops, knowing which specific vulnerabilities they wanted to focus on. They then assessed how the planting materials given and the new farming practices taught addressed those vulnerabilities. Rose, a food-secure farmer in Kamuli, explained that gaining household food security was her main attraction to VEDCO: 'When I discovered that VEDCO's major objective was to fight food insecurity and then poverty, such that a farmer can have a lot of produce and then a surplus for sale, that was the best seducing factor for me' (Interview, 1 June 2005).

In contrast to Rose, market linkages were the most important consideration for Erasto:

> There were two organizations [already in the area]: ADRAK and AMREF. I wasn't involved in AMREF because they deal with orphans and I don't have one. But ADRAK normally gives out fruits and coffee, then boar goats. But then they just educate. They don't go out and look for a market for you. VEDCO, when they sensitize you, they go ahead to look for a market – that's where VEDCO beats the rest of the organizations ... that is why I wanted to work with VEDCO. (Interview, 7 July 2005)

The use of social networks

In both study areas, adoption of technologies and practices was supported by farmers' social networks in two ways:

1   acquiring important resources, such as financial support and labour; and
2   spreading innovations by exchanging related information and practices.

The spread of information and practices affected social status in farmer networks.

Information transfer
When agricultural innovations are introduced to an area, how information is spread is an important indicator of the way in which social networks are organized and change (German et al, 2006). Information transfer in the research areas emerged as an important dimension in innovation adoption and sustainability. Figure 8.1 shows from whom the farmers interviewed in Luwero District first heard about VEDCO (information for three farmers is missing).

The information paths show that social networks played a crucial role in determining which farmers had access to information and, consequently, how innovations were spread. Five farmers indicated that they were introduced to VEDCO by Mathias, two by Isaac and none by Aisha, indicating that the span of an individual farmer's network had a discernable influence on innovation. While local leaders played an important role in organizing initial meetings for farmers when VEDCO started up in Luwero, other farmers were responsible for most of the subsequent spread of innovations. Approximately two-thirds of the farmers had been introduced to VEDCO by volunteer rural development extensionists (RDEs) or friends. RDEs are farmers who receive training

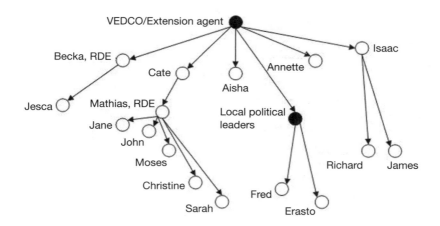

Figure 8.1 *Information transfer among farmers in Luwero District*

and planting material from VEDCO and, in turn, disseminate these to other farmers. A majority of those who had been introduced to VEDCO by other farmers were food insecure or moderately food secure. The remaining one-third of farmers indicated they had been introduced to VEDCO's innovations by an extension agent or by local political leaders. Farmers who had first heard of VEDCO from an extension agent tended to be moderately food-secure and agricultural-trade farmers. Only two farmers had first heard of VEDCO from local political leaders: Fred and Erasto, who were already involved in regular domestic trade prior to their involvement with VEDCO. This pattern contrasted sharply with that in Kamuli, where the majority of farmers had first heard of the innovations from a local political leader or an extension agent. This difference may be primarily attributable to the different lengths of time that VEDCO had been operating in each district.

As the innovation diffusion processes unfolded, the type of information that was shared changed. Initially, information shared between farmers was on plant and animal varieties, as well as agricultural practices introduced by VEDCO. It included the constraints likely to be faced before and after adoption and the benefits that farmers could expect after adoption. In later stages, the information shared was principally related to sustaining adopted varieties and practices. This was in three general areas:

1 information on farming practices, including mulching, pruning and harvesting;
2 educative information, such as pesticide use and how to achieve market prices; and
3 information on the success or failure of field experiments conducted by farmers.

While information on the first two areas originated from VEDCO's workshops, experimentation information came primarily from farmers.

Exchange of farming practices and experimentation
In both Kamuli and Luwero, social networks were also used extensively to exchange farming practices learned through VEDCO. While some of the farmers reported sticking to the crops and practices as taught by VEDCO, the majority reported experimenting further with these methods. This included using the techniques learned on other crops and animals not introduced by VEDCO and modifying the practices/methods to suit their situation. For instance, a female farmer who started out rearing an improved breed of pigs provided by VEDCO switched to a traditional breed after realizing that the former was more susceptible to disease. However, she continued to use the practices taught by VEDCO to manage the animals. Other farmers kept the crops introduced but completely eliminated a cropping method that they had learned. Fred, an agricultural-trade farmer, explained how he eliminated some methods learned:

> One of the things that VEDCO showed us was planting in straight lines, mulching and then raising beds. But I realized that Uganda is very fertile land; even when I don't mulch, I can get a good harvest. The land is still fertile. (Interview, 7 July 2005)

Individual farmers then shared the results of their experiments with others, thus enabling mutual learning.

Changes in social status
Ridgeway (2003) describes social status as an evaluative hierarchy that exists within social groups, such as farmers and traders, or between individuals within a social group. This evaluative hierarchy is structured by a wide range of characteristics such as age, gender and class (Goffman, 1951). Indicators of social status include esteem, respect, likeability and belonging (Triandis et al, 1996). Social status is determined by a wider range of factors than just economic wealth or class. Instead, status is a distribution of social prestige that develops in groups and individuals who regularly interact with each other, but is also recognized by those who are socially distant (Ridgeway, 2003). Because status is an evaluative hierarchy, it informs expectations of a person's or group's ability to perform tasks. These expectations consequently influence access to resources and the willingness of other social actors to cooperate with individuals or groups (Berger et al, 1978). Berger et al (2002) also note that social status is the result of negotiations by actors in the social system. Thus, the categorization of actors according to status changes over time. Transformations in social status may happen as a result of technical or economic changes. In both Luwero and Kamuli, interviews with farmers indicated that innovation adoption and its impacts upon the vulnerability

context had a profound effect on the social status of individual farmers, as well as farmers as a social group. For farmers in all four categories, interactions based on sharing of practices and information became a more important delineator of the farmers' social networks as opposed to spatial or temporal boundaries. Aisha illustrated this when talking about her interactions with other farmers:

> I have many new friends not from within [the village] but, let's say, from Kawanda, Ibero and Katuka: basically, those who come to check on my innovations. I have become very good friends and the relationships are being upheld. (Interview, 12 July 2005)

For farmers, this increased interaction provided what Anderson (2003) refers to as 'settings of sociability', in which farmers were able to transform their status. Changes in individual status were associated with the increase in household food security and income. Farmers across all four categories reported an increase in food security and income after they had adopted improved varieties and practices. This was true even for those farmers who only sold surplus produce periodically. This extra income was first invested in an asset that would secure livelihoods, then in large livestock and thereafter in children's education. Aisha explained how she had used her increased income:

> It has really changed for us. We had no bicycle before. But after this intervention, we managed to work on some innovations and managed to buy a bicycle. It was followed by having a cow. We now have animals we keep [chickens] and we produce milk as well. Our income levels have risen and we can now take our children to school. We have money to pay off our school fees. (Interview, 12 July 2005)

Another farmer, James, explained how he invested his income:

> I have changed in that I had no hoe. I had to borrow one for cultivating. I have my own hoes now. And the clothes, compared to what I was putting on before, it's better. So this is a big change. (Interview, 8 July 2005)

As these responses show, while increased income was invested in tangible assets to secure or improve their livelihoods, the investments also had a social value in the way in which farmers thought of themselves. Fred, a Luwero farmer involved in selling okra for export, explained how increased food production and income had changed how he felt about himself:

> For me, as somebody who is now earning an income, I can meet my domestic demands. If I don't have paraffin or salt, I can sell some of my produce and meet those demands. I don't have to go to my neighbours saying, 'Help me! Help me!' This is the difference. (Interview, 7 July 2005)

The interviews indicate that this increased ability to meet household needs transformed the social esteem and respectability of farmers. Mathias explained how working with VEDCO had changed his status:

> I have changed a lot because I am respected by people. Apart from respect, my standard of living has changed. I am changed in that, altogether, I command respect. I have changed in that, if you look at me, I am a youth but I am like an elder … After joining VEDCO, I became famous. …yeah, it's different to be a youth and to be called *Mzee* [elder]. (Interview, 13 June 2005)

While instrumental gains played a large part in transforming the farmers' status, information and practices acquired in the process of innovation adoption also shaped social status. Information and practices acquired a social value in the process of becoming knowledge and skills possessed by individual farmers and farmer groups. Erasto explained how agricultural practices had become part of his repertoire of skills:

> When I feel like I want to graft a mango sapling, *I can*. When I feel like I want to make compost manure, *I can*. When I feel like I want to plant bananas correctly, *I can*. It is because I have the skills now. (Interview, 7 July 2005, emphasis added)

Possession of knowledge and skills was especially important for female farmers, all of whom indicated that their status had been transformed not only by the fact they could now earn money 'as women', but that they could now teach other women and men in the village. Betty, a food-insecure farmer and a representative of a disabled women's group in Kamuli, explained it this way:

> We are emancipated. We have gained confidence to represent ourselves. We used to say, 'Let one of us represent us'; but we would be afraid. But now we can represent ourselves as women. We now have the skills we need to fight hunger. (Interview, 1 June 2005)

Rose, a food-secure female farmer in Kamuli, explained further:

> One thing is that fame is not always found with a Msoga woman. Being known by very many people, rendering all those skilful facilities, I see myself as a really changed woman. (Interview, 1 June 2005)

Status transformations were not limited to female farmers. Among Luwero farmers, social status had improved as they could now collectively act for the benefit of the group. Fred talked about how adopting okra as an export crop had changed the relationships between farmer neighbours:

> I want to reflect on the relationships with neighbours. Now we are organized as a group in the village. Every week, farmers get 60,000 Ugandan shillings [approximately US$36] out of their sales. I have changed and the neighbours have also changed in that they have something that is generating income for them daily. I see that the village is growing and, if this growth continues, we will develop more. (Interview, 7 July 2005)

Information was also a key leverage point in helping farmers to change their position within the agricultural sector in regard to other actors, especially traders. At the time of the interviews, farmers had access to weekly market prices from VEDCO. This meant that farmers, even those who traded on the spot market, were in a better position to negotiate with traders. Because of the availability of market price information and the consequent change in the power dynamic during price negotiations, farmers as a group viewed themselves as more legitimate actors in the agricultural sector.

Status transformations, however, were not uniform for all farmers. An important development was the emergence of high-status individuals who were central in sustaining the farmers' social networks. These individuals, particularly RDEs, had become central in transferring information and, consequently, shaped how innovations were sustained. For instance, RDEs who attended workshops to learn new farming practices then passed on these skills to other farmers in their group or neighbourhood. These farmers were also inspirational individuals who served a cohesive role in the social network. If they were no longer able to interact with other farmers, this affected not only the transmission of information and skills to other farmers, but also the ability of other farmers to maintain their social network. This point is demonstrated by the case of one women's group in which the RDE 'quarrelled' with her husband. Her husband decided that he did not want her to use part of his land as a demonstration garden. As a result, farmers in this group lost not only a space to experiment and a formal gathering space where they could routinely share information, but also the 'glue' of the group; eventually, the group broke up. Annette, who had been a member of this group, put it this way:

> The chairperson had a misunderstanding with her husband, which later led to the husband saying: 'I no longer want you in VEDCO.' She was inspirational. She could mobilize and advise us and, when she left, members lost hope and neglected group activities. So we are now working on our own again. (Interview, 11 July 2005)

While there were positive outcomes for some farmers, others within the same group could not sustain the varieties or practices that they had adopted. The case of Emily, a 50-year-old widow in Luwero District who was raising three grandchildren, illustrates this. Emily had been receiving assistance from VEDCO for three years at the time of the interview. Although agriculture was now her only source of livelihood, she had previously been involved in a variety of non-agricultural businesses, with monetary support and advice from her eldest son. Because of the illness and death of this son, she was unable to continue her business.

Shortly before her eldest son passed away, she was approached by Mathias to join a VEDCO farmer group. She adopted improved banana varieties promoted by VEDCO. During this period, her second son fell ill and she spent most of her time nursing him. As a consequence, most of the banana plants died because of neglect. Her last child is also now sick and Emily is afraid that she, too, will die. When asked to comment on the main reason why she hasn't been able to grow successfully the improved banana varieties adopted from VEDCO, Emily mentioned a lack of time and labour and the fact that her social network had shrunk as each of her children died.

CONCLUSIONS

In this chapter we have examined how crop production and animal-rearing practices adopted from an NGO affected the individual and group status of 26 small-scale farmers in Luwero and Kamuli districts in Uganda. All of the farmers involved in the study relied on farming as their major source of livelihood. Farmers in both districts faced major constraints in adopting innovations, which included environmental changes, labour, finances, access to transportation and markets. In both areas, farmers viewed the adoption of new varieties and practices as a means of addressing these challenges. Farmers used their social networks extensively to support adoption. This included the transfer of information and practices adopted. Increases in household food security and income, as well as the transformation of information and practices into knowledge and skills contributed to improving the farmers' status. Most farmers, especially women, mentioned improvement in indicators of social status such as social esteem, social respect and 'fame' based on knowledge and skills.

While these changes in social status were reported at an individual level, there was a group dimension noted in both Luwero and Kamuli. Here, farmers reported that their status as a social group within the agricultural sector had changed with regard to other actors, particularly traders. A final major finding was that the transformation of social networks, as evidenced by status changes, had differential outcomes for individual farmers. While some farmers became high-status individuals with enormous influence on the well-being of social networks, others were unable to take advantage of innovations and the social benefits that they offered.

As pointed out previously, studying non-instrumental roles of in-novation adoption is important in order to identify major leverage points and bottlenecks and to understand social niches in which technologies fit best. Within the study, those individuals who have access to information and continuous training occupied a central location in farmer networks and could potentially serve as focal points to support innovation sustainability. The case of Emily, on the other hand, shows how bottlenecks that extend beyond deficiencies inherent in agricultural innovations can prevent farmers from benefiting. With regard to social niches, the results indicate that male and female farmers engaged in agricultural trade were more comfortable with different innovations. Women tended to choose livestock, while men chose export crops. This was intriguing since farmers denied that these differences had anything to do with traditional cultural divisions in agricultural labour or the prestige of export crops versus domestic animals.

Consequently, several questions emerged from this study. A persisting question in innovation adoption and sustainability is whether there is transformation or persistence of status distinctions among farmers who share innovation spaces. Related to that is a concern with how status distinctions contribute to differential benefits. Research conducted in Tanzania, for instance, found that – despite targeting women as the initial beneficiaries – men benefited most during the 'spill-over' stage, when information on innovations was transferred between farmers (German et al, 2006). The persistence of old status distinctions may mean that initial benefits from interventions that targeted certain social categories may revert back to those who are privileged by status and structural conditions. Future research on innovation and status, therefore, should consider how innovation in the second and third 'spill-over' stages is patterned by social niches that reflect new status traits formed during initial innovation processes and those in existing status categories.

ACKNOWLEDGEMENTS

Thanks to the entire staff of VEDCO and the farmers in Kamuli and Luwero districts. Special thanks to Ibra Mbadhi and Dorothy Masinde. Funding source: Iowa State University Council for International Programs grant: Roles of Social Capital in Technology Adaptation for Food Security in HIV/AIDS-Impacted Communities.

REFERENCES

Anderson, E. (2003) *A Place on the Corner*, University of Chicago Press, Chicago, IL

Berger, J., Fisek, M., Norman, R. and Zelditch, M. (1978) 'Status construction and social interaction', *Contemporary Sociology*, vol 7, no 1, p90

Berger, J., Ridgeway, C. and Zelditch, M. (2002) 'Construction of status and referential structures', *Sociological Theory*, vol 20, no 2, pp157–179

Fairhead, J. and Leach, M. (2005) 'The centrality of the social in African farming', *IDS Bulletin*, vol 36, pp86–90

German, L., Mowo, J. and Kingamkono, M. (2006) 'A methodology for tracking the "fate" of technological interventions in agriculture', *Agriculture and Human Values*, vol 23, no 3, pp353–369

Glaser, N. B. and Strauss, A. L. (1967) *The Discovery of Grounded Theory: Strategies for Qualitative Research*, Aldine, Chicago, IL

Goffman, E. (1951) 'Symbols of class status', *British Journal of Sociology*, vol 2, no 4, pp294–304

Lindkvist, L. (1998) 'Knowledge communities and knowledge collectivities: A typology of knowledge work in groups', *Journal of Management Studies*, vol 42, pp1189–1210

Ridgeway, C. (2003) 'Status construction theory', in P. Burke (ed) *Contemporary Social Psychological Theories*, Stanford University Press, San Francisco, CA, Chapter 13

Sanginga, P. C., Best, R., Chitsike, C., Delve, R., Kaaria, S. and Kirkby, R. (2004) 'Enabling rural innovation in Africa: An approach for integrating Farmer Participatory Research and market orientation for building the assets of rural poor', *Uganda Journal of Agricultural Sciences*, vol 9, pp942–957

Scott, R. (1986) *Social Processes and Structures: An Introduction to Sociology*, Holt, Reinhart and Winston, Chicago, IL

Triandis, C., Vasiliou, V. and Thomanek, E. (1996) 'Social status as a determinant of respect and friendship acceptance', *Sociometry*, vol 29, no 4, pp396–405

Wennink, B. and Heemskerk, W. (2006) Farmers' Organizations and Agricultural Innovation: Case Studies from Benin, Rwanda and Tanzania, Bulletin 374, Royal Tropical Institute, Amsterdam, The Netherlands

From Participation to Partnership: A Different Way for Researchers to Accompany Innovation Processes – Challenges and Difficulties

Henri Hocdé, Bernard Triomphe, Guy Faure and Michel Dulcire

INTRODUCTION

Top-down approaches to innovation development are still frequent or even dominant in many circles. However, such approaches have long ceased to be the only paradigm for designing and delivering the inventions needed to help farmers adapt to a rapidly evolving environment. Agriculturalists moving away from top-down approaches were following the steps of social scientists such as Lewin (1946), who embarked on research conducted in close interaction with local actors. During the 1970s, farming systems approaches were developed both in the English- and French-speaking spheres (see, for example, Norman et al, 1982; Jouve and Mercoiret, 1987) and were soon followed by the emergence of participatory approaches: from participatory rural appraisal (PRA) (Chambers et al, 1989) via participatory technology development (PTD) (Ashby and Sperling, 1995; Veldhuizen et al, 1997) to participatory learning and action research (PLAR) (Scoones et al, 1994). This gradual evolution reflected a growing awareness by researchers that it was crucial to better involve the farmers in the research process and to empower them in the process of doing so. Today, many researchers are engaged in refining such approaches and methodologies in order to further improve the way in which research works with an array of stakeholders of the rural sector in the hope that this will speed up the innovation process, increasingly seen from an innovation systems perspective (World Bank, 2006).

Even after several decades, the shift away from top-down approaches is far from complete and has not happened without resistance. The first steps towards introducing farmer participation into the research process are relatively painless because researchers keep a fair degree of control over the process. The subsequent steps, leading to the development of full-fledged partnerships, are much more difficult because researchers

have to reassess many conventional research methodologies and scrutin-
ize their deeply held individual and institutional values and mechanisms
for decision-making.

This chapter focuses on gaining a better understanding of such difficulties. Drawing from ten projects conducted by researchers working with
farmers in a variety of contexts over the past decade, it provides insights
and lessons on five major issues:

1   the conditions that led to encounters between individual researchers
    and farmers, including an analysis of the 'breakaway' from conventional
    modes of operation of individuals and institutions;
2   the non-linearity and low predictability of the project trajectories over
    time;
3   the diversity and specificities of the project set-up;
4   the role of farmers and farmer organizations in the partnership; and
5   the types of results obtained.

The chapter concludes by outlining some specific challenges for researchers
involved in action research in partnership.

MATERIALS AND METHODS

This chapter is based on an investigation developed within the context of a
research project called Construction of Innovation and Role of Partnership
(CIROP) conducted in 2005 to 2007 by the French Agricultural Research
Centre for International Development (CIRAD). CIROP addressed two
interrelated questions:

1   What types of partnership are required to strengthen the capacity of
    rural societies to innovate?
2   Which methods derived from action research are required to do so?

CIROP focused on innovation processes and partnerships. Innovation
processes were considered in their technical, social and organizational
dimensions, whereas partnerships were defined as the set of formalized
linkages established among actors in a given territory to federate means
(material and immaterial) around projects or programmes constructed
jointly to achieve shared objectives (Lindenperg, 1999). While CIROP inter-
venes directly in two ongoing action research projects in West and Central
Africa, it was also involved in making a comparative study of past and
ongoing research projects in which local actors were involved to differing
degrees and in different ways in innovation processes.

The case studies

Table 9.1 lists the ten experiences selected for the comparative study. Six of them took place in Latin America, two in Africa and two in France. CIRAD researchers were directly involved in seven of them. Issues addressed were highly diverse. They ranged from focused technical interventions (e.g. plant breeding for sorghum and durum wheat, dissemination of plantain transplants and conservation agriculture) to strengthening the adaptive capacity of farming systems to drought, supporting farmer-level decision-making, structuring the cocoa supply chain, community-based land management, and the creation of future scenarios for smallholder agriculture. Four of these ten experiences are still ongoing. In all cases, research and farmer organizations were the main stakeholders involved. In half of the cases, they were joined by extension services. Other stakeholder types were involved in about half of the cases: agro-industry (Ecuador), education (Brazil Cerrados), land-use planning agency (Reunion Island) and non-governmental organizations (NGOs) (northeast Brazil). Across all projects, researchers expressed a desire to 'do research differently' in close cooperation with farmers, but only one of them (Costa Rica) explicitly claimed to be doing action research.

The ten cases can be divided into three main clusters according to the source of initiative in launching the project: research (seven cases), stakeholders other than research (two cases) and research and farmers jointly. While diverse, these cases do not constitute a representative sample of the breadth of existing partnerships in research. Selection biases include, among others, an over-representation of research-led projects, with priority given to research–farmer relationships.

Case study framework and comparative analysis

A common framework was developed to analyse the case studies by applying concepts related to action research (Liu, 1992). Four principles form the heart of action research:

1 an equilibrium between a will to change and a research purpose;
2 a dual objective aimed at resolving a problem and producing new knowledge;
3 collaborative work in a mutual learning process; and
4 an ethical framework devised by all participants.

Putting action research into practice involves implementing three main overlapping phases: the initial phase, including defining the problem, objectives and commitments; the realization phase, including diagnosis, planning, applying potential solutions and evaluation; and the disengagement phase. Recently, Chia et al (2005) introduced the notion of 'action

Table 9.1 *Selected characteristics of the ten case studies*

Site	Main focus	Major stakeholders involved*				Period of operation	References
		Research	*Farmer organizations*	*Extension*	*Miscellaneous*		
Central/south Cameroon	Diffusion of plantain transplants	XXX		X		1997–2002	Temple et al (2006)
Nicaragua	Participatory sorghum breeding	XXX	X	X		2003–present	Trouche et al (2005)
North Cameroon	Design of farm management advisory method	XXX		XXX		1999–2003	Djamen Nana et al (2005)
Reunion Island, France	Negotiation between stakeholders on land-use management	XXX	X		X	1999–2000	Dulcire et al (2006)
Mexico	Design of multi-stakeholder platform for conservation agriculture and irrigation	XXX	XX	XXX	X (input supplier)	2000–2004	Triomphe et al (2006)
Northeast Brazil	Participatory innovation to cope with drought	XXX (NGO)	XXX	X		1992–2003	Sabourin et al (2006)
Brazil Cerrados	Strengthening sustainable development in the agrarian reform sector	XXX	XX	X	X (education)	2002–present	Scopel et al (2005)
Ecuador	Implementation of quality cocoa supply chain	X	XX		XXX (agro-industry)	2000–present	Dulcire and Roche (2007)
Southern France	Participatory organic durum wheat breeding	XXX	XX			2003–present	Chiffoleau and Desclaux (2006)
Costa Rica	Farmer organizations imagining the future of smallholder agriculture	XXX	XXX	X		2004-2005	Faure et al (2007)

Note: * Importance of involvement is qualified on a scale ranging from some (X), medium (XX) to very strong/leading role (XXX).

research in partnership' to add and emphasize the importance of associating multiple stakeholders in the action research process.

The framework included five main headings: overall context; description of the various phases of the experience; analysis of the results and impacts; synthesis of the most outstanding features of the experience; and major doubts and questions. It was applied in several stages. First, a short bibliographic review was produced on each experience. Then, one or two outsiders conducted semi-structured interviews with researchers closely involved in the experiences and, whenever possible, with other stakeholders. They developed an iterative version of a written document on each case study, based on feedback from the initial interviewees and other CIROP team members. The aim was to produce a final document of about 15 pages for each case.

This chapter presents the results of the initial comparative analysis of the ten cases, focusing specifically on three main issues: the types of results obtained; formalization of commitments and relationships; and operational and governance set-up.

SELECTED LESSONS LEARNED FROM THE CASE STUDY ANALYSIS

Encounters among key individuals and breakaways

In most case studies, research initiated the participatory process by proposing to its would-be partners to help them solve problems that they faced (supply-driven process). In other cases, the non-research partners took the initiative and contacted research as a valuable contributor to solving a previously identified problem or constraint (demand-driven process). In only one case did supply and demand coincide at a given time.

All case studies showed that, rather than institutions, it was individuals with specific skills and historical trajectories who initiated the encounter. Institutions did play a role by granting such individuals a certain degree of freedom or by giving them an actual mandate to 'do things differently', in the best cases. In several cases, however, it was up to the individuals themselves to create the space they needed to operate.

Reasons abound to explain why such individuals were eager to work with other actors. Some of them, keenly aware of the problems and dead-ends associated with more conventional approaches, were actively seeking more pertinent and efficient approaches to innovation development and diffusion (Mexico, Nicaragua and central/south Cameroon). Others were simply convinced that they could not reach their objectives without dialogue and cooperation with other stakeholders (Mexico again, Costa Rica, Reunion Island and southern France). Still others wanted to give the same importance to scientific and social objectives, implying that research had to identify organizations with a strong social and political legitimacy

(Brazilian cases). In all cases, the discovery of the need to work with other actors was gradual: indeed, most individuals had actually started their journey towards 'the other' quite some time before, when they embarked in a reflective, iterative process of reassessing and reorienting their approaches and methodologies. They usually engaged in such novel collaborative processes in an empirical *ad hoc* manner, accepting the challenge of learning by doing and adapting the approach as they went along.

How research partners, and particularly farmers, reacted to these novel approaches was variable and evolved over time. Some watched intently without strong involvement (central/south Cameroon), others instantly adhered (Nicaragua), others started by watching 'over the hedge' the behaviour of these unconventional researchers before involving themselves actively (Nicaragua again). Some were quite ready to take over responsibilities previously assumed routinely by research (north Cameroon), but without accepting the entirety of what researchers wanted them to do (e.g. active involvement of women). In the cases where farmers took the initiative, their requirements and expectations *vis-à-vis* research were clearer ('we are the ones who set the criteria for what we want to do research on'). This led to an immediate definition of terms of reference for the farmer–research interaction, which were gradually refined (southern France). Things played out quite differently in Ecuador: the researcher assumed the role of mediator between the agro-industry and the farmer organizations.

In summary, our case studies illustrate the key role played by individual researchers who embarked on a long-term professional trajectory, seeking novel ways of conducting research. This eventually led them to work closely with stakeholders whom they did not necessarily know in advance. In doing so, they could usually count on the discrete benevolence of their institutions. They also revealed to farmers during the course of action a new unusual face of research and of themselves as individuals. This encounter, in turn, created favourable conditions for engaging in fruitful dialogue and negotiating objectives and modalities for joint work. Much the same analysis applies to other stakeholders as they work with research. The story did not end with the encounter. Other challenges soon followed, such as how to mobilize the resources needed to implement joint activities. But a key lesson so far is that the conditions under which these initial encounters take place bear in themselves powerful ingredients for the eventual shape and success of the project.

Non-linear pathways of partnership projects

There was nothing automatic in the actual trajectory of the projects, which tended to take non-linear, highly unpredictable pathways for several reasons. For one, most projects were the result of highly personalized interactions and negotiations at the local level, with only limited efforts made to scale up and institutionalize the corresponding agreements and approaches. Under such circumstances, the course of the project tended to change matter-of-factly as soon as the need for this was perceived and agreed upon. In southern France, for example, farmers decided to put emphasis on legislation problems, and research followed. In other cases, stark differences in the core interests of the various participants – at both individual and institutional level – emerged over time, modifying the initial agreement on objectives and activities. Another reason for non-linearity had to do with the limited knowledge that research usually had about prevailing social interactions and power relationships among stakeholders, which furthermore evolved over time. In Mexico, for example, none of the stakeholders was willing to challenge decisions made by the powerful state representative, when he grabbed control of the project from the researchers who initiated it. The unequal ability (and at times, willingness) of the various participants in a multi-stakeholder project to follow agreed-upon rules for project operation may also play a key role. In Costa Rica, for instance, the farmers did not realize all of the consequences of the rules agreed upon when the project started.

In conclusion, non-linearity and unpredictability appear to be key intrinsic features of multi-stakeholder projects. Providing or negotiating enough time for partnerships to evolve is critical. These take shape gradually and mature thanks to the mutual knowledge and learning gained by individuals and institutions adjusting to each other's vision and behaviour. But this happens if and when they are given enough time to periodically revisit not only their activities, but also strategic aspects such as objectives, modes of operation, and roles and responsibilities of each partner (Liu, 1992). Unfortunately, these much-needed adjustments do not necessarily occur smoothly and gradually, but rather in crisis mode and at unexpected times. Crises may produce negative consequences, but can also provide the opportunity to address issues not properly tackled at earlier stages of the project. When properly managed, crises also allow various partners to take ownership of the project. Thus, it is wise to devise from the start some mechanisms to pick up early and still weak signals of impending tensions and crises, and to manage them adequately once they emerge in order to minimize collateral damage.

Diversity of set-ups and governance mechanisms

Tasks and responsibilities are distributed among participating stakeholders at two distinct levels: operational set-ups and governance mechanisms.

The *operational set-ups* are designed whenever stakeholders jointly decide to carry out an agreed-upon activity, with the willingness to 'do it together', whether or not clear 'rules of the game' have been formalized between them. Our case studies illustrate the diversity of operational set-ups designed for conducting diagnosis, monitoring and evaluation, training courses, exchange visits, group development, assessing results, experimenting on station or on farm, planning activities, etc. (see Table 9.2). Some of these activities rely on rather conventional research methods and tools. Other set-ups are less conventional, such as when experiments are managed directly by farmers (Mexico, Nicaragua, north Cameroon, northeast Brazil, southern France).

The shift towards co-piloting of these set-ups is strong. Recurring questions pop up in the debates taking place during joint planning sessions, such as 'up to what point should such and such an activity be implemented by research or by its partners?' However, the actual effect of co-piloting remains unclear: in what form and to what degree does it contribute to strengthening the partnership spirit, to the quality of the problem-solving solutions and to the generation of new knowledge?

In terms of *governance*, most case studies did not develop specific mechanisms for deciding jointly on strategic project orientation or for managing conflict among stakeholders. Project coordinators tended to be much more accountable to their own hierarchy rather than to other stakeholders. In Brazil, researchers brought other actors into the steering and leadership of the project. The projects in Costa Rica and Mexico decided from the start that setting up a formal inter-institutional governance system was, in itself, among the explicit objectives of the project. In Mexico, this led to the rapid formalization of an inter-institutional entity, with a clear mandate to plan, conduct and assess the joint activities.

In conclusion, our case studies illustrate the importance of formalizing governance mechanisms and rules as they bring an added capacity to partnerships to solve problems jointly over the medium to long term. Rules evolve dynamically during the project life, reflecting the accumulated learning and evolving power relationships among stakeholders. Thus, rules can be considered as much a product as a starting point of an action-research process.

Involving farmers and their organizations

Effective involvement of all stakeholders is a crucial issue in any partnership process. Indeed, the identification of objectives and set-ups, and the lessons and conclusions drawn from the experiences, depend greatly upon the capacities of each partner to carry out agreed-upon activities and to negotiate with other stakeholders. Our case studies illustrate how difficult it is to move away from token participation and to ensure a strong, balanced involvement of all participants, especially the farmers.

Diversity and difficulties related to farmers' participation
Involvement of farmer organizations depends greatly upon the genesis of the project. When, as in most cases, research took the initiative of launching the process, strong farmer participation was more difficult to achieve. In the remaining three cases, farmer organizations played a key role throughout the process, as well as in driving it. Involvement also depends

Table 9.2 Components of operational set-up in the ten case studies

Site	Surveys* Diagnostic	M&E	Training	Exchange visits	Trials, experiments and nurseries** On-station	On-farm	Farmer innovation	Participatory experiments	Nurseries	Farmer focus groups	Workshops Data analysis	Result assessment	Planning activities
Central/south Cameroon	X	X	XX			X			X				
Nicaragua			X	XX			XXX	X			XXX	XX	X
North Cameroon	XX		XX	X			X				XX		
Reunion Island, France	XXX	X		X								XX	XXX
Mexico	X	X	X	XX	X	X	(X)	XX			X	X	XXX
Brazil Cerrados	X	X	XX	X	X		X	XX		XX	XX		XXX
Northeast Brazil	XX	X	X	XX		X	XX			XX	XX	XX	XX
Ecuador	X		XX	XXX	X		X	XX	X		X	XX	X
Southern France	X			X	X	X	X	X			X	XX	
Costa Rica	XX		X	X							XXX		XX

Notes: Importance of the component throughout the project: X = some; XX = medium; XXX = very important

*Types of surveys: diagnostic (comprehensive farming system or thematic survey); M&E (monitoring and evaluation survey).

**On-farm: research-designed experiments on farmers' fields; farmer innovation: experiments conducted autonomously by farmer innovators; participatory experiments: trials jointly designed and managed by farmers and researchers.

upon who represents the farmers, whether they participate as individuals (Cameroon), as representatives of relatively young or weakly organized farmer organizations (Costa Rica, Nicaragua, Mexico, Brazil Cerrados and Reunion Island) or of strongly organized and politically vocal farmer organizations (Ecuador, southern France and northeast Brazil).

The size of the farmer organization appears less important than its capacity to organize its activities and to establish relationships with other stakeholders. Organizations of poor small-scale farmers tend to lack adequate financial resources, limiting their representatives' ability to take part in events or to carry out activities and limiting their motivation to involve themselves intensively throughout the process. There are also competing requirements for investing time in the organization's activities versus on the farm.

Representativeness and legitimacy
When farmer organizations are absent, the representativeness of farmers in a partnership is an issue that other stakeholders need to especially take into account. When farmer organizations are involved, farmers' selection may be the result of a mostly internal process (Mexico) or, more often, of an interaction with other stakeholders (Costa Rica and Reunion Island). Because of their considerable experience in interacting with the outside world, elected representatives of farmer organizations have political legitimacy in the eyes of their fellow farmers (Mercoiret and Berthomé, 1995) and thus are often expected to represent their organizations in multi-stakeholder projects. However, the array of skills required for such collaboration may be quite different from those typically brought by elected farmer leaders. In several cases (Nicaragua, Brazil, south Cameroon and Ecuador), technical skills of farmers are indeed key to ensure that innovation and new knowledge are produced, especially when the project involves a strong component of farmer-managed experimentation. Altogether, the role of personal characteristics and social status, the willingness to participate, the technical and interpersonal skills (facilitating a meeting, reaching consensus) and the legitimacy inside the farmers' world are more important than representativeness *per se*.

Building trust and reaching clear commitments

Relationships between stakeholders
The capacity to establish adequate relationships between the worlds of the farmers, the technicians and the researchers is yet another critical issue. All of the case studies insist on the importance of trust among farmers, technicians and researchers, and propose different ways of achieving this. But it takes time to build trust. In north Cameroon, a full year was needed to establish trust between technicians and farmers and to start working on topics of real interest to the farmers. In addition to developing farm management capacities, there was a need to develop activities specifically

geared to allow stakeholders to come to know each other and to build a common language (e.g. field visits and training events). This critical time factor explains why rapid participatory rural appraisal approaches are severely limited in their ability to generate adequate relationships between stakeholders and, hence, in providing a sufficient basis for solving problems though participatory processes.

Internal communication among stakeholders is also critical. Managed poorly, it may become a source of frustration to participants. In most case studies, large meetings were organized at key moments in the project life to discuss, validate and disseminate results. Circulating reports about such meetings cannot, however, be the only means of ensuring effective communication. Dissemination strategies should ensure that the information reaches beyond those individuals who take part directly in project activities and should involve institutional decision-makers. Costa Rica illustrates the importance of holding regular meetings between grassroots organizations and their representatives in the project to improve the proposals and facilitate gradual appropriation of the results.

Nature of the commitments
Another key issue has to do with the nature of the commitments and responsibilities of each stakeholder within the partnership. Some commitments are strategic (e.g. quality management in cocoa production through producing and marketing suitable varieties in Ecuador). Others are more tactical or operational (e.g. management of field trials in Nicaragua). Some are global and influence the whole process (in Costa Rica the first six months were dedicated to defining the objectives and methodology), while others are partial and involve only specific stakeholders (in Brazil Cerrados, where separate agreements were reached between NGOs and farmer organizations). Interestingly, none of the ten case studies had any procedures in place for monitoring the various stakeholders' commitments and for enforcing sanctions when potentially counter-productive deviations were observed.

Formalizing commitments is a different issue than establishing and keeping them. Usually, technicians and researchers trust in written agreements based on negotiations among stakeholders and/or specially established governance and technical committees. While some case studies (Mexico and Ecuador) illustrate such a situation, in most others, commitments and rules remained informal. Beyond the issue of whether unwritten agreements and commitments may or may not be formal, what appears critical is to use forms of engaging and committing stakeholders that cohere with their prevailing values and culture. While some experienced farmer organizations may trust written agreements and formal committees, others may prefer a commitment expressed in a special place or in front of a respected moral authority.

Overall, the case studies illustrate four major results related to trust and commitment:

1   Partnerships largely depend upon the trust progressively established among stakeholders.
2   Trust, in turn, results from establishing effective learning processes.
3   Commitments should be linked to the actual capacities of stakeholders to fulfil them.
4   Commitments need not be formalized in written documents or through the establishment of formal committees.

Main types of results achieved in partnership mode

The ten case studies produced three types of results: knowledge generation, learning processes and empowerment, and problem-solving.

Knowledge generation
Unsurprisingly, the knowledge produced on the biophysical processes or on farming systems related closely to the specific topics addressed in each project (e.g. five academic reports were published on farming systems and on the plantain supply chain and three on the process of disseminating new seeds in south Cameroon). Original knowledge was also generated in the innovation process and in the strategies of the different stakeholders.

All cases also generated research information useful for the stakeholders during the course of the participatory process. This is in contrast to what happened with the purely scientific products targeted at the scientific community (e.g. papers at congresses and articles undergoing peer review). These were relatively few compared to the production observed in conventional research and tended to appear after the end of the participatory process, thus effectively preventing an efficient use of these products by the stakeholders (e.g. the first scientific presentation about Reunion Island was made two years after the project ended and the first scientific article five years after). This illustrates the difficulties the researchers engaged in such projects face in finding enough time to distance themselves from the pressure of action-related commitments.

Learning processes and empowerment
Learning is a key product of partnership processes. It derives from the dynamic exchange of experiences, knowledge and know-how among the different stakeholders while working together. Learning takes diverse dimensions. Participants build knowledge about new technologies (e.g. new germplasm in Nicaragua and France, and conservation agriculture in Mexico). They also learn about organizational issues (e.g. farmer experimenter groups in northeast Brazil and Nicaragua), designing new projects (cocoa supply chain in Ecuador) and developing capacities to negotiate with other stakeholders (with the ministries in Costa Rica and Mexico). The learning process is mutual as it involves all who take part in it. For example, in the case of Nicaragua, farmers learned about the resources and conditions (time and risks) required to create new varieties, while the researcher learned about relevant criteria to create 'ideal' varieties.

The learning process is usually complex as it is embedded in different activities, mixing:

- access to knowledge and know-how through classical training;
- strengthening capacities during the whole process; and
- developing skills by putting acquired knowledge and capacities into action.

While the first two are present in all case studies, the last one occurred in only a few cases. Defining precisely the nature of the learning processes remains difficult: it would require identifying a set of unambiguous criteria and assessing the corresponding impacts in and outside the group of participants.

Problem-solving
Last, but not least, results were related to solving problems in the form of technical, organizational and institutional innovations. In south Cameroon, technical innovations were derived from knowledge held and developed by both researchers (new techniques to grow young plantain plants) and farmers (e.g. termite control and material needed for building nurseries, etc.). In north Cameroon, the main result was a new method for providing farm management advice with the assistance of a public institution. In other cases, the innovation was more of an institutional nature (definition of a contract for community-based land management between farmers and the Ministry of Agriculture in Reunion Island, and the creation of a regional institution for promoting conservation agriculture in central Mexico).

All of the case studies showed, however, that the problem initially identified by the stakeholders was not completely solved by the end of the project, even though relevant results were achieved. Most of the time, results were partial because, at the start of the participatory process, the definition of objectives was either imprecise or overly ambitious. The objectives may also evolve during the project because of an evolution both of the problem and the stakeholders. Fortunately, the process of negotiation and the search for new solutions tend to continue even after the project ends.

CONCLUSIONS, CHALLENGES AND PERSPECTIVES FOR RESEARCH

The comparative analysis of the case studies confirms that researchers who engage in collaborative processes have distinct characteristics and professional trajectories that lead them away from conventional approaches typically used in their institutions. Along the way, researchers discover that they need to face strong partners. This gives feedback to the *ex-ante* design of action research projects and to the importance of dedicating enough time and efforts to the initial stages, during which the foundation for the partnership is laid.

Our study also stresses the need to invest more thinking into a number of areas essential for improved performance of partnership projects, such as facilitation and negotiation, conflict prevention and resolution, developing the rules of the game, etc. In all such endeavours, researchers have to think hard how they can learn and deploy new skills without losing their professional identity and becoming little more than technical advisers and facilitators of social processes. Engaging in participatory processes gives rise to new roles and functions for researchers (e.g. as facilitators, communicators, negotiators and mediators) (Chia et al, 2005), and as catalysers of unpredictable and non-linear innovation processes. However, it appears to be difficult to allocate time to these functions without diverting it from the time required for generating new knowledge and capitalizing it in forms acceptable to the academic world. A solution could, of course, be to share these functions more equally among stakeholders; but how to strike the correct balance remains unclear.

As roles evolve over time in any dynamic partnership process, the balance among partners in steering and coordinating the project has to be readjusted periodically to take into account the often-claimed willingness of researchers to contribute to the gradual autonomy and empowerment of their weakest partners, frequently farmers and their organizations. This leads to critical reassessment by research of the level of control that it must and can share over the partnership process with other actors, without fearing that less control will translate into failure to produce legitimate, useful science and knowledge. This also relates to how the various stakeholders commit themselves to the partnership rules and work plans, and how they formalize such commitments with mechanisms providing both sanctions and incentives.

The uncertainties about achieved results oblige researchers to carefully negotiate their place and status within their institutions in order to avoid marginalization and loss of status because mainstream researchers perceive work in partnership as lacking in scientific legitimacy. Questions and challenges abound in this respect. For example, how may researchers accept taking a back seat in order to contribute better to the empowerment of their weaker partners? How can they find the time necessary for more self-critical assessment of what they do or should do, when pressure to deliver ever more ambitious results and impacts in ever decreasing time-frames is mounting? How can researchers pursue the necessary systematization of results and lessons obtained within the context of partnership processes, and with whom?

Ideally, a solution would be to somehow find a way of readjusting in-stitutional signals and incentives and of investing significant efforts to provide adequate training and learning opportunities to many researchers, as well as to some institutional decision-makers on the principles, approaches and practices of action research in partnership.

ACKNOWLEDGEMENTS

The authors wish to thank the stakeholders of the different case studies for their multifaceted contributions, and the many CIRAD colleagues who helped to shape the ideas and perspectives presented in this chapter.

REFERENCES

Ashby, J. and Sperling, L. (1995) 'Institutionalizing participatory, client-driven research and technology development in agriculture', *Development and Change*, vol 26, pp753–770

Chambers, R., Pacey, A. and Thrupp, L. A. (1989) *Farmer First: Farmer Innovation and Agricultural Research*, Intermediate Technology Publications, London

Chia, E., Dulcire, M. and Hocdé, H. (2005) 'Comment favoriser les apprentissages collectifs d'un groupe de chercheurs?', Sixième Congrès Européen de Sciences des Systèmes, 19–22 September, Paris, France

Chiffoleau, Y. and Desclaux, D. (2006) 'Participatory plant breeding: the best way to breed for sustainable agriculture', *International Journal of Agricultural Sustainability*, vol 4, pp119–130

Djamen Nana P., Djonnéwa, A. and Havard, M. (2005) 'Co-construction d'une démarche de conseil aux exploitations agricoles familiales du Nord-Cameroun', in Revue Scientifique 2005 de l'IRAD, La Recherche au Service des Acteurs du Monde Rural, 25–28 July, Yaoundé, Cameroon

Dulcire, M. and Roche, G. (2007) 'Chercheurs – agriculteurs – industriel: Co-construction d'une filière de cacao fin et "bio" en Équateur', Third Conference on Living Knowledge, Communities Building Knowledge – Innovation through Citizens' Science and University Engagement, Paris, 29 August–1 September

Dulcire, M., Piraux, M. and Chia, E. (2006) 'Du contournement de la LOA au détournement des CTE: Les exploitations agricoles à l'épreuve de la multifonctionnalité dans les DOM insulaires, la Réunion et la Guadeloupe', *Cahiers Agricultures*, vol 15, no 4, pp363–369

Faure, G., Hocdé, H. and Meneses, D. (2007) 'Réflexions sur une démarche de recherche-action: Les organizations paysannes du Costa Rica construisent leur vision de l'agriculture familiale', *Cahiers Agricultures*, vol 16, pp205–211

Jouve, P. and Mercoiret, M. R. (1987) 'La recherche-développement: Une démarche pour mettre les recherches sur les systèmes de production au service du développement rural', *Les Cahiers de Recherche-Développement*, vol 16, pp8–15

Lewin, K. (1946) 'Action research and minority problems', *Journal of Social Issues*, vol 2, pp34–46

Lindenperg, G. (1999) 'Les acteurs de la formation professionnelle: Pour une nouvelle donne', Rapport au Premier Ministre, Paris, France

Liu, M. (1992) 'Présentation de la recherche-action: Définition, déroulement et résultats', *Revue Internationale de Systémique*, vol 6, no 4, pp293–311

Mercoiret, M. R. and Berthomé, J. (1995) 'Les organizations paysannes face au désengagement de l'Etat', in *Introduction aux Travaux, Atelier International, Mèze*, CIRAD, France, pp15–19

Norman, D. W., Simmons, E. M. and Hays, H. B. (1982) *Farming Systems in the Nigerian Savanna: Research and Strategies for Development*, Westview Press, Boulder, CO

Sabourin, E., Hocdé, H., Tonneau, J. P. and Sidersky, P. (2006) 'Production d'innovations en partenariat: Une expérience dans l'Agreste de la Paraiba, Brésil', in Agronomes et Innovations: Troisième Edition des Entretiens du Pradel, Actes du Colloque des 8–10 Septembre 2004, L'Harmattan, Paris, France, pp191–206

Scoones, I., Thompson, J. and Chambers, R. (eds) (1994) *Beyond Farmer First: Rural People's Knowledge, Agricultural Research and Extension Practice*, Intermediate Technology Publications, London

Scopel, E., Triomphe, B., Goudet, M., Valadares Xavier, J. H., Sabourin, E., Corbeels, M. and Macena da Silva, F. A. (2005) 'Potential role of conservation agriculture in strengthening small-scale farming systems in the Brazilian Cerrados, and how to do it', in Proceedings of the Third World Congress on Conservation Agriculture, Nairobi, Kenya, 3–7 October 2005

Temple, L., Kwa, M., Fogain, R. and Mouliom Pefoura, A. (2006) 'Participatory determinants of innovation and their impact on plantain production systems in Cameroon', *International Journal of Agricultural Sustainability*, vol 4, no 3, pp233–243

Triomphe, B., Hocdé, H. and Chia, E. (2006) 'Quand les agronomes pensent innovation et les institutions transfert: Des malentendus sur la forme ou des visions différentes sur le développement? Le cas du Bajio guanajuatense (Mexique)', in Agronomes et Innovations: Troisième Edition des Entretiens du Pradel, Actes du Colloque des 8–10 Septembre 2004, L'Harmattan, Paris, France, pp247–266

Trouche, G., Hocdé, H. and Aguirre, S. (2005) 'Sélection participative des sorghos au Nicaragua', in J. Lançon, E. Weltzien and A. Floquet (eds) Partenaires pour Construire des Projets de Sélection Participative, Atelier de Recherche, 14–18 March 2005, Cotonou, Bénin

Veldhuizen, L., Waters-Bayer, A. and de Zeeuw, H. (1997) *Developing Technology with Farmers*, Zed Books, London

World Bank (2006) *Enhancing Agricultural Innovation: How to Go Beyond the Strengthening of Research Systems*, World Bank, Washington, DC

III

Policy, Institutional
and Market-Led Innovation

CHAPTER 10

Participatory Analysis of the Potato Knowledge and Information System in Ethiopia, Kenya and Uganda

Peter Gildemacher, Paul Maina, Moses Nyongesa, Peter Kinyae, Gebremedhin Woldegiorgis, Yohannes Lema, Belew Damene, Shiferaw Tafesse, Rogers Kakuhenzire, Imelda Kashaija, Charles Musoke, Joseph Mudiope, Ignatius Kahiu and Oscar Ortiz

INTRODUCTION

Potato is important for smallholders in Kenya, Uganda and Ethiopia as both a cash crop and a food-security crop. Potato production has tripled in ten years since the mid 1990s in sub-Saharan Africa, almost exclusively because of area expansion (FAOSTAT, 2006). With its cultivation restricted to the highlands and its ever-increasing consumption in cities, potato is the cash crop of the future for the densely populated eastern and central African highlands. To satisfy the growing demand from urban centres for cheap food, there is room for additional growth in potato production. Further area expansion will, however, put a strain on natural highland forests in eastern Africa. Producing potatoes at lower altitudes in the equatorial tropics is not feasible because of pest and disease pressure and physiological limitations of the crop. The only option for increased potato production is, therefore, raising crop productivity.

In Kenya, potatoes are the second most important food crop after maize (FAOSTAT, 2006) while, in Ethiopia, potato production can fill the gap in food supply during the 'hungry months' before the grain crops are harvested. In southwestern Uganda, potato production is crucial in supporting the income and food security of the rural population. Average potato yields for 2005 in Kenya, Uganda and Ethiopia were estimated at 7.7, 6.9 and 10.5 tonnes per hectare, respectively (FAOSTAT, 2006), while progressive farmers in these countries attained yields of 25 tonnes per hectare under the same rain-fed conditions in the same period.

This yield gap can be explained by poor management of late blight, bacterial wilt and viruses, low soil fertility and drought stress. Interventions to improve crop husbandry of poor potato farmers by increasing their knowledge could have considerable impact upon their livelihoods in terms of both improved food security and increased income.

To promote technological and methodological innovations successfully, it is important to understand the current agricultural knowledge and information system related to the potato crop (AKIS–potato). AKIS–potato can be defined as a group of individuals, public organizations (governmental and non-governmental) and the private sector who exchange information and knowledge related to potato management, processing and trade (Engel, 1997).

Understanding this system, its components and the way in which they interact is the essential first step towards a more efficient innovation system (Lundvall et al, 2002; Hall et al, 2004). Understanding the AKIS–potato system will allow research and development organizations to coordinate interventions in a way that makes use of the comparative advantages of each stakeholder. As part of a larger project on farmer participatory research, the AKIS of the potato sector in Ethiopia, Kenya and Uganda, and the interactions between stakeholders in the sector were analysed. The objectives of the study were to:

- identify bottlenecks in interaction between the different stakeholders;
- highlight priorities for intervention in the potato sector; and
- draw conclusions on how to improve the flow of information in the system.

METHODOLOGY

Multi-stakeholder workshops were organized to identify constraints and opportunities in the potato sector, with specific focus on improving the AKIS–potato in Kenya, Uganda and Ethiopia. In Ethiopia, the workshop was a two-day event which brought together representatives of potato-related organizations and farmers from Alemaya, Galessa, Jeldu and Degem districts. In Uganda, it was a one-day workshop with potato stakeholders from Kabale District. In Kenya, two one-day stakeholder workshops were conducted in both Bomet and Nyandarua districts.

Workshop participants were grouped together according to stakeholder categories such as ware-potato farmers, seed-potato farmers, public extension, non-governmental organizations (NGOs), processors, transporters and agricultural-input suppliers. Stakeholder categories represented at the workshops varied by country, depending upon their responses to the invitations. All groups analysed their own role and the role of other stakeholders in the potato chain and constructed a matrix of interactions, following a method described by Biggs and Matsaert (2004). First, each stakeholder group identified its interactions with other stakeholders in the potato chain. Then, the groups identified constraints in these interactions. The complete matrix of interactions was then put together by the workshop facilitators and the opinions of the different stakeholder groups about the others were presented in plenary and discussed.

In Kenya, in both Bomet and Nyandarua, the problems identified in the first workshop were prioritized in the second workshop. Each participant ranked the five most important constraints, with every constraint receiving points (5 to 1) according to importance. Solutions to the most important constraints were subsequently discussed in mixed groups of stakeholders and reported back in plenary for further elaboration.

RESULTS AND DISCUSSION

Kenya

The main stakeholders of the AKIS–potato in Kenya were the Kenya Agricultural Research Institute (KARI), the public extension service of the Ministry of Agriculture (MoA), agricultural-input dealers, the Kenya Potato Growers and Marketing Association (KPG&MA), local government, potato transporters, traders, brokers and middlemen, seed-potato producers and consumption-potato producers. All were represented at the meetings, except for the brokers and middlemen, who were invited but did not attend. NGOs were notably absent.

Almost all stakeholders at the workshop complained about the so-called 'extended bag', which is a very large packing unit of 150kg to 200kg. According to farmers and extension workers, this results in low prices. Even the traders acknowledged that the extended bags were not optimal, but forced upon them by market brokers in Nairobi. There are, however, some efforts to standardize the bag used for ware-potato marketing at 110kg. The participants agreed that a price per kilogram would be ideal, but realized that this required a certain level of community organization to obtain communal weighing scales.

Many participants cited the exploitation of farmers by brokers as a point of concern; but the brokers' counter-arguments could not be heard as they did not attend the meeting. Producers, however, acknowledged that field-level brokers were members of their communities and fulfil a role in the marketing chain. They suggested a fixed commission instead of one that varies on the speculation skills of the broker.

As a result of the involvement of many different interim handlers, the transaction costs between producer and consumer are relatively high (Kirumba et al, 2004). The dilapidated road network pushes down farm-gate prices even further. Prices at the farm gate fluctuate widely, and no price information is exchanged between farmers. Farmers' access to price information could enhance their bargaining power and increase the price they get from traders (Bakis, 2002). At the level of market brokers, who mediate between transporters and wholesalers, there are unnecessary transaction costs as a result of cartel formation.

The long marketing chain is a barrier to the flow of information on both product quality and market prices. Low-quality farm-gate produce – as a result of no grading on tuber size and quality, immature harvesting and mixing of varieties – led to high losses in the transport, marketing and processing chain, as indicated by traders and processors. There is, however, no feedback from the market to the farmer about the quality of the produce, and there is hardly any price incentive that stimulates farmers to deliver better-quality potatoes.

An important problem identified in both districts is the lack of high-quality seed potato. The need for certification was stressed by farmers, extension workers and the KPG&MA, who claim that farmers are cheated by poor-quality potatoes sold as seed. However, the seed growers state that farmers are not willing to pay extra for good-quality seed.

The lack of information transfer between research, extension and farmers was another concern raised. Research is considered slow in responding to problems raised by the extension staff. Extension staff are blamed for not delivering new technology, reacting slowly to farmers' needs, not being visible and not leaving their offices. In the opinion of farmers (potato growers, seed farmers and the KPG&MA), research and extension are also to blame for the inadequate supply of high-quality seed potato. The lack of credit facilities was also mentioned as a shortfall of the extension service.

Farmers do not consider 'change agents' in research and development as messengers of information only, but have wider expectations from them as service providers. Extensionists stand between research and farmers in the agricultural knowledge system and are easily blamed for inadequate communication. On the one hand, they have to live up to high expectations from the side of the farmers, even under poorly resourced conditions. On the other hand, research expects them to communicate 'new information' to farmers, who are not necessarily receptive to, or interested in, this information.

Figure 10.1 clearly illustrates how the outcomes of this analysis provide insight into the interrelations and perceptions of the actors in the potato value chain. It presents the opinions and the intensity of interactions between agricultural-input dealers, farmers and extension workers. Extension workers noted the low attendance of input dealers in their training efforts as a constraint, while the input dealers identified the bad timing of meetings by extension staff as a problem. Potato producers noted that the extension workers lack knowledge on new technologies, while the extension workers accused farmers of resisting new technologies. The input dealers felt that they could play a role in information transfer and advice regarding the use of agrochemicals. The extensionists, however, did not recognize such a role for input dealers and accused them of misinforming farmers. In reality, these dealers do give advice to farmers, but complain that farmers do not follow the advice regarding the use of chemicals. Farmers complained that the dealers sell them adulterated products. Looking at Figure 10.1, there seem to be opportunities to improve the flow of information in the triangle by enhancing the linkage between the extension staff and agricultural-input dealers, who already have strong contacts with farmers. Mistrust by both farmers and extension workers towards the dealers stands in the way of such communication. Moreover, the dealers indicated that farmers are not willing to learn, an opinion they share with the extension workers.

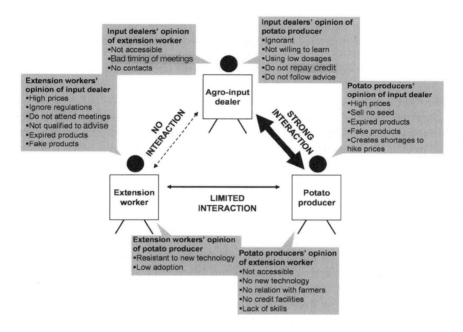

Figure 10.1 *The level of interaction between agricultural-input dealers, extension workers and potato producers and their perception of each other, Bomet and Nakuru districts, Kenya, 2005*

Table 10.1 ranks the problems in the potato value chain and possible solutions suggested by the stakeholders. When analysing the suggestions for improvement, the need for farmer organization became clear. For almost any intervention, a certain level of farmer organization is required. The KPG&MA appeared to be the obvious forum that could support further intervention. The farmers, however, indicated a general reluctance to join such initiatives as a result of a long history of failed organizations and dishonest leadership. Participants indicated that community leaders with track records of failed communal projects should be left out of any new organizational initiatives to reduce the level of mistrust among potential members. The setting-up of study groups and common-interest groups was suggested as a possible option to improve farmer–extension–research linkages, test new technologies, receive training and multiply seed.

Interestingly, the different actors were very aware of the need for quality improvement at the farm-gate level and suggested size grading, purity of variety and the proper hardening of tuber skin for this purpose. Higher-quality seed is also required as part of quality improvement. To improve the bargaining power of small-scale farmers, on-farm storage or communal storage was suggested. Contract farming by a crisp (chips) processor was also indicated as an option to ensure higher and stable farm-gate prices. Resistance to some of these changes could be expected from the side of brokers, transporters and traders, which would require support and enforcement of change by the local administration.

Ethiopia

The stakeholder workshop in Ethiopia identified 14 AKIS–potato stake-holders, including researchers, farmers, potato traders, consumers, district bureaux of agriculture, transporters, casual labourers, NGOs, farmer co-operatives, brokers, store owners, the media, agricultural-input suppliers and supermarkets. Marketing was identified as the activity with most interaction between stakeholders. The main providers of information to farmers were identified as research, extension and agricultural-input suppliers.

The analysis of constraints in interaction showed that researchers were particularly disappointed in the uptake of technologies by farmers, in spite of much-increased efforts to involve farmers in technology development. The flow of information from trained farmers to others in the community was also considered to be limited. With few public extension workers in the district bureaux of agriculture, working under time constraints, collaboration with researchers was said to be difficult.

The farmers indicated the low quality of agricultural inputs to be a constraint. The extension staff shared this opinion and blamed the suppliers for low-quality products at inflated prices. Farmers also identified low potato prices and dishonest brokers as problems. It was noted that extension staff also sought their own interests in activities undertaken with farmers.

The traders indicated low-quality produce at farm-gate level as their main problem and identified this as the reason for low prices offered to farmers. Furthermore, they saw the absence of large buyers as a constraint. The product is retailed in small quantities, which takes longer to sell the stock, with higher risk of spoilage.

The public extension workers indicated a lack of good interaction with researchers. NGOs indicated a slow response from the side of research to requests from practice, resulting in outputs not reaching the end users in time. Extension staff felt that farmers ignored advice given to them and

Table 10.1 *Constraint ranking and suggested solutions for potato production and marketing in Kenya*

Constraint	Suggested solutions
Lack of high-quality seed	Train seed multipliers Teach positive selection Farmer-group seed multiplication
Minimal contacts between market and knowledge chain actors	Use church gatherings and other meetings to introduce new technology Demand-driven technology that does not require capital investment Initiate study groups with farmers and extensionists to improve interaction and provide a platform for technology testing
Extended bags	Standardization (by the time of the second workshop, efforts for standardization were being initiated)
High prices/low use of fertilizers and chemicals	Credit scheme to be run by KPG&MA
Minimal exchange of price information between farmers	Improve price communication between farmers through the formation of common-interest groups
Low prices for potatoes	Conduct research into simple ware-potato storage Contract farming for the crisp (chips) industry Better timing of production on the basis of price information supplied by the MoA Improve quality of potatoes (see suggested solutions in Table 10.2)
Poor roads	Community road maintenance paid through levies collected by local government Setting up levy collection points by communities on feeder roads
Bacterial wilt	Train seed multipliers Teach positive selection (selection of healthy-looking mother plants in ware-potato farmers' fields as a source of seed for the next season)
Low quality of potatoes offered to market	Harmonize size grading to standardize prices with the assistance of KPG&MA, the Community Development Agency (CDA) and the local administration Improve quality of ware potatoes by using high-quality seed and limited training Harvest crop when mature (hardened skin)
Lack of credit facilities	Credit scheme to be run by KPG&MA
Low yields	Improve seed quality; credit scheme to be run by KPG&MA to increase fertilizer and fungicide use
Lack of storage facilities at farm level	Farmer-managed research into simple on-farm storage

did not adopt newly introduced technologies despite being trained. NGOs believed that farmer participation in different development activities was below expectation. The extension service acknowledged a limited interaction with NGOs.

Similarly to Kenya, different types of innovations are needed to improve the AKIS–potato in Ethiopia. Table 10.2 summarizes the most important

Table 10.2 *Constraints and suggested solutions for potato production and marketing in Ethiopia*

Constraint	Suggested solutions
Limited interaction between research, extension, NGOs and farmers	Enable researchers to transfer information faster Existing stakeholder forum should be strengthened and new forums set up Improve training to transfer more information to farmers Create a desk at the agricultural office for exchange between research and extension Make leaflets, manuals and other training materials available to development agents and farmers Cultivate a culture of collaboration among development organizations
Low prices for ware potatoes at farm gate	Strengthen farmer organizations Joint marketing Improve exchange of price information Encourage farmers to construct improved ware-potato stores
Bad roads	District and zonal councils should repair roads
Unavailability of inputs	Open more input-supply shops in rural areas Farmer unions could play a role in the supply of agrochemicals Train farmers on alternative low-input management strategies
Low-quality products	Introduce federal control of the quality of chemicals
Limited adoption and further dissemination of technology by farmers	Improve training Select early adopters to assist in facilitating innovation Develop demonstration sites Collaboration between researchers and extension staff in training farmers Research should develop cost-effective innovations
Limited skills of extension staff	Train extension staff continuously and increase the cadre
Low quality of potatoes	Train farmers on how to improve quality, especially on harvesting (at maturity) Set quality standards for potato production
Weak credit schemes	Raise awareness about credit and payback mechanisms Extend the periods of loans
Lack of high-quality seed potato	Train and list reliable seed-potato producers

constraints identified and the solutions proposed. Discussions among the participants revealed that the linkages between many of the stakeholders in the potato production and marketing system are weak. This hampers the flow of information and development of knowledge in the system.

Creative ways of improving the interaction between farmers, agricultural extension providers and researchers should be sought. Organizational innovation on the part of farmers is identified as crucial if the system is expected to be enhanced as a whole. The strengthening of farmer organizations was widely recognized as imperative for improving linkages with farmers in terms of technology dissemination, as well as for improving input supply and output marketing. The lack of a forum for exchange between all stakeholders in the potato innovation system was noted. Such a forum could be an instrument to improve linkages between stakeholders and could help in improving the flow of information through the system. This would assist in increasing production and improving the marketing chain of potatoes in Ethiopia. The lack of high-quality seed potatoes featured prominently in the discussion. Training and promoting specialized seed producers was suggested as a solution.

Uganda

In Uganda, the interactions between stakeholders in the AKIS–potato were mapped out (see Figure 10.2).

The different knowledge system interactions in the potato value chain were ranked according to their current importance in managing information. The mass media were considered to play the smallest role, while the farmers and the national research and extension institutions were ranked highest.

The type of innovations needed to improve the potato sector in Uganda (see Table 10.3) were similar to those required in the cases of Kenya and Ethiopia. The highest priority was given to improved interaction between stakeholders in the potato chain and mechanisms for better coordination of interventions. Inappropriate packaging of information was identified as a major problem, especially the language in which information material was produced. Moreover, most of the material was considered inappropriate for illiterate people. A limited flow of information was noticed between the wealthy and poor sectors of the communities.

The local mass media (radio) are poorly connected to information suppliers (researchers and public extension services). Input dealers are not considered as information suppliers by extension and research, while they are considered an important source of information by farmers. NGOs and extension appreciated the research organizations for their participatory research activities, but considered their outreach limited. The outreach of the NGOs was also considered to be limited. Privatized extension (National Agricultural Advisory Services, or NAADS) was noted as having a wider reach, but with limitations in terms of agricultural extension skills.

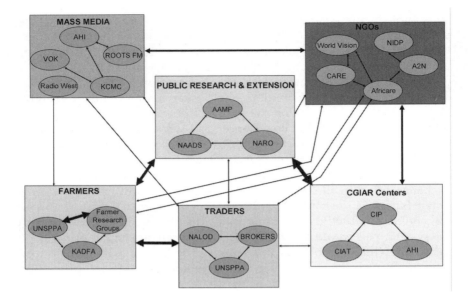

Figure 10.2 *Interaction of AKIS–potato actors in Kabale, Uganda, 2004*

Notes: The thickness of the arrows indicates the strength of the linkages and information exchange.

A2N = Africa 2000 Network; AAMP = Area-Based Agricultural Modernization Programme; AHI = African Highlands Initiative; CIAT = International Centre for Tropical Agriculture; CIP = International Potato Centre; KADFA = Kabale District Farmers' Association; KCMC = Kachwekano Community Multi-Media Centre; NAADS = National Agricultural Advisory Services; NALOD = NAMLOD Perfect Consult Ltd; NARO = National Agricultural Research Organization; NIDP = Nangara Integrated Development Project; UNSPPA = Uganda National Seed Potato Producers Association; VOK = Voice of Kibwezi.

Interestingly, farmers were said to provide limited feedback to development organizations. These organizations also complained that farmer-group continuity is unsatisfactory and that farmers show little initiative in seeking information. Farmer-group formation and cohesiveness were aspects that needed attention.

Suggestions for improving the flow of information were: capacity-building for research and extension in the development of appropriate training materials; and improving collaboration between research, NGOs and private service providers to use the higher skills available in NGOs and research organizations within the larger NAADS programme and the public extension service.

Table 10.3 *Constraints and suggested solutions for potato production and marketing in Uganda*

Constraints	Solutions proposed
Inappropriate packaging of training materials	Capacity-building in creating training materials for research and extension staff
Outputs of research and NGOs do not reach many farmers	Collaborate closer with NAADS service providers and public extension service Use radio Involve agricultural-input dealers
Some incompetent contractors in the National Agricultural Advisory Services (NAADS)	Collaborate closer with NGOs and research
Reluctance of farmers to be involved in learning new ideas; group sustainability weak	More focus on sustainable farmer-group formation
Adulterated inputs sold	–
Lack of credit facilities for input dealers	–
Limited funds for radio stations	Collaborate more closely with researchers, extensionists and NGO staff

CONCLUSIONS

The participatory workshops proved very effective in identifying AKIS bottlenecks and options for intervention. It was a rare opportunity for the different stakeholders of the potato innovation system to come together and discuss issues. This was, in itself, a key output. The matrix of interactions was an appropriate tool to identify constraints in the AKIS–potato. The construction of the matrix led to a better understanding on the perceptions of different stakeholders about each other and improved mutual understanding. This could be the beginning of a process to improve collaboration between stakeholders in the potato value chain. The workshops alone, however, are not enough to spark action and to induce positive change towards a more effective innovation system through improved collaboration. Further follow-up and facilitation would be required to continue the process.

Although the exercise set out to map imperfections in the information flow, the workshops eventually identified potato value chain constraints in a wider sense, especially in Kenya and Ethiopia. Giving special attention to knowledge flows is not easily accepted by the different stakeholders as it is just one factor in the innovation system that cannot be separated from other interactions. In a conceptual sense, this is possible for a researcher; but, in practice, the distinction of the information and knowledge system from the wider potato production and marketing system is artificial and, thus, not practically useful in a multi-stakeholder setting. Especially when a relatively large number of farmers are engaged in the process, the direction of discussions will naturally be geared towards solving practical problems in the value chain, rather than focusing on information exchange.

The limited presence of extension (both governmental and non-governmental) is a major impediment to the effective flow of information, and clear strategies need to be developed by the different stakeholders to mediate this and to improve extension coverage. In the first place, agrochemical dealers need to be considered as agents for delivering information to farmers. They have close contacts with farmers and could serve as hubs for providing written and oral information on improved technologies to farmers. Second, research organizations have to engage more in developing mass dissemination strategies for their information and developing communication materials in collaboration with extension partners. Research organizations need to gain specific expertise for this purpose. The mass media, especially radio, are underutilized in all three countries. It may not necessarily be the best tool to improve knowledge and induce change in farming practices, but it can arouse the interest of farmers and change agents in new technology.

More research is required on how to improve farmer-to-farmer flow of information, which is an important form of exchange. Information on innovations from trained farmers to the rest of the community does not flow automatically, as is often assumed. Farmer facilitators or farmer organizations could be used as agents to transmit information as an alternative to formal extension workers.

The study of the AKIS–potato in the three countries gives clear direction on how the potato-related innovation system can be made more dynamic, efficient and responsive to the needs of the different value chain actors. In the first place, it can be concluded that improved organization of farmers will allow them to become more active actors in the innovation system. This, in turn, would provide the other stakeholders with stronger and better-defined feedback on opportunities, needs and constraints in the potato value chain. As a result, research institutions and both governmental and NGO extension services could become more responsive to farmers' needs.

Second, the meetings showed a clear need for building a more durable forum for information exchange and collaboration towards technological, methodological and organizational innovation in the potato sector in all three countries. A potato stakeholder forum would ensure a more holistic and coordinated effort in potato-sector innovation. It would provide research, extension, producers, trade and industry with the much required arena for closer interaction and would create synergies through combining the comparative strengths of different stakeholders. Increased intensity of interaction would improve information flow between potato stakeholders and make them aware that they are part of the same system and that their actions are interlinked. This would enhance the capacity of the potato sector to innovate effectively. In short, the potato stakeholder forum could serve as a catalyst for the better functioning of the potato-related innovation system.

The question arises as to who should champion the establishment of such a potato stakeholder forum. National research institutes may be best positioned to initiate this forum in spite of the fact that they have, in the past, shown reluctance to shift from the old linear mode of research and extension to innovation systems thinking. Making this shift towards an innovation systems perspective in agricultural research and development will hopefully be facilitated through their involvement in building the forum. Compared to the national extension services, research organizations are better able to draw in expertise from different disciplines internally, bridging between social, organizational and technical sciences. The national research organization will be more sustainable than NGOs, which often operate for shorter periods and are more susceptible to shifts in the priorities of donors. Furthermore, research would be more impartial than extension, the staff of which are more directly involved with all other stakeholders and, as mentioned earlier, are in the difficult position of being in the middle between research and the farming community.

ACKNOWLEDGEMENTS

The research project upon which this chapter is based was part of work in Uganda and Ethiopia sponsored by the International Fund for Agricultural Development (IFAD) and work in Kenya sponsored by the Organization of the Petroleum Exporting Countries (OPEC) Fund for Development.

REFERENCES

Bakis, R, Saudolet, E. and de Janvry, A. (2002) 'Transaction costs and the role of bargaining and information: Evidence from Peru', Working Paper, Agriculture and Resource Economics, University of California, Berkeley, CA
Biggs, S. and Matsaert, H. (2004) 'Strengthening poverty reduction programmes using an actor-oriented approach: Examples from natural resources innovation systems', Agricultural Research and Extension Network Paper 134, Overseas Development Institute, London
Engel, P. (1997) The Social Organization of Innovation: A Focus on Stakeholder Interaction, Royal Tropical Institute (KIT), Amsterdam, The Netherlands

FAOSTAT (2006) FAO Database of Agricultural Statistics, www.fao.faostat.org/
    faostat.
Hall, A., Mytelka, L. and Oyeyinka, B. (2005) 'Innovation systems: Implications
    for agricultural policy and practice', ILAC Brief 2, Institutional Learning and
    Change Initiative, Rome, Italy
Kirumba, W., Kinyae, P. and Muchara, M. (2004) 'Potato market survey Kenya',
    Internal Report to the Promotion of Private-Sector Development in Agriculture
    Programme, GTZ/MoA, Nairobi, Kenya
Lundvall, B., Johnson, B., Andersen, E. S. and Dalum, B. (2002) 'National systems
    of production, innovation and competence building', *Research Policy*, vol 31,
    pp213–231

Enabling Rural Innovation: Empowering Farmers to Take Advantage of Market Opportunities and Improve Livelihoods[1]

Susan Kaaria, Jemimah Njuki, Annet Abenakyo, Robert Delve and Pascal C. Sanginga

BACKGROUND

Agricultural markets can play significant roles in reducing poverty in poor economies, especially in countries that have not achieved significant agricultural growth. Dorward et al (2005) highlight three broad mechanisms through which agricultural growth can drive poverty reduction:

1 the direct impacts of increased agricultural productivity and incomes;
2 the benefits of cheaper food for both the urban and the rural poor; and
3 agriculture's contribution to growth and the generation of economic opportunity in the non-farm sector.

However, experience has shown that markets can fail the poor, especially the poorest and marginalized groups, including women. In his review on how to make market systems work better for the poor, Johnson (2005) argues that, in remote rural areas, markets may fail because they are too 'thin', or the risks and costs for poor people to participate may be too high, or there may be social or economic barriers to participation.

Other factors can also influence the role of agricultural markets in reducing poverty in poor economies. For instance, market-oriented production may result in the capture of new economic opportunities that were previously undertaken by the poor (DFID and OPM, 2000) or create a privileged group of farmers with access to a new technology. Evidence also shows that, in some instances, increased access to market opportunities can open up competition by other producers, driving local producers out of production (Dorward et al, 2003).

Women face many constraints as they endeavour to engage with market systems. Empirical studies on intra-household gender dynamics in Africa have shown that, when a crop enters the market economy, men are likely to take over from women, who therefore do not benefit (Braun and Webb, 1989; Kaaria and Ashby, 2001). In some instances, women's social and cultural roles may assign productive and reproductive roles to men and women that limit the latter's access to markets (OECD, 2004). Women's role in household provisioning versus the men's role in providing the cash requirements of the household affects women's ability to participate in markets.

Many approaches for linking smallholder farmers to markets are based on commodities and cash crops and use arrangements such as contract farming and out-grower schemes that link smallholders to large-scale growers. Such arrangements, while indeed linking the smallholders to regional and domestic markets, also leave them vulnerable because they lack capacity to engage effectively in markets and to analyse and negotiate with these markets. In a review of case studies, Bingen et al (2003) found that investment in human capital formation could determine the ability of rural communities to participate effectively in markets. They argue that, although human capital investments can be slow, the skills in marketing often determine the ability of a community to access inputs and to market produce beyond the life of a project.

Together with partners and communities in eastern and southern Africa, the International Centre for Tropical Agriculture (CIAT) is testing and evaluating a participatory approach for linking farmers to markets: enabling rural innovation. The approach integrates specific strategies to encourage and promote participation by the poor and by women, and builds their capacity to engage effectively in markets in a more sustainable manner. This chapter presents preliminary lessons from applying this innovative approach, using two country case studies: Malawi and Uganda.

ENABLING RURAL INNOVATION

The enabling rural innovation (ERI) initiative is a research for development framework that uses participatory research approaches to strengthen the capacity of research and development (R&D) partners and rural communities to access and generate technical and market information in order to improve farmers' decision-making. The aim is to create an entrepreneurial culture in rural communities, where farmers 'produce what they can market rather than trying to market what they produce' and are encouraged to invest in natural resources rather than depleting them for short-term market gain (Best and Kaganzi, 2003; Ferris et al, 2006). This initiative emerged from three main streams of CIAT's experiences over the last 20 years:

1   farmer participatory research (FPR) (Ashby et al, 2000);
2   rural agro-enterprise development; and
3   natural resource management (NRM).

The initiative seeks to use the most effective elements from these three approaches when working with rural communities to build more robust livelihood strategies.

CIAT has been implementing this approach in partnership with rural communities, national agricultural research and extension services, and non-governmental organizations (NGOs) for the past five years in Africa. Emphasis is on developing and testing innovative partnerships that bring together stakeholders with complementary skills and expertise along the resource-to-consumption and policy continuum.

This section highlights some of the key aspects of the ERI approach:

● the resource-to-consumption conceptual framework within which it is being tested and evaluated;
● the enterprise development approach, including a participatory research process; and
● the key steps in the ERI process.

Conceptual framework for enabling rural innovation: From resource to consumption

ERI applies a resource-to-consumption conceptual framework that emerged from a review of experience on what has worked or not in different approaches to benefit women through technological change (Kaaria and Ashby, 2001). This framework builds positive backward and forward linkages from the resources or assets of a community (natural, human, social, physical and financial) to production, post-harvest handling and processing, market opportunities and household consumption (see Figure 11.1). It expands conventional production to consumption or commodity-chain approaches by explicitly basing decisions about new productive activities on the combination of community assets that will best meet the dual needs of household food production and income generation.

The resource-to-consumption framework is based on the following principles:

● It takes a 'beneficiary' rather than a 'commodity' starting point for technology development. Research objectives are defined by assessing the market, community interests and their assets.
● Technology development is driven by a comprehensive beneficiary diagnosis to identify differences in intra-household allocation and control over resources and responsibilities in order to understand constraints and opportunities to technology adoption and reinvestment in NRM.

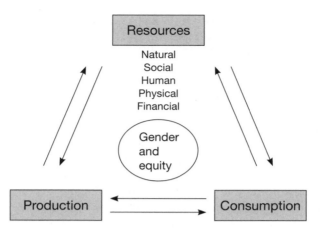

Figure 11.1 *The resource-to-consumption conceptual framework*

Source: Kaaria et al (2005)

- Gender and stakeholder differentiation of roles and perceptions is explicit and integrated within the process of technology development in order to ensure equity in access to technologies and distribution of benefits.
- It takes a 'territorial' rather than a 'commodity' focus in developing market opportunities. The approach builds community skills in identifying and analysing the opportunities for new or existing products, matching market opportunities with their asset base.

Strategy for agro-enterprise development and participatory market research

The approach builds the skills and knowledge of communities, local service providers and farmer organizations to engage effectively in markets. It emphasizes a market orientation that enables smallholders to link themselves successfully to potential markets, with support from R&D partners. It builds on CIAT's approach to rural agro-enterprise development as described by Ostertag (1999), Best and Kaganzi (2003), Lundy et al (2003, 2006) and Ferris et al (2006). ERI recognizes that risk assessment plays an important role in the strategy of a smallholder farmer. Therefore, when selecting products and new business options, it is crucial to assess the appropriate level of risk that a client group can handle. Tools such as cost–benefit analysis and the Ansoff matrix (see Table 11.1) are used to categorize risk options by comparing types of products and markets. Market opportunity analyses of products based on demand and profitability tend to bias results towards higher-risk options, and enterprise groups need

Table 11.1 *The Ansoff matrix for risk assessment*

	Existing products	New products
Existing markets	Market penetration (lowest risk)	Product development
New markets	Market development	Diversification (highest risk)

Source: Ostertag et al (1999)

to be aware of the risks, costs and benefits of such higher-profit options. For groups with more experience in marketing, higher-risk strategies are likely to be more attractive.

Once the group has selected the most appropriate option, the farmer organization or group then follows a stepwise approach to developing sustainable enterprises. The process begins with a participatory diagnosis to assess community assets and market opportunities, and constraints based on these assets. The group elects an enterprise planning committee to undertake market studies on the group's behalf. Participatory market research builds farmers' skills in analysing markets and understanding them better, in consolidating relationships with traders and in negotiating for better prices for their produce. Enterprise selection is based on the analysis of sound technical and economic information, as well as community criteria. Business plans of the best enterprise options are designed and tested for collective marketing (for further details, see Best and Kaganzi, 2003; Ferris et al, 2006).

The key steps in the enabling rural innovation process

Various steps are involved in establishing the ERI process with communities. Groups are facilitated by the partner organization and are supported at critical moments by CIAT. Figure 11.2 shows the key steps in implementing the ERI process:

- engagement of R&D partners and communities;
- participatory diagnosis to assess community assets, finances, current income opportunities, potential options, access to services, skills base, degree of cooperation, access to new technologies and organizational structures;
- formation of farmer research and market research groups, and building the groups' capacity to participate actively in selecting, testing and evaluating technology options and marketing strategies;
- participatory market analysis to identify market opportunities for competitive products that will increase household income and employment;

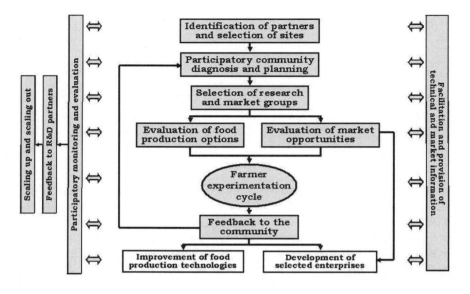

Figure 11.2 *Key steps to enabling rural innovation*

Source: Kaaria et al (2005)

- prioritization of opportunities and selection of agro-enterprise options based on social differences, including gender and wealth;
- planning and implementing experiments by farmer research groups to support enterprise and food security options;
- feedback of results to the community and R&D organizations, and identification of further research questions;
- participatory monitoring and evaluation that are useful to both communities and their service providers; and
- scaling up (expanding) of participatory research results and the process of community enterprise development.

METHODOLOGY AND DATA COLLECTION

Objectives of the study

ERI is being tested and evaluated with a variety of R&D partners and communities to assess the feasibility and outcomes of applying it within ongoing development processes or projects. Three case studies were conducted to assess the benefits of the initiative at household level in the countries where ERI was first tested: Uganda and Malawi.

The specific objectives of the study were to:

- assess the effectiveness of the ERI approach in promoting pro-poor market linkages;
- assess other tangible and non-tangible benefits (empowerment, capacity-building, gender dynamics, social and human capital build-up) of the ERI approach and how these differ by gender; and
- identify key gaps and areas that need strengthening, as well as potential opportunities.

Data collection

Case study methodology
Case studies were established to derive an understanding of the dynamics between pro-poor market linkages and benefits in inter- and intra-household dynamics and farmers' investment decisions. Case studies emphasize detailed contextual analysis of a limited number of events or conditions and their relationships (Yin, 1984; Stake, 1995). According to these authors, when systematically implemented, the case study methodology can establish reliability and generality of findings even with a small number of cases. To assess change, this study used a before-and-after impact model, where respondents were asked to compare the current situation with the situation three years ago. Although recalling data has some disadvantages in that it depends upon farmers' memory, it was useful in providing a frame of reference for assessing change.

Sampling
Groups and communities for the study were selected from the initial ERI countries (Uganda and Malawi). An important criterion was that the community or group had earned significant income over several years from the community agro-enterprises. At least 50 per cent of group members were interviewed (see Table 11.2). The sample was normally stratified.

Formal survey
The formal questionnaire focused on collecting information on:

- characterizing the agro-enterprise;
- investments in natural resource management;
- gender and intra-household dynamics;
- social and human capital;
- income and assets;
- food and nutrition security; and
- household characteristics.

Empirical model for assessing impacts on income
A multiple regression analysis was used to understand the variables that influence income from the enterprises. Table 11.3 provides an overview of

Table 11.2 *Groups and communities surveyed*

Country	Name of group/community	Type of agro-enterprise	Size of group	Sample size
Uganda	Nyabyumba Farmers' Group	Potato	120	72
Malawi	Katundulu village	Pig	36	26
	Chinsewu village	Bean	75	34

the variables used in the analysis. The analysis focused on the determinants of income from the potato enterprise in Nyabyumba in Kabale District, Uganda. Because of data limitations, the Malawi cases were not included in this part of the analysis.

Background of the communities

Nyabyumba Farmers' Group, Kabale District, Uganda
The Nyabyumba Farmers' Group was formed in 1998 with 40 members. The group, supported by the NGO Africare, focused on producing improved potatoes from clean seed provided by the National Agricultural Research Organization (NARO). In 2000, the Nyabyumba group formed a farmer field school (FFS) to improve its technical skills in potato production and to increase yields. In 2003, equipped with the necessary skills for producing a high quality and quantity of potatoes, the group

Table 11.3 *Variables used in the multiple regression analysis*

Variable	Units/codes
Total revenue from potatoes in 2005	Ugandan shillings (USh)
Price offered by Nandos	Ugandan shillings per kilogram (USh/kg)
Price offered if sold as seed	USh/kg
Price offered by other buyers	USh/kg
Marital status of head	1 = married; 0 = other
Level of formal education	0 = none; 1 = some
If seller of potatoes is wife	1 = yes; 0 = no
Year of membership in a farmer field school (FFS)	Year
Land allocated to potatoes	Acres
Size of household	Number of individuals

wanted to increase their commercial sales and requested support from Africare, NARO, the Regional Potato and Sweet Potato Improvement Network in Eastern and Central Africa (PREPACE) and CIAT. Through this consortium of partners, the Nyabyumba Farmers' Group received training in identifying and analysing market opportunities and in developing a viable business plan for the potato enterprise. From the market study, the group identified Nandos (a fast-food restaurant in Kampala) and the local wholesale markets in Kampala as potential market outlets. The group set up a series of committees to manage, plan and execute their production and marketing.

In order to increase the competitiveness of production, the group conducted research supported by NARO to determine the most suitable nutrient levels of nitrogen, phosphorus and potassium (NPK) fertilizer and the time of de-haulming potato plants to produce larger tubers with higher organic content, firm skin and higher yields as required by the buyers. The Nyabyumba Farmers' Group has expanded to a membership of 120 members, 80 of whom are women. They have supplied 190 tonnes of potatoes to Nandos over a four-year period, bringing them an income of about US$50,000 (for further details, see Ferris and Kaganzi, 2005).

Tikolane Farmers' Club, Chinsewu village, Kasungu District, Malawi
The Tikolane Farmers' Club has a membership of about 75 households, of which 18 are female headed. In 2003, CIAT, in partnership with Plan-Malawi, began activities by taking the community through the ERI steps of participatory diagnosis, identifying market opportunities and selecting an enterprise, farmer participatory research, and gender and HIV/AIDS awareness. The community selected six committees: the main committee plus committees for participatory market research, farmer participatory research, livestock, production, and savings and credit.

The vision of the Tikolane Farmers' Club is to have member households with enough food at all times, permanent houses, decent clothing and adequate money to pay for necessary services in their daily lives by the year 2010.

The major enterprises of the group are bean-seed production and goat production. Farmers were trained in multiplying bean seeds and have been doing a number of experiments, including participatory variety selection, identifying options for effective pest and disease control, and identifying technologies to improve soil fertility.

Tigwirane Dzanja Club, Katundulu village, Lilongwe District, Malawi
In May 2003, ERI was established in Katundulu village in partnership with the Department of Agricultural Research Services and the Lilongwe Agricultural Development Division. The village has 36 households, all smallholder farmers. The community formed a group called the Tigwirane Dzanja Club, which literally means 'Let us hold each other's hand', with the prime purpose of alleviating poverty through group action.

This community, like the Chinsewu community, was taken through the ERI steps. It then selected committees: the main one (which also served as the participatory monitoring and evaluation committee) plus committees for participatory market research, farmer participatory research and livestock.

Katundulu village selected a pig enterprise and started with ten sows and two boars. The ten sows were distributed to ten of the 36 households and it was agreed that, after farrowing, two female piglets would be passed on to two different households as a form of repayment. Research focused on identifying the most cost-effective feeding ration using soybeans, lime, premix, salt and maize bran. The pig enterprise has grown quickly and become a profitable venture. One of the most successful farmers now has 14 pigs, has sold more than 10 and has opened a bank account with 9000 Malawi kwacha from the proceeds.

RESULTS AND DISCUSSION

Characterizing the households in the three communities

Each Nyabyumba household had an average landholding of 11 acres (4.45ha), of which 7.6 acres (3ha) were cultivated; the average Katundulu household had 2.9 acres (1.2ha) and cultivated almost all of it (2.8 acres); and the average Chinsewu household had 4.7 acres (1.9ha) and cultivated 4.1 acres (1.7ha).

All farmers surveyed were relatively young, with an average age of household head of 44 years for Nyabyumba and Chinsewu, and 36 years for Katundulu. In all of the communities, most households were male-headed: 88 per cent in Nyabyumba, 83 per cent in Chinsewu and 88 per cent in Katundulu. In terms of educational level, in Nyabyumba, 38 per cent of the household heads had no formal education, 54 per cent had primary education and 9 per cent had secondary education and higher. Chinsewu and Katundulu were similar in terms of educational level: most of the respondents were literate (72 per cent), 75 per cent had primary education, few had no formal education (13 per cent in Chinsewu and 4 per cent in Katundulu), and a significant proportion had attained secondary education or higher (12 per cent in Chinsewu and 24 per cent in Katundulu).

There were significant differences between houses in Uganda and Malawi. Most of the Nyabyumba households had iron-sheet roofs (65 per cent), about 26 per cent had semi-permanent houses and only a few (8 per cent) had grass-thatched houses. This is an indicator of wealth, as was highlighted during informal meetings with group members who said that there are few poor households in the group because of the income from potato sales. On the other hand, most houses in Chinsewu were made with mud walls (72 per cent), a few had used unburned bricks (19 per cent),

9 per cent had used burned bricks, and only 9 per cent had a cement floor. Katundulu had poorer households: 88 per cent of the houses were made with mud walls, and only 8 per cent had a cement floor.

In terms of livestock ownership, Nyabyumba households owned the largest diversity of livestock. Livestock ownership by households was as follows: in Nyabyumba 33 per cent of households had goats, 24 per cent poultry and 19 per cent local cattle; in Chinsewu 42 per cent had chickens and 39 per cent goats; and in Katundulu 33 per cent had chickens and 33 per cent rabbits.

These differences in land holdings, types of houses and livestock ownership between Ugandan and Malawian farmers are typical of the differences in well-being of communities in the two countries. A baseline study of comparable ERI sites in Malawi and Uganda found that average-income households in Kabale, Uganda, still had higher incomes than those regarded as wealthy households in Dedza, Malawi (Sanginga, 2006).

Impacts of the ERI approach

Social and human capital benefits of farmer organizations for the community
Proponents of participatory approaches argue that applying 'empowering' types of participatory research can build human and social capital in various ways. It can:

- enhance the innovative capacity of farmers to experiment with new agricultural practices; and
- strengthen farmers' general analytical abilities, problem-solving skills, and ability to initiate and sustain innovation without external facilitation.

These arguments are supported by Johnson et al (2003), who found that these types of human capital benefits occurred when empowering participation was used: participating farmers could advise their neighbours on agricultural problems and help them to negotiate with traders for better prices.

In this study, social and human capital impacts were measured using self-assessments of farmers' capabilities over the past three years. Results from both Nyabyumba (Uganda) and Chinsewu (Malawi) showed that farmers' abilities had changed significantly (see Figure 11.3). Most farmers indicated that their abilities to help other farmers solve agricultural problems were currently very good to good, whereas very few felt like this three years ago. Similarly, when asked about their abilities to conduct their own experiments to test new varieties, without external facilitation, most of the members felt that they were very good to good, while only a few felt like this three years ago. However, when asked about capabilities to bargain with traders, although both communities had increased their abilities, farmers in Nyabyumba indicated they had improved more than the Chinsewu farmers felt they had.

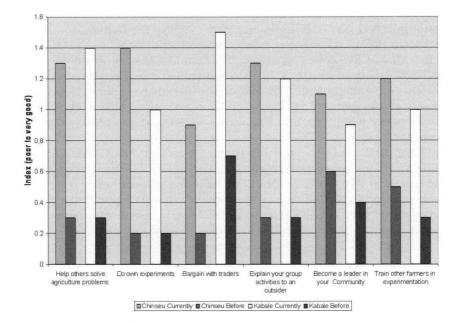

Figure 11.3 *Changes in social and human capital in Nyabyumba (Kabale)*
and Chinsewu

Similar results were found for the Katundulu farmers in Malawi, who
were asked to assess their capabilities. Most felt that their abilities to help
other farmers solve their problems related to pig production, to explain
their group activities or plans to an outsider, to become a leader in their
group and to keep their own farm records had been greatly enhanced.

However, in comparing between female and male members in
Nyabyumba, Uganda, results were mixed. Although women had
improved their skills overall, the results showed that, in various areas,
men had improved significantly more than had the women. Results
showed significant differences in abilities to: bargain with traders to get
better prices (p = 0.0557), become a leader in the group (0.0015), become a
leader in the community (0.0012), train other farmers in experimentation
(0.0077) and keep their own records (0.0001).

An additional analysis was made to compare differences between
ordinary group members' and committee members' levels of ability. The
results show that there were significant differences between group and
committee members in terms of their ability to understand and apply
production-oriented activities (p = 0.001), market-oriented activities (0.01)
and community-oriented activities (0.001). These results imply that there
is a significant difference in skills gained (and therefore in human capital)
between ordinary members and committee members.

Similar findings showing inequity in the distribution of benefits from social capital have emerged in other empirical studies. For example, Gotschi et al (2006) found that gender was a key variable in determining group members' abilities to generate supportive relations and benefit from social capital. However, her study also found that group position was important in increasing social capital benefits, and that women are more likely to obtain help and access information when they are leaders instead of mere members of the group.

Women's empowerment in decision-making and control of income
Gender equity and empowerment of women are central to the ERI process. Therefore, one of the key research questions was whether market orientation benefits women. These aspects were integrated in various ways by:

- ensuring that at least 30 per cent of the members of any committee are women;
- selecting enterprise options based on the extent to which both men and women can benefit and that the enterprise will not adversely affect women and the poor; and
- building community capacities in group development, leadership, conflict management, group relations, social integration with emphasis on gender, and HIV/AIDS awareness.

In this study, we assessed gender equity in two ways by:

1 asking who keeps the income from the sale of the enterprise; and
2 assessing changes in decision-making patterns in the household.

Our results revealed that, in the potato enterprise, 46 per cent of the respondents indicated that income was kept by women. On the other hand, in the pig enterprise, all respondents (100 per cent) indicated that income was kept by men. In informal discussions, women farmers in Katundulu, Malawi, indicated that they could access these benefits indirectly through sale of surplus maize (on which fertilizer purchased with earnings from pigs had been applied).

In this study, we hypothesized that increasing income under the control of women would have significant implications on intra-household decision-making and that household decision-making would become more shared. Changes in decision-making patterns in the household were assessed by asking who made decisions on where to plant, which markets to go to, and how income from the sales was used. Figures 11.4 and 11.5 show changes in intra-household decision-making in Nyabyumba (Uganda) and Chinsewu (Malawi). In both instances, there was a significant reduction in decisions made by men alone, and a corresponding increase in decisions made by men and women jointly. These results are supported

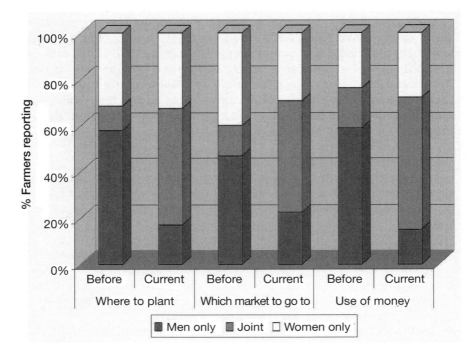

Figure 11.4 *Changes in intra-household decision-making in Nyabyumba, Uganda*

by literature on intra-household dynamics in resource allocation and decision-making; Ulph (1988), Aldermann et al (1995) and Doss (1996) argue that household decisions often reflect the bargaining power of its different members. Putting income in the hands of women can increase their bargaining power. The Chinsewu example extends this argument by demonstrating that increasing women's bargaining power can have spill-over benefits into other household decisions; in this case, it influenced decisions on the use of income from tobacco.

Factors that influence income disparities and gender differences in households

Multiple regression analysis was used to understand the variables that influence income from potatoes (see Table 11.4). Income from potato sales in 2005 was used as a proxy for income.

The results indicated that prices offered – price offered by Nandos (p = 0.020), price offered if sold as seed (0.001) and price offered by other buyers (0.079) – were all statistically significant. This was expected. However, the difference in the order of importance was surprising. The results revealed that the price offered if potato was sold as seed was more

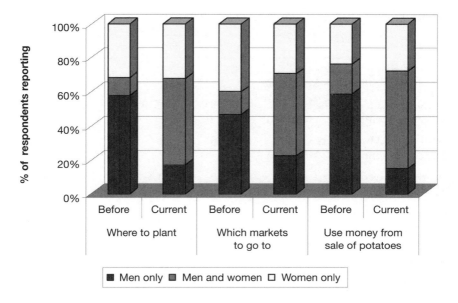

Figure 11.5 *Changes in intra-household decision-making in Chinsewu, Malawi*

significant than the price offered by Nandos. In focus group discussions, farmers in Nyabyumba had given priority to their partnership with the Nandos fast-food restaurant in terms of income from sales; but the results of the multiple regression analysis indicate that farmers earned more when they sold their potatoes as seed versus selling their potatoes to Nandos. The seed potato is smaller than that demanded for making potato chips, which is what Nandos required.

Other interesting results were the gender implications of the farmer-to-market linkages. The highly significant, negative coefficients if the respondent was female (p = 0.029) and if the wife sold the potatoes (0.001) indicate that women members or wives who sold potatoes received lower prices for them. This finding is validated by the earlier results on human capital benefits, which showed that men's abilities increased significantly more than women's in terms of ability to negotiate for good prices. Thus, although women had been involved in the enterprise (e.g. 80 of the 120 members of the Nyabyumba group were women), they still received lower prices than their male counterparts when they sold potatoes.

Other variables such as land allocated to potatoes (p = 0.577), age of household head (0.260), size of household (0.109) and level of education (p = 0.060) were not significant. The land allocated to potatoes was not significant probably because, with commercialization, the farmers are intensifying production and gaining higher yields from the same land area. On the other hand, the finding that the level of education was not significant was remarkable. In many instances, market opportunities are

Table 11.4 *The determinants of income in 2005 from the potato enterprise*

Variable	Coefficient	Standard error	t	P>|t|
Price offered by Nandos (Ugandan shillings per kilogram)	0.440	0.184	2.39*	0.020
Price offered if sold as seed (Ugandan shillings per kilogram)	0.659	0.195	3.38*	0.001
Price offered by other buyers (Ugandan shillings per kilogram)	0.348	0.195	1.79*	0.079
Marital status of head (1 = married; 0 = otherwise)	2.355	1.177	2.00*	0.050
Sex of respondent (1 = female; 0 = male)	−2.611	1.165	−2.24*	0.029
Seller of potato is wife (1 = yes; 0 = no)	−4.008	1.192	−3.36*	0.001
Land allocated to potatoes (acres)	−0.808	1.439	−0.560	0.577
Age of household head	−1.537	1.352	−1.140	0.260
Size of household	1.461	0.898	1.630	0.109
Constant	4.910	4.354	1.13	0.264

Note: * t-tests are significant at the 90% confidence level.

captured by the more educated and younger community members; but this did not happen here. One reason may be because the ERI approach invests in building skills and expertise of farmers to analyse and understand markets using simple tools and methods that are also appropriate for farmers without any formal education.

LESSONS LEARNED

This study highlighted the benefits of applying the ERI approach through case studies in Malawi and Uganda. However, the results also identified various areas where there is a need to adjust the ERI approach to ensure that women can benefit equitably. Several lessons can be derived:

● Although the ERI approach takes specific measures to integrate gender considerations in the distribution of benefits, there are clear gaps. The results highlighted significant gender differences in the distribution of social and human capital benefits, which translated into significant differences in income by men and women. Therefore, ERI needs to strengthen the gender component.

- These results imply that, if the approach aims to promote pro-poor market linkages with gender equity, then the choice of enterprise matters. For example, in the potato enterprise in Nyabyumba, 46 per cent of the women kept the money and decided on their use. On the other hand, in the pig enterprise in Malawi, women did not have direct access to the money from sales. However, from the limited dataset, it would be incorrect to make an overall general conclusion. More case studies will need to be conducted before further conclusions can be drawn.
- The study showed that, using the ERI approach, groups can make significant increases in income – for example, the Nyabyumba Farmers' Group with a membership of 120 members made US$51,136 (90 million Ugandan shillings) in four years. The challenge is how to scale up these impacts to more groups and more communities. This may involve working with farmers at a higher level, such as second-level associations of farmers. However, to do this will require significant adaptations to the approach, which currently focuses on household and group level.

CONCLUSIONS

This chapter presents lessons from applying a novel approach for linking smallholder farmers to markets. ERI aims to strengthen social organization and entrepreneurial capacity in rural communities, encouraging farmers to produce what they can market rather than try to market what they produce.

This is the first in a series of case studies that CIAT is conducting to assess the benefits of applying ERI at the household level. The results indicate that linking farmers to markets led to significant increases in household income. The study found that, when the enterprises were potatoes and beans (food crops that can be sold), women were able to keep the income and make key decisions on the enterprise. However, although there were substantial changes in social and human capital among all group members, men's abilities improved significantly more than women's, leading to lower income gains for women.

ACKNOWLEDGEMENTS

The authors would like to thank the following people who have contributed to this chapter in various ways as researchers, collaborators and implementers in the field: Winnie Alum, Flavia Asimwe, Julius Barigye, Rupert Best, Colletah Chitsike, Diiro Gracious, Elly Kaganzi, Ignatius Kahiu, Peace Kankwatsa, Robert Muzira, Grace Nalukwago, Samson Kazombo Phiri, Noel Sangole and Linda Soko.

NOTE

1   This chapter is a modified and expanded version of an article published in *Natural Resources Forum* (Kaaria et al, 2008). The research was conducted while the first author was with the International Centre for Tropical Agriculture (CIAT) in Uganda.

REFERENCES

Aldermann, H., Hoddinott, J., Haddad, L. and Udry, C. (1995) 'Gender differentials in farm productivity: Implications for household efficiency and agricultural policy', Food Consumption and Nutrition Division Discussion Paper 6, International Food Policy Research Institute (IFPRI), Washington, DC

Ashby, J. A., Braun, A. R., Gracia, T., Guerrero, M. P., Hernandez, L. A., Quirós, C. A. and Ro, J. I. (2000) *Investing in Farmers as Researchers: Experience with Local Agricultural Research Committees in Latin America*, CIAT, Cali, Colombia

Best, R. and Kaganzi, E. (2003) 'Farmer participation in market research to identify income-generating opportunities', *CIAT in Africa Highlight*, vol 9, CIAT-Africa, Kampala, Uganda

Bingen, J., Serrano, A. and Howard, J. (2003) 'Linking farmers to markets: different approaches to human capital development', *Food Policy*, vol 28, pp405–419

Braun, J. von and Webb, P. (1989) 'The impact of new crop technology on the agricultural division of labor in a West African setting', *Economic Development and Cultural Change*, vol 37, no 3, pp513–534

DFID (UK Department for International Development) and OPM (Oxford Policy Management) (2000) 'Making markets work better for the poor: A framework paper', DFID and OPM, London

Dorward, A., Poole, N., Morrison, J. A., Kydd, J. and Urey, I. (2003) 'Markets, institutions and technology: Missing links in livelihoods analysis', *Development Policy Review*, vol 21, no 3, pp319–332

Dorward, A., Kydd, J., Morrison, J. and Poulton, C. (2005) 'Institutions, markets and economic co-ordination: Linking development policy to theory and praxis', *Development and Change*, vol 36, no 1, pp1–25

Doss, C. R. (1996) 'Testing among models of intrahousehold resource allocation', *World Development*, vol 24, no 10, pp1597–1609

Ferris, S. and Kaganzi, E. (2005) 'Linking farmers to markets: The Nyabyumba potato farmers in Uganda', *CIAT in Africa Highlight*, vol 22, CIAT-Africa, Kampala, Uganda

Ferris, S., Kaganzi, E., Best, R., Ostertag, C., Lundy, M. and Wandschneider, T. (2006) *A Market Facilitator's Guide to Participatory Agro-Enterprise Development*, CIAT, Cali, Colombia

Gotschi, E., Delve, R. J. and Freyer, B. (2006) 'The "wrong" gender: Is social capital more accessible to men?', Conference on International Agricultural Research for Development, Tropentag 2006, 11–13 October, Bonn, Germany

Johnson, A. (2005) 'Making market systems work better for the poor (M4P)', Asian Development Bank (ADB) Discussion Paper 9, ADB–DFID Learning Event, ADB, Manila, www.markets4poor.org

Johnson, N. L., Lilja, N. and Ashby, J. A. (2003) 'Measuring the impact of user participation in agricultural and natural resource management research', *Agricultural Systems*, vol 78, pp287–306

Kaaria, S. K. and Ashby, J. A. (2001) 'An approach to technological innovation that benefits rural women: The resource-to-consumption system', Working Document 13, Participatory Research and Gender Analysis Programme, Cali, Colombia

Kaaria, S., Chitsike, C., Best, R., Delve, R., Ferris, S., Kirkby, R., Kaganzi, E., Sanginga, P., Njuki, J. and Roothaert, R. (2005) *Enabling Rural Innovation in Africa: A Program that Empowers Communities to Improve Livelihoods*, ERI brochure, CIAT-Africa, Kampala, Uganda

Kaaria, S., Njuki, J., Abenakyo, A., Delve, R. and Sanginga, P. (2008) 'Assessment of the Enabling Rural Innovation (ERI) approach: Case studies from Malawi and Uganda', *Natural Resources Forum*, vol 32, pp53–63

Lundy, M., Gottret, M. V., Cifuentes, W., Ostertag, C. F. and Best, R. (2003) *Design of Strategies to Increase the Competitiveness of Production Chains with Small-Scale Producers*, Field Manual, CIAT, Cali, Colombia

Lundy, M., Ostertag, C. F., Best, R., Gottret, M. V., Kaganzi, E., Robbins, P. and Ferris, S. (2006) 'Territorial based approach to agro-enterprise development', Strategy Paper, Rural Agro-enterprise Development Project, CIAT, Cali, Colombia

OECD (Organisation for Economic Co-operation and Development) (2004) *Accelerating Pro-Poor Growth through Support for Private Sector Development*, Development Assistance Committee Network on Poverty Reduction, OECD, Paris, France

Ostertag, C. F. (1999) *Identifying and Assessing Market Opportunities for Small Rural Producers,* Series Tools for Decision Making in Natural Resource Management, CIAT, Cali, Colombia

Sanginga, B. N. (2006) 'Gender and socio-economic differences in rural livelihoods and natural resources management in ERI sites in Uganda, Malawi and Tanzania: A synthesis report', CIAT-Africa, Kampala, Uganda

Stake, R. E. (1995) *The Art of Case Study Research,* Sage, Thousand Oaks, CA

Ulph, D. (1988) 'A general noncooperative Nash model of household consumption behaviour', Mimeo, University of Bristol, Bristol, UK

Yin, R. K. (1984) *Case Study Research: Design and Methods*, Sage, Newbury Park, CA

Doing Things Differently: Post-Harvest Innovation Learning Alliances in Tanzania and Zimbabwe

Brighton M. Mvumi, Mike Morris, Tanya E. Stathers and William Riwa

INTRODUCTION

Conventional approaches to technology transfer within small-scale farming systems have frequently failed. Household food security remains precarious for many smallholder farmers, and food production levels show little or no increase. Post-harvest service provision and research have focused on technology development, with less attention being given to understanding delivery-system constraints, to distinguishing between the needs and priorities of different households or to exploring farmers' own research capabilities. Recent approaches to scaling up technologies – both products and processes – point to the dependence of up-scaling on the activities and interactions of a diversity of key players and organizations, all together referred to as the innovation system (see Figure 12.1) (Arnold and Bell, 2001; Hall et al, 2003; Lundy et al, 2004). Here, institutional arrangements (or institutions) refer to 'the mechanisms, rules and customs by which people and organizations interact with each other' (North, 1990). The key challenge to effecting impact is perceived less in terms of devising new technologies – *doing different things* – and more in terms of improving the working of the innovation system – *doing things differently* – to overcome institutional constraints. Translating ideas into social and economic use requires appropriate technologies (hardware innovation), compatible mindsets (software innovation) and favourable institutional settings (system-ware innovation).

The idea that extension services need to be demand led, client oriented, farmer empowering, etc. has gained widespread acceptance (MoAFS, 2003; MoFA, 2003; Leeuwis and Van den Ban, 2004). Reaching this point, however, has involved many adherents in a tortuous learning path, and one which has not yet ended.

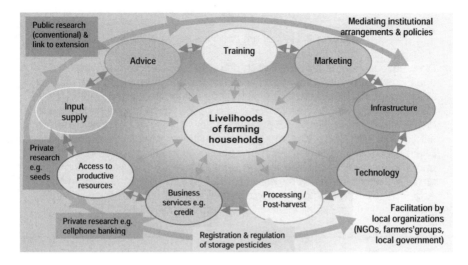

Figure 12.1 *Innovation system from farmers' perspectives*

Source: adapted from a diagram presented by Ian Goldman at the UK Department for International Development (DFID) workshop Improving the Productivity of Smallholder Farmers in Southern Africa, Harare, Zimbabwe, 27–29 September 2005

UNDERLYING PROBLEMS AND THE SEARCH FOR SOLUTIONS

The common denominator for the core research team was initially the problem of insect pests in grain storage. Only after much work – and frustration – was it appreciated that, for the research to have widespread impact, it was also essential to identify and address shortcomings in the institutional context. Food security and income opportunities of many rural households in sub-Saharan Africa are seriously undermined by storage insect pests. Many small-scale farmers rely on imported organophosphate-based pesticides to protect stored grain; but farmers and various authorities are increasingly questioning the safety and efficacy of synthetic pesticides (PAN and CTA, 1995; Marange et al, 1997; Arthur, 2002). Households that follow the traditional method of using ash, botanicals and sand to control storage insect pests are faced with inconsistent and often poor results (Tran and Golob, 1999; Stathers et al, 2002a).

A set of research projects was commissioned by the Crop Post-Harvest Programme of the UK Department for International Development (DFID) from the mid 1990s through to 2006 to explore the efficacy of diatomaceous earths (see Box 12.1) in order to counter this problem.

Box 12.1 *What are diatomaceous earths?*

Diatomaceous earths are soft whitish powders formed from the fossils of tiny aquatic plankton. After processing, these powders can be mixed with grain to kill insect pests. When diatomaceous earths come into contact with insects, they absorb the wax from the cuticle of the insect, which then dehydrates and dies. Diatomaceous earths have extremely low toxicity to mammals and are safe to mix with food. They are used in industry as filters for beverages and pharmaceuticals, as fillers in paints and plastics, as coating agents in fertilizers, etc. and are also mixed into feed to combat internal parasites in domestic animals. Diatomaceous earths are registered as grain protectants in America, Asia, Australia and Europe. In addition to commercial diatomaceous earths, there is potential to exploit local African diatomaceous earths.

These research projects established that diatomaceous earths were efficacious as grain protectants in a range of agro-ecological zones in Zimbabwe and Tanzania. They also established that the technology was readily usable by diverse smallholders and that food stocks (maize, sorghum, bean and cowpea) were successfully protected for periods of more than eight months (Stathers et al, 2002b; Stathers et al, 2008). Trials were done on imported commercial diatomaceous earths (Protect-It® and Dryacide®) and samples from a few of the many local deposits found throughout sub-Saharan Africa that showed insecticidal efficacy. Although local diatomaceous earths will probably represent more economically sound (i.e. to the state) and financially viable (i.e. to business and to farmers) options in the longer run, further work is required to establish and implement safety, extraction and processing protocols.

GETTING DIATOMACEOUS EARTHS INTO USE

Establishing the efficacy, safety and usability of diatomaceous earths as protectants of stored grain proved to be the start rather than the end of the journey. Getting diatomaceous earths (or any seemingly appropriate technology) into economic use amongst rural households that are currently treating their food stocks with potentially dangerous pesticides requires various changes in the institutional setting. Farmers and extension staff need to understand the limitations and dangers of existing protectants, advisers need to make informed recommendations for policy and regulation changes, registration authorities need to 'buy into' and support research findings, and the business community needs to champion registration processes – all of which involves many more stakeholders than conceived as active participants in the original research initiatives.

In addition to hardware (i.e. diatomaceous earth technology) and software (i.e. skills and knowledge required to use the technology) issues, the challenge was also to change the in-country post-harvest systems (system ware) and, in particular, constraints within and between different organizational stakeholders at all levels. Rising to this broader challenge, the Crop Post-Harvest Programme funded a further study to explore how national innovation systems could be better mobilized to sustain the uptake and adoption of crop post-harvest knowledge for the benefit of poor farmers. The research propositions were as follows:

- *Proposition 1:* undertaking action research within an alliance of practitioners, researchers, policy-makers and activists will lead to greater impact and facilitate scaling up through, amongst others, development of broader ownership of concepts and processes, enhancement of local capacity (particularly for adaptive management) and the emergence of locally appropriate solutions or innovations.
- *Proposition 2:* current practices in statutory post-harvest service provision and supporting initiatives are failing to distinguish between the needs and priorities of different households and are therefore unable to meet their diverse requirements.
- *Proposition 3:* researchable constraints and opportunities exist at the current interface of supply and utilization, and the planned insights into these could help facilitate improvement in terms of 'shaping' and delivery of post-harvest information by the range of post-harvest knowledge management organizations.

THE RESEARCH APPROACH

Learning alliances (LAs) (see Box 12.2) were to be both the means of testing proposition 1 and the vehicle within which research would examine propositions 2 and 3. Moriarty et al (2005) define LAs as a series of connected stakeholder platforms, created at key institutional levels (typically national, intermediate and community), and designed to break down barriers to both horizontal and vertical information-sharing and, thus, to speed up the process of identification, development and uptake of innovation.

With respect to proposition 1, implementation involved inviting key post-harvest stakeholders in Tanzania and Zimbabwe to join and initiate Post-Harvest Innovation Learning Alliances (PHILAs). The LAs would explore new ways of working together, with the aim of scaling up post-harvest innovations appropriate to the needs, priorities and circumstances of diverse farmers. The LAs were not only to improve sharing and adoption of existing ideas, but also to provide creative spaces in which institutional constraints could be identified and addressed, adaptive management could be encouraged and local ownership of emerging solutions could thrive.

Box 12.2 *Typical characteristics of learning alliances*

Learning alliances (LAs):

- are groups of individuals or organizations with a mutual interest in solving an underlying problem and scaling up solutions;
- bring together a wide range of partners with capabilities in implementation, regulation, policy and legislation, research and learning, documentation, dissemination, etc.;
- represent part of the bigger whole and capture aspects of the organizational complexity that constitutes the day-to-day realities of the innovation system;
- comprise partners typically clustered at different 'administrative' (e.g. national, regional and district) levels – stakeholder platforms – within the innovation system;
- aim to identify and break down barriers constraining learning – across platforms (i.e. horizontally) and between platforms (i.e. vertically); and
- promote flexible and adaptive working practices, share responsibilities, costs and benefits.

Source: adapted from Moriarty et al (2005)

With respect to propositions 2 and 3, a series of collaborative research initiatives (case studies) were undertaken by newly formed partnerships of alliance members. These case studies were designed to explore current post-harvest service provision and opportunities and constraints at the interface of supply and demand.

To stimulate new ways of working, the research management team sought to build iterative action–reflection cycles into all processes. Case study methodologies typically involved triangulation between sources, and the findings and conclusions were interactively peer reviewed at workshops.

Propositions 2 and 3 subsequently evolved along the lines of supply- and demand-side issues, with output objectives reflecting these two aspects. The supply-side objective was to develop practical 'insights' from current working practices and to generate 'improved practice' recommendations. Supply-side case studies included:

- *case study 1:* interface analysis of public service provision and public-sector research;
- *case study 2:* interface and comparative analyses – public service providers and farmers; farmer-centred organizations and farmers;
- *case study 3:* district 'nodal' studies – multi-stakeholder workshops to establish what works well and what does not.

The demand-side objective was to explore and improve the ability of farmers and of commercial enterprises to access and use relevant post-harvest information. Case studies to explore and improve the demand side included:

- *case study 4:* 'people-focused' studies – studying a small number of agencies who use an 'empowerment' approach;
- *case study 5:* household 'enquiry visits' – learning to listen and listening to learn from farmers;
- *case study 6:* exchange visits between farmers and frontline extension staff;
- *case study 7:* interface analysis of commercial enterprises and service providers.

The overall output objective, which was to be applicable at the national level, was to generate and promote recommendations for policy and implementation strategies that would improve the performance of post-harvest service providers and researchers and enhance related decision-making by farmers and commercial enterprises. Activities to generate policy and implementation strategy recommendations included:

- *case study 8:* literature review on post-harvest policy experience, advice and formulation;
- *case study 9:* the agro-processing industry – opportunities and constraints for small-scale farmers.

The relationship of the project with different end-user groups would be determined by the LA approach, in general (LAs are a microcosm of the whole system), by PHILA activities, in particular, and by the focus of the research activities. Given the limited resources and time-frame, it was felt that the studies would be most insightful if focused at district level. Subsequent delays in accessing funds reduced the time to complete this complex multi-stakeholder undertaking to less than ten months.

Two districts, exhibiting contrasting characteristics, were identified and selected by the PHILA membership: Manyoni and Singida districts in Tanzania, and Buhera and Binga districts in Zimbabwe.

Engagement with end users was to be through the PHILA project's collaborative research, its internal information-sharing or strategic engagement activities. Raising awareness and extending its members' individual and organizational capacity was to be achieved through their involvement in the collaborative case studies, either as lead researchers or as participants. The case studies would also involve and benefit other potential end users (e.g. case study 6: farmers involved in the exchange visits; case study 3: district councillors and administrators who participate in the district workshops) (Morris et al, 2006).

PROJECT ACHIEVEMENTS

Post-Harvest Innovation Learning Alliances established

The aim of the PHILA project was to advance the understanding and effectiveness of LAs as agents of change. The PHILA approach, with its underpinning innovation systems perspective and inclusive principles, provided a safe space for diverse stakeholders from multiple organizations to work and learn together in strategic pursuit of a common purpose. Achievements include the following:

- PHILAs were established in Tanzania and Zimbabwe; membership at time of writing totals more than 40 organizations from the public, voluntary and private sectors.
- The PHILA approach and activities raised awareness and understanding on matters of institutional learning and change among members and other post-harvest system stakeholders.
- New working patterns and approaches tested by the PHILA project were documented; these are generating fresh understanding and suggesting new modes of working. Many members have adopted tools and techniques introduced to them through PHILA.
- The PHILA project's performance against its core activities (i.e. collaborative research, information-sharing, engagement and management) was assessed by the membership in both Tanzania and Zimbabwe, and lessons were noted (Morris et al, 2005a; Mvumi et al, 2005).
- Lessons learned were documented and shared at two innovation symposia; in Delft, The Netherlands (June 2005), and Kampala, Uganda (November 2006).
- Members' assessment of PHILA's performance roundly endorsed the use of, and emphasis placed on, information and communication technologies (ICTs), while acknowledging the disadvantage that those lacking access suffered.
- A PHILA website was established (www.nri.org/PHILA) with a members-only working area.

Practical insights developed and improved practices recommended

Based on practical insights developed from current working practices, recommendations for 'improved practice' were generated. Achievements here included the following:

- A body of critical information on the interface between public service providers and public-sector research in Tanzania was generated, shared and reviewed.

- A body of critical information comparing and contrasting how public service providers and farmer-centred organizations work with farmers in two dissimilar districts of Tanzania and one district in Zimbabwe was generated, shared and reviewed.
- Past, present and potential post-harvest service provision were explored, appraised and documented in pairs of contrasting districts in both Tanzania and Zimbabwe; participatory agricultural development planning exercises were initiated, and associated multi-stakeholder processes, tools and techniques were developed and shared with the respective district personnel and PHILA members.
- Practical ideas for increasing responsiveness to farmers' demands were developed and documented.
- A case study was made on learning by farmers and extension staff through exchange visits, and learning materials were generated, including a film based on participants' 'video diary' entries.
- Extension staff from two districts in Tanzania and one in Zimbabwe were trained in the enquiry-visit approach, a portfolio of techniques and tools for facilitating understanding of farmers' diverse circumstances, and responsiveness to their decision-making.

Access to and use of post-harvest information better understood

The ability of farmers and commercial enterprises to access and use relevant post-harvest information was explored and improvements were proposed.

Farmers
More than 260 farmers in Kongwa and Babati districts in Tanzania and in Buhera and Binga districts in Zimbabwe learned about, and independently assessed, a range of grain-protection treatments, including diatomaceous earths. At least 70 farmers in Tanzania and Zimbabwe who had been involved in researcher- and farmer-managed grain storage trials continued their own experimentation with diatomaceous earths. In Tanzania, farmers and extension staff from Singida and Manyoni districts visited farmers in Kongwa district, who had been running grain protection trials for three years. Some of the visitors had been conducting grain protection trials with botanicals themselves. Farmers and extension staff in Zimbabwe testified to the efficacy of diatomaceous earths, demanding that the products be made available to the farming community through the normal market channels. Case studies on empowerment initiatives in Tanzania and Zimbabwe were undertaken, and lessons with the potential for transfer were identified.

Commercial enterprises
PHILA members undertook, documented and reviewed a case study on the interface between agrochemical companies and public service provision and research in Tanzania.

Recommendations for policy and implementation strategies generated

Achievements with regard to policy and implementation strategies included the following:

- Detailed presentations of earlier findings on the use of diatomaceous earths as grain protectants were made to a range of post-harvest stakeholders at inception workshops in Zimbabwe and Tanzania.
- A meeting was held with key representatives from the agricultural-input private sector in Tanzania in March 2005. The research findings on the use of diatomaceous earths as grain protectants in Tanzania and Zimbabwe were shared to stimulate them to champion the registration process in Tanzania (Stathers et al, 2005).
- Case studies on the formulation, implementation and bearing of diverse policies on the post-harvest situations in Tanzania and Zimbabwe were undertaken and, in Tanzania, were shared and reviewed.
- A set of recommendations based on multi-stakeholder reviews of the case studies associated with improving responsiveness to farmer demand was generated (Morris et al, 2005b; Mvumi et al, 2005).
- Steps for improving both policy formulation and policy implementation in Tanzania and Zimbabwe were identified and drawn up by PHILA members at the review workshops (Morris et al, 2005b; Mvumi et al, 2005).
- Literature on the requirements of small to medium agro-processing systems for effective manufacturing and marketing and on agro-processing service provision, research and extension linkages in Zimbabwe was reviewed and synthesized (Mhazo et al, 2005).
- Participatory agricultural development planning exercises were initiated in two districts in each country. Associated multi-stakeholder processes, tools and techniques were developed and shared with the respective district personnel and PHILA members. Singida District Council has since adopted key elements of this approach in its district agricultural development planning.
- Key stakeholder types with which PHILA needs to build and foster relationships were identified (see Box 12.3) and the process of engagement was introduced into strategic planning.
- PHILA's strategic action plans for Tanzania, drawn up by the membership, envisage PHILA playing a mainline role in the workings of the Ministry of Agriculture, Food and Cooperatives and/or the ministry continuing to play a major role in PHILA's future.

Box 12.3 *Key stakeholder types with whom the Post-Harvest Innovation Learning Alliance needs to build and foster relationships*

It is important that PHILA fosters sound relationships with the following key stakeholder types:

- those who make decisions or effect changes in policy and practice (e.g. policy-makers, district councillors, service providers and innovative farmers);
- those who can influence these decision-makers directly (e.g. members of parliament and private-sector companies);
- those in civil society who can bring pressure to bear on decision-makers;
- those who can support, reinforce and strengthen PHILA's recommendations (e.g. training, academic and research organizations, as well as financial organizations);
- those in the media who provide a means by which PHILA can reach the public; and
- the donor community, who can finance and support PHILA's activities.

Some of these stakeholders may already be members of the alliance; but many others are still outside the alliance, suggesting that either enrolment or engagement initiatives are required.

DISCUSSION OF LESSONS LEARNED

The system-ware approach

Innovation systems approaches stress that research takes place within a socio-economic, political and environmental context, and that a wide range of organizations and institutions play a role in translating ideas and knowledge into widespread social and economic use. The rationale for adopting an LA approach was to improve the working of the innovation system through 'doing things differently' in order to overcome constraints in the way in which people and organizations interact with each other (i.e. a system-ware approach).

Morris et al (2005b) identified two levels of institutions, based on Williamson's (2000) hierarchical classification scheme, which could be influenced by LAs in the short to medium term. The first relates to rules defining governance structures, incentive structures, business contracts, etc. – and we include here policy (e.g. agricultural extension and pesticide registration) and research and development programmes. Williamson, who refers to this level as 'the play of the game', suggests it leads to the building of organizations and networks, and has a frequency of change of one to ten years. The game of influence, however, plays two ways; while LAs might influence governance structures, incentives, etc., the LAs themselves, their constitution and processes are, in turn, influenced by the prevailing play and its players.

We interpret Williamson's second level, which relates to allocation mechanisms and affects adjustments in prices, outputs and incentive alignments, to include rules governing staffing arrangements, and access to and use of ICT. Williamson suggests that their frequency of change is short term and continuous.

Inter-organizational working

The LAs not only brought different stakeholders from multiple organizations within the respective agricultural systems together, but also provided a safe and effective space for them to work and learn together, improving inter-organizational relationships. While there was already some cooperation between public and non-state service providers, there were also rivalries and tensions. Working together in an alliance allowed complementarities to be developed, reduced conflicts and inefficiencies, and proved mutually beneficial to all stakeholders. Action research by local partners built or strengthened local capacity for adaptive management. This was exemplified by the district nodal studies (case study 3), which were, in effect, participatory planning exercises and have been followed up by the respective districts.

The approach also embraces sharing information on products and processes and planning for strategic engagement with key stakeholders (see Box 12.3), who may not be part of the alliance. The sharing drew attention to resource and capability gaps associated with access to and use of modern ICT, and efforts and resources were applied to address these shortcomings. In the exchange visits (case study 6), farmers were given disposable cameras and introduced to video diaries, both to facilitate the direct learning experience and to provide for later sharing with neighbours. All told, the LA experience has led to a much more realistic understanding, both of the essential inputs and costs, and of the challenges and opportunities associated with the establishment of multi-stakeholder, multi-agency, multi-level LAs.

Policy implications

Many sub-Saharan African countries have recently reformed their agricultural extension policy objectives, typically under 'encouragement' from donors, and now place emphasis on pluralism, farmer empowerment, and client-focused and demand-led services (MoAFS, 2003; MoFA, 2003). Implementation strategies have also been drafted with donor help; but operationalization is still very much 'work in progress', with donor commitment less evident.

The policy review in Tanzania (case study 8) generated concern amongst members about the apparently isolated nature of policy formulation, the predominance of senior government officials in the process and, conversely, poor levels of engagement with, and involvement of, civil society. Similar findings were reported for Zimbabwe. District staff complained that there is no process in place to explain to them, let alone to farmers, the implications of agricultural or other policies to their own activities, which caused them to function oblivious of recent policy initiatives. Members are also concerned about the complexity of policy language and argue that existing policy needs to be made more accessible (e.g. through the use of farmer networks, radio programmes, etc.). With regard to policy implementation, members suggest posting local-language summaries on village notice boards; multi-stakeholder strategic planning initiatives (e.g. case study 3); regular evaluations of policy implementation and sharing of the findings widely; and training on policy issues. The PHILA project has been lobbying for improvements in existing policy processes, but is hampered by resource and time constraints. However, a minor success has been scored in bringing about improvements to the Plant Protection Act in Tanzania.

Commercialization and product registration

Engagement with agri-business has had mixed success. While diatomaceous earths offer a considerably safer grain protectant than synthetic pesticides, agri-business – the main suppliers of these pesticides – is not convinced that imported diatomaceous earth products would be more profitable for them. This situation could change if the authorities responded to the conclusive knowledge about the dangers of organophosphates and/ or should agri-business become more responsive to its corporate social responsibility. The real benefit could then come from local production of diatomaceous earths, which would, in turn, probably lower market prices. Local production, however, seems a long way off, not least because of the cost of mining, regulation and production. Meeting these costs might be speeded up if significant markets were identified, either through increasing the use of imported diatomaceous earths or through the growth of other industrial uses such as production of paints and filters (see Box 12.1).

In this case, neither good science nor the adoption of an LA approach has succeeded, as yet, in addressing the barriers to registration and commercialization in Zimbabwe and Tanzania. This institutional or system-ware failure may not, however, typify all innovation systems; parallel entrepreneurial initiatives in Zambia and Mozambique, prompted by the earlier diatomaceous earth studies and using local diatomaceous earths, are already more commercially advanced.

Vulnerability and adaptive capacity

If LAs are to accommodate grassroots linkages and if service provision is to be more responsive, then ways have to be found to learn about and develop the capacity to address those factors that constrain farmers' circumstances. Inadequate and erratic rainfall typically leads to poor harvests and food shortages in the project districts, with more vulnerable households and individuals suffering the most. The enquiry tool (case study 5), which enables extension staff (and others) to listen to and learn from farmers, provides for systematic gathering of disaggregated information on production and post-harvest practices, which can counter prevailing ignorance, including that resulting from the private and often secretive nature of grain storage practices. A clearer understanding of the causes and manifestations of poverty amongst rural households will ensure that proffered solutions better match needs and priorities. The extent to which this learning will be institutionalized within the respective structures remains to be seen; but continued interest and use of the enquiry tool suggest room for optimism.

Learning about learning

Much of the PHILA experience has been documented and is available on the PHILA website (www.nri.org/PHILA).

PERSISTING CHALLENGES

LAs provide alignment for the three key components of innovation:

1   the hardware that enables technological innovation;
2   the software that, through action research, develops compatible mind-sets and adaptive capacity; and
3   the system ware in which constraints in the institutional environments, outside the remit of conventional research approaches, can be systemically addressed.

As such, LAs offer a strategic approach for developing the much-sought-after empowering, client-oriented, demand-led services. However, challenges persist:

● Establishing LAs – typically building on existing networks – involves high front-end transaction costs. Donors who are happy pushing at the forefront of policy development should equally consider stepping up to meet the associated implementation challenges and costs. The costs may be high; but the costs of not doing so are even higher – 'invest now or pay later' (Barnett, 2006).

- Meaningful innovation is fundamentally about changing institutional/ social relationships and developing more effective ways of learning. Technological aspects still seem predominant, with 'information' often misrepresented as 'knowledge', and ideas on 'knowledge manage- ment' confined to 'technology uptake'.
- Conflict is inherent in change. Those benefiting from the *status quo* are happy to continue dictating play but are unlikely to concede voluntarily to changes of rules by others. LAs need good facilitation to draw stakeholders together and to enhance negotiation and conflict management because:
  - existing elites will tend to exclude some stakeholders, deliberately or otherwise – 'ineffective communication' may otherwise be 'politically effective';
  - private-sector players with competing interests and busy schedules do not readily appreciate inclusive participatory approaches; and
  - representation of farmers frequently excludes more vulnerable households and minority groups.
- LAs would generally be built on existing stakeholder platforms or network arrangements for reasons of legitimacy and/or to avoid creating parallel structures; but learning may not be central to these existing platforms. Incorporating institutional learning and change within the existing structures, without appearing to be subversive, requires trust to be built and consolidated – a time-consuming process.
- LAs are about changing the dynamics between and within organiza- tions. But this largely depends on the skills and energy of individuals. LAs are about processing ideas rather than peddling products. Learning about and relating to a process approach remain challenging. Process documentation is still at an early stage of development.
- If LAs are to have sustainable impact, then they need both to influence and to win over policy-makers and other stakeholders who can facil- itate their formalization within the system. Currently used advocacy tools include policy briefs and outcome mapping, but other methods and tools are needed.
- Staff turnover or transfers associated with policy or organizational changes can be disruptive and frustrating after a large investment of resources (Mvumi et al, 2003). Sharing of information within each partner institution and between partners is required to strengthen institutional memory and to reduce the negative impact of staff turnover.

CONCLUSIONS

The literature on knowledge networks – precursors to LAs – suggests that the formation period typically takes from one to three years (Creech and Ramji, 2004). The formation of PHILAs and understanding of the associated processes may be on course for success; but much remains to be learned. This will require continued support from the research and development communities – for example, the Post-Harvest Forum for Research and Development in Eastern, Southern and Central Africa, the Association for Strengthening Agricultural Research in Eastern and Central Africa (ASARECA), the Southern African Development Community (SADC), the DFID, the World Bank, the African Development Bank, the European Union, etc. – and greater commitment from the statutory authorities to promote this challenging approach. It is also essential to ensure the extended involvement of the private sector and the widest representation and participation of farmer organizations and networks, including representatives of more vulnerable groups and minorities, if research outputs are to be effectively taken up.

ACKNOWLEDGEMENTS

This chapter is an output of the Post-Harvest Innovation Learning Alliances in Tanzania and Zimbabwe. Funding for presenting the paper upon which this chapter is based at the Innovation Africa Symposium in Kampala, Uganda, was provided by the World Bank through the International Centre for Tropical Agriculture (CIAT), for which we are grateful.

REFERENCES

Arnold, E. and Bell, M. (2001) 'Some new ideas about research for development', in *Partnership at the Leading Edge: A Danish Vision for Knowledge, Research and Development*, Danish Ministry of Foreign Affairs, Copenhagen, Denmark
Arthur, F. H. (2002) 'Survival of *Sitophilus oryzae* (L.) on wheat treated with diatomaceous earth: Impact of biological and environmental parameters on product efficacy', *Journal of Stored Products Research*, vol 38, pp305–313
Barnett, A. (2006) *Journeying from Research to Innovation: Lessons from the Department for International Development's Crop Post-Harvest Research Programme 'Partnership for Innovation'*, The Policy Practice Limited, Brighton, UK
Creech, H. and Ramji, A. (2004) 'Knowledge networks: Guidelines for assessment', Working Paper, International Institute for Sustainable Development, Winnipeg, Canada
Hall, A. J., Yoganand, B., Sulaiman, R. V. and Clark, N. G. (eds) (2003) *Post-Harvest Innovations in Innovation: Reflections on Partnership and Learning*, DFID Crop Post-Harvest Programme South Asia, Patancheru, India/Natural Resources International, Aylesford, UK
Leeuwis, C. and Van den Ban, A. (2004) *Communication for Rural Innovation: Rethinking Agricultural Extension*, third edition, Blackwell Science, Oxford, UK
Lundy, M., Gottrett, M. V. and Ashby, J. (2004) 'Learning alliances: An approach for building multi-stakeholder innovation systems', ILAC Brief 8, Institutional Learning and Change Initiative, Rome, Italy

Marange, T., Mvumi, B. M., Chinwada, P., Mushayi, P., Mautsa, L. and Mhunduru, J. (1997) A PRA Survey of Kawere Ward in Mutoko District, Crop Post-Harvest Research Programme Report, Harare, Zimbabwe

Mhazo N., Mvumi, B. M., Nazare, R. M. and Nyakudya, E. (2005) 'The status of the agro-processing industry in Zimbabwe with particular reference to small and medium enterprises', PHILA Review Workshop, 23–25 November, Morogoro, Tanzania

MoAFS (Ministry of Agriculture and Food Security) (2003) The Agricultural Sector Development Programme: Agricultural Extension Reform, Draft Main Report, MoAFS, Dar es Salaam, Tanzania

MoFA (Ministry of Food and Agriculture) (2003) Agricultural Extension Policy, Directorate of Agricultural Extension Services, MoFA, Accra, Ghana

Moriarty, P., Fonseca, C., Smits, S. and Schouten, T. (2005) 'Learning alliances for scaling up innovative approaches in the water and sanitation sector', in S. Smits, C. Fonseca and J. Pels (eds) Proceedings of Symposium on Learning Alliances for Scaling up Innovative Approaches in the Water and Sanitation Sector, 7–9 June 2005, IRC International Water and Sanitation Centre, Delft, The Netherlands

Morris, M., Mvumi, B. M., Stathers, T. E. and Riwa, W. H. (2005a) 'Post-harvest innovation to improve food security in Tanzania and Zimbabwe: Learning Alliance lessons', in S. Smits, C. Fonseca and J. Pels (eds) Proceedings of Symposium on Learning Alliances for Scaling Up Innovative Approaches in the Water and Sanitation Sector, 7–9 June 2005, IRC International Water and Sanitation Centre, Delft, The Netherlands

Morris, M., Stathers, T. E., Mvumi, B., Gasana, D., Riwa, W. and Mathias, D. (2005b) Post-Harvest Innovation Learning Alliance (PHILA): Review Workshop – Tanzania, Report of a workshop organized by the PHILA management team, Plant Health Services (Tanzania), the Natural Resources Institute (UK) and the University of Zimbabwe, 23–25 November 2005, Muslim University of Morogoro, Morogoro, Tanzania

Morris, M., Mvumi, B., Stathers, T. and Riwa, W. (2006) Post-Harvest Innovation: Enhancing Performance at the Interface of Supply and Utilization, Project Final Report, Project R8460, Crop Post-Harvest Programme (CPHP) of Natural Resources International Ltd, for DFID, Aylesford, UK

Mvumi, B. M., Morris, M. J., Stathers, T. E. and Riwa, W. (2003) 'Partnership in research and development of diatomaceous earth technology for small-scale grain storage', Paper prepared for Global Forum for Agricultural Research meeting 22–24 May, Dakar, Senegal

Mvumi, B. M., Stathers, T., Mhazo, N., Soroti, Z., Mwanga, J., Marongwe, L. S. and Morris, M. (2005) Post-Harvest Innovation Learning Alliance (PHILA): Review Workshop – Zimbabwe, Report of a workshop organized by the PHILA management team, the University of Zimbabwe, the Natural Resources Institute (UK) and Plant Health Services (Tanzania), 13–14 December 2005, Harare, Zimbabwe

North, D. C. (1990) Institutions, Institutional Change and Economic Performance, Cambridge University Press, Cambridge

PAN (Pesticides Action Network) and CTA (Technical Centre for Agricultural and Rural Cooperation) (eds) (1995) Pesticides in Tropical Agriculture: Hazards and Alternatives, Margraf, Weikersheim, Germany

Stathers, T. E., Chigariro, J., Mudiwa, M., Mvumi, B. M. and Golob, P. (2002a) 'Small-scale farmer perceptions of diatomaceous earth products as potential stored grain protectants in Zimbabwe', *Crop Protection*, vol 21, pp1049–1060

Stathers, T. E., Mvumi, B. M. and Golob, P. (2002b) 'Field assessment of the efficacy and persistence of diatomaceous earths in protecting stored grain on small-scale farms in Zimbabwe', *Crop Protection*, vol 21, pp1033–1048

Stathers, T. E., Riwa, W. H., Mvumi, B. M. and Morris, M. (2005) Engaging the Private Sector in the Promotion of Diatomaceous Earths, Report of a meeting to engage the private sector and other stakeholders in the registration and promotion of diatomaceous earths as grain protectants in Tanzania, 22 March, Plant Health Services, Dar es Salaam, Tanzania

Stathers, T. E., Riwa, W., Mvumi, B. M., Mosha, R., Kitandu, L., Mngara, K., Kaoneka, B. and Morris, M. (2008) 'Do diatomaceous earths have potential as grain protectants for smallholder farmers in sub-Saharan Africa: The case for Tanzania?', *Crop Protection*, vol 27, pp44–70

Tran, B. M. D. and Golob, P. (1999) Improvements in the Storage and Marketing Quality of Grain Legumes, Final Technical Report, DFID Crop Post Harvest Programme, Project R6503, Natural Resources Institute, Chatham, UK

Williamson, O. E. (2000) 'The new institutional economics: Taking stock, looking ahead', *Journal of Economic Literature*, vol 38, pp595–613

CHAPTER 13

Alternative Funding Mechanisms for Local Innovation Systems

Willem Heemskerk, Ninatubu Lema, Bertus Wennink and Henriette Gotoechan-Hodounou

INTRODUCTION

The perceived need to strengthen the demand side for agricultural service provision for enhancing innovation and the call for a separation of responsibilities for policy-making, funding and implementation have resulted in innovations such as alternative funding mechanisms for agricultural research for development (R4D) at national and local levels (NEPAD, 2002; FARA, 2006). These institutional innovations aim at enhancing multi-stakeholder resource control, increasingly involving research clients and the end users of agricultural production and processing technology in decisions concerning the allocation of research staff, money and infrastructure (Carney, 1998; Chema et al, 2003; Heemskerk and Wennink, 2006). It is envisaged that these reorganized funding mechanisms for agricultural innovation will combine greater efficiency in resource management with improved effectiveness in innovation development through stronger client control, thus better addressing the agricultural and natural resource management needs, particularly of small-scale farmers and processors.

Although agricultural knowledge and information systems (AKIS) and agricultural innovation systems (AIS) (Engel, 1997; Hall et al, 2001; Rivera et al, 2005) often emphasize technological innovation(s) in any component of the value chain, putting these systems to work requires substantial organizational and institutional innovation.

Stakeholder involvement and client empowerment have led to a de-concentration of funding mechanisms for agricultural innovation. The roles of the various groups of stakeholders in an agricultural innovation system are changing rapidly. The state increasingly emphasizes its policy and regulatory functions through which it also tries to stimulate the effectiveness and efficiency of R4D service provision. For reasons of transparency and effectiveness, the functions of financing, planning and budgeting, as well as providing services in an AIS, are all being separated. Ideally, planning and budgeting is a multi-stakeholder and client-driven activity: the actual R4D financing is provided either through the state (including donors), through jointly managed funds (i.e. by clients and providers) or through public–private partnerships (PPPs) (Gill and Carney, 1999; Hartwich et al, 2003; Spielman and von Grebmer, 2004). Implementation is mostly through specialized agencies, including research centres (public or private), universities, non-governmental organizations (NGOs), etc. In local innovation systems, the need for interaction with and ownership by smallholders and their organizations (Wennink et al, 2003) is crucial.

In sub-Saharan Africa, it has become increasingly difficult to mobilize financial resources for AKIS from the public sector, while donors prefer to channel more of their funds to demand-side requests. The resulting pressure on resources calls for alternative financing mechanisms to generate incremental funds and to use these more effectively, while the issue of increasing state financing for R4D needs to be addressed simultaneously (Tabor et al, 1998; NEPAD, 2002).

This chapter, based on Heemskerk and Wennink (2006), examines some experiences with local alternative funding mechanisms in Tanzania and Benin, describing and analysing the performance of stakeholder-controlled funding mechanisms such as competitive grant systems (CGSs) and public–private sector matching funds.

ISSUES AND CHALLENGES

Innovative alternative funding mechanisms require far-reaching institutional innovations, such as enhanced client control over priorities and resources, and expanding the range and skills of service providers, as well as organizational change within the various stakeholder organizations, not only in the public sector, but also with regard to farmer organizations (FOs) and the private sector (Echeverría, 1998). Stakeholders from both the supply and demand sides must have the capacity to participate meaningfully in the AIS, in general, and in its funding mechanisms, in particular, in order to gain the desired effectiveness and efficiency. Innovative funding mechanisms must be designed in such a way that they contribute to strengthening R4D partnerships and other learning alliances, and become vehicles for attracting funding from both public and private sources. Local funding mechanisms face the challenge of combining enhanced stakeholder participation with long-term sustainability. De-concentrated funds are more likely to have better stakeholder participation than national funds, but will also have larger overheads, which are influenced by economies of scale. These funds focus on adaptive research and pre-extension and therefore demand greater stakeholder participation for interactive learning. Local competitive funds for agricultural innovation financed by public financing mechanisms (national budget or levies and taxes) need to be matched with other funds to become sustainable. This means establishing local stakeholder ownership and integrating different priorities and perspectives.

CASES: KEY FEATURES

The following cases were selected on the basis of reported experience with local funding mechanisms (Tabor et al, 1998; Gill and Carney, 1999; Blackie et al, 2003; Matthess and Arodokoun, 2005; Rivera et al, 2005).

National Agricultural Research Fund in Tanzania

In Tanzania, the National Agricultural Research Fund (NARF) is a competitive funding mechanism that pools resources for all agricultural research priorities. The NARF has been established as a transparent mechanism for funding highly innovative and applied (priority) agricultural research and development initiatives that would also facilitate collaboration among allied research institutions, notably the universities. A multi-stakeholder committee manages the fund, which can be accessed by various actors in the National Agricultural Research System (NARS).

Zonal Agricultural Research Funds in Tanzania

Complementary to the NARF, there are seven sub-national Zonal Agricultural Research Funds (ZARFs), which concentrate on adaptive research and dissemination and address zonal research priorities established by local stakeholders. Local ownership is stronger within the ZARFs, partly because local district governments contribute to the zonal funds (Blackie et al, 2003). All seven research and development zones in Tanzania formed such a fund, managed by multiple stakeholders, during the period of 1998 to 2005. The total financial allocation amounted to approximately US$2 million for the given period.

District Agricultural Research Funds in Tanzania

In Tanzania, some districts have established their own competitive grant mechanisms for outsourcing research and extension services, seen as priorities by the stakeholders in the district. In 4 of the 22 districts of the eastern zone of Tanzania, District Development Funds were established that had a special allocation for contracting research and development services. A total of US$200,000 annually was available to the four districts (Lema et al, 2003).

Competitive funds for zonal programmes in Benin

In Benin, the National Agricultural Research Institute (INRAB) manages a national competitive fund – a consolidated funding mechanism (CFM) – which has been de-concentrated to two zonal agricultural research centres. In each zone, a multi-stakeholder-driven competitive funding mechanism was instituted in 2001 under the overall supervision of INRAB. A multi-stakeholder Zonal Research and Development Committee (ZRDC) administers the fund. Research proposals are screened and resources allocated by the ZRDC's Project Appraisal Committee. These zonal competitive funding mechanisms for adaptive research and dissemination can be accessed by all NARS member organizations, as well as the public-sector agricultural extension service (Matthess and Arodokoun, 2005).

Tanzania Coffee Research Institute

The coffee sector in Tanzania has established a coffee R4D fund that is financed through coffee export levies. In March 2001, the existing public coffee research centres, as well as relevant donor funds, were transferred to the newly established, privatized Tanzania Coffee Research Institute (TaCRI). It is now a membership-based organization and is managed by a multi-stakeholder management board through a management team. The coffee fund is managed by the Tanzania Coffee Board and made available almost exclusively to the privatized TaCRI.

Public and private funding of agricultural extension in Benin

In Benin, the Ministry of Agriculture has agreed with the Cotton Association (AIC), representing most cotton-sector stakeholders, to establish a common fund based on cotton export levies to finance cotton research and extension support services. Private parties (i.e. cotton producers and ginners) have contracted the public-sector extension service to provide agricultural extension services. After the freezing of staff recruitment for public agricultural extension, the organization began to decline, causing national cotton stakeholders to involve the private sector in cotton-sector extension. In 2000, private entities such as the AIC and an input-supply company, and the Ministry of Agriculture, contributed to the establishment of a common fund for cotton-sector support services. The national extension service was contracted by this common fund, financed through cotton levies, to provide cotton advisory services at provincial and district level.

REVIEW

The following criteria were used for analysing the above-mentioned cases:

- stakeholder participation in planning, monitoring and evaluation, leading to joint resource allocation and interactive learning;
- institutional change to ensure the involvement of all stakeholders;

- efficiency and sustainability of the mechanism in terms of upward and downward accountability, transparency, good governance and stakeholder ownership; and
- effectiveness and relevance of the research and development services.

Stakeholder participation

In most of the cases, some form of multi-stakeholder committee has been established with widely varying degrees of authority to allocate resources. Tanzania's various management committees at national, zonal and district level – NARF, ZARF and the district Agricultural Research Fund (ARF), composed of representatives of the primary stakeholders – make decisions on research priorities, call for proposals, screen these through external reviewers and give final approval. The committees involve farmers in monitoring during field visits (see Table 13.1). Similar committees in Benin only establish priorities, while INRAB implements the entire process (calls, screening and contracting). All research providers are qualified to submit proposals to these funds. The commodity funds in Tanzania (coffee) and Benin (cotton) organize local-level consultations for setting research priorities. The multi-stakeholder coffee and cotton boards consolidate the priorities and then contract out the adaptive research proposals to the privatized Coffee Research Institute in Tanzania or to the public extension services in Benin (see Table 13.1).

Institutional change

Although the stakeholders have an established role (see Table 13.1), further institutional change is needed to strengthen this role and to develop true partnerships in prioritizing and monitoring resource allocation for R4D and innovation. In Tanzania, stakeholder representatives were mainly from NARS institutions, with little representation of established FOs. The fund did not attract closer collaboration between research service providers, although this was one of its main objectives. At zonal level, the public and private advisory services' influence over the research agenda improved; but farmers still had little influence. Because of stakeholder pressure, funds became more autonomous and focused more on R4D activities, leading to a better balance between research and development. Farmer groups in Tanzania influence the priority setting at district level and participate in district-level farmer forums twice a year, contributing to further integrated district planning. The management committees in Benin, in which farmer representatives and commodity groups have their capacity developed, are further empowered in research proposal screening and resource allocation. Researchers in the committees are rotated rapidly in order to institutionalize the demand-driven approach and the required attitudinal change. The TaCRI board, with coffee-sector

Table 13.1 *Stakeholder participation in planning, monitoring and evaluating the mechanism*

National Agricultural Research Fund (NARF) Tanzania	Zonal Agricultural Research Fund (ZARF) Tanzania	District Agricultural Research Funds (ARFs) Tanzania	Consolidated funding mechanism (CFM) Benin	Tanzania Coffee Research Institute (TaCRI)	Public–private partnership (PPP) extension Benin
Lema and Kapange (2006a)	Lema and Kapange (2006b)	Akulumuka and Lugeye (2006)	Gotoechan-Hodounou et al (2006)	Kapange and Lema (2006)	Sogbohossou et al (2006)
Multi-stakeholder involvement in priority setting, open calls, fund management and field visits for monitoring and evaluation (M&E)	Multi-stakeholder involvement in priority setting, open calls, fund management and field visits for M&E	Multi-stakeholder involvement for designing tenders based on village plans and proposal reviews; M&E by village	Annual multi-stakeholder establishment of priorities; open call by research	Coffee innovation system actors establish board plan and approve proposals based on strategic needs	Village sets priorities; districts consolidate; Multi-stakeholder endorsement of programme

stakeholders and government representation, has full autonomy to set priorities. Although the stakeholders' voice – in particular, that of the members – has been strengthened considerably, the smallholder is still not sufficiently represented. TaCRI supports a capacity development process, also for priority setting (see Table 13.2). The AIC in Benin forms the public–private platform in which all cotton stakeholders participate, including the farmer organization Fédération des Unions de Producteurs du Bénin (Federation of Producer Unions in Benin; FUPRO). Performance-based annual management contracts are signed between the AIC and public extension. Performance assessment of extensionists by FOs is being developed.

A general institutional change is the establishment of multi-stakeholder platforms. Although various changes to stimulate stakeholder interaction in R4D resource allocation have been made, the voice of the FOs remains weak, while the private sector is largely absent other than in the chain-based (coffee and cotton) mechanisms. Although changing, researchers are still inclined to emphasize strategic and applied research, rather than adaptive research and pre-extension. Major attitudinal changes are therefore required to pursue the change process (see Table 13.2).

Efficiency and sustainability

The main factors that contributed to more efficient research programme development were the use of peer reviews of the submitted proposals and the emphasis on local priority setting, validated at higher levels. Financial transparency was improved through the application of direct costing and activity-based budgeting in the required formats, although this aspect needs further development in Benin. The reviewing and accounting mechanisms of the funds have improved transparency and confidence to the extent that earmarked funds of projects have started to follow similar procedures. At the district level, however, the transparency and downward accountability of the funds need further improvement. Disbursement in the Tanzanian system was considered more efficient compared to inflexible on-budget public procedures. The widespread communication on the programme has increased 'competition' for resources and, hence, efficiency in Benin. Downward accountability to members has improved; but not all coffee-producer organizations are members (see Table 13.3).

The sustainability of the funds has improved as a result of financial diversification, as well as through district-level contributions and fund-raising activities for zonal funds, triggered by matching funds. The financial contribution to zonal funds by districts has been considerable; but the flow of funds is stagnating. Funding for agricultural development mainly came from the agricultural sector programme and only limited funds were provided by the districts' core budgets. The sustainability of the funding mechanisms is at stake as some overhead costs (transaction costs were still over 10 per cent in Tanzania), such as multi-stakeholder

Table 13.2 *Institutional change and alternative funding mechanisms*

NARF *Tanzania*	ZARF *Tanzania*	District ARFs *Tanzania*	CFM *Benin*	TaCRI *Tanzania*	PPP Extension *Benin*
NARS actors are primarily involved; limited farmer organization (FO) representation; poor interaction between National Agricultural Research Institute (NARI) and universities	Funds more autonomous and development oriented; farmers' influence still limited	Village farmer groups and district farmer forums involved; integration of sectors and district budgets and procedures	Empowerment of stakeholder committee in resource allocation; rapid rotation in approval committee to maintain accountability	TaCRI privatized and owned by coffee stakeholders; emphasis on social capital development for farmer empowerment	Public–private platform with all cotton stakeholders; performance-based annual management contracts

meetings and workshops, are often financed separately by donor-funded projects. The issue is being addressed through cost-sharing and integration of overheads in research proposals. In Benin, the overhead costs and time allocated by partners could be reduced by signing two-year contracts. TaCRI is financed by both public (government and donor) funds and private-sector funds (coffee cess revenues) and through revenue-generating activities such as contract research and extension. The amount of collected coffee and cotton levies fluctuates with production and requires some stability. The conflicting interests between large coffee and cotton traders can threaten the partnership with the public sector. Contribution from the state budget will, for reasons given, still be required to balance the funds. The level of capacity in fund management and fundraising is a threat, particularly to the local funds. Nevertheless, all mechanisms have improved transparency, contributing to decreased overhead costs, greater efficiency and trust and, hence, sustainability, but need further improvement by strengthening PPPs (see Table 13.3).

Effectiveness and relevance

In general, research proposals have become more relevant because of competition, the screening process and the rejection of weak proposals, as illustrated by the percentage of rejected proposals (see Table 13.4). The project formulation leads to well-defined outputs, which can be monitored and evaluated; but effects and impacts are yet to be shown. Out of the 107 proposals taken up for review by NARF, 54 considered to be addressing zonal issues were sent back to the zones. Of the 53 reviewed proposals, 31 did not meet the set criteria. Meanwhile, because of funding problems and limited capacity, NARF could not address all of the submitted proposals. ZARFs funded 120 research projects over the referred period with grants of roughly US$7000 to $25,000 per project, which amounted to 6 to 15 per cent of total zonal research funding. The approval rate for submitted proposals varied between zones from 41 to 93 per cent. With the district funding, only about 40 per cent of the funds could be used because of lack of implementation capacity. Districts had an average of five proposals for each of two annual calls – hence, an approval rate of 20 per cent. Benin's CFM research proposals were better focused on needs and the problems of producers. Proposals were rejected mainly because of lack of strategic relevance. The relatively open character of the ZRDC has improved the flow of information and knowledge, while participatory research methods have been scaled out; but information on the rate of adoption and, hence, innovation system performance is still limited.

TaCRI has shifted its emphasis from a research focus to a coffee chain innovation focus and, as such, links up with not only international know-ledge providers, but also with producers through the farmer field school approach. Capacity development and information and knowledge management receive more emphasis. Although cotton yields in Benin have

Table 13.3 *Efficiency and sustainability of alternative funding mechanisms*

NARF *Tanzania*	ZARF *Tanzania*	District ARFs *Tanzania*	CFM *Benin*	TaCRI *Tanzania*	PPP Extension *Benin*
Peer review and the use of both national and zonal priority-setting mechanisms leads to efficiency	Enhanced transparency; activity-based budgeting; high transaction costs addressed by fund diversification; fundraising; matching funds	Transparency and downward accountability to be improved; merging sector and district budgeting; enhanced disbursement	Competition not yet leading to more R4D; improved transparency and trust; cost-sharing and activity-based budgeting for sustainability	Public and private funding not stable; downward accountability to FOs to be improved	Public and private funding not stable; two-year contracts and activity-based budgeting to reduce overhead costs; partnerships at risk

Table 13.4 *Effectiveness and relevance of alternative funding mechanisms*

NARF *Tanzania*	ZARF *Tanzania*	District ARFs *Tanzania*	CFM *Benin*	TaCRI *Tanzania*	PPP Extension *Benin*
20% of projects approved, 50% referred to ZARFs; proposal writing not funded; limited capacity for reviewing	41–93% approved; ZARF funding amounts to 6–15% of total research funding	20% approved; only 40% of funds used	More producer-relevant needs addressed; knowledge management improved	Pluralist knowledge sources; knowledge management capacity strengthened	Involvement of non-public service providers limited; impact studies required

improved, little is known about the effects and impact of the extension services. Involvement of other service providers could enhance competition and quality; but the number of qualified providers is limited. Competitive funding mechanisms lead to the selection of well-focused and relevant proposals by the multi-stakeholder committee and the rejection of others, and, simultaneously, to interactive learning to make such decisions. Knowledge management, in general, needs attention in all of the funding mechanisms because of both the competitive element as well as the larger number of stakeholders (see Table 13.4).

LESSONS LEARNED

The Tanzanian NARF brought about a clearer research focus on key priorities, but contributed little to closer collaboration between stakeholders within NARS, although this was one of its main objectives. Two key clusters of NARS actors, the ministerial research departments and the agricultural universities, need to greatly strengthen collaboration at research project level. This weakness was partly caused by inadequate monitoring and evaluation (M&E), a responsibility of the NARF management. Another problem was the erratic flow of funds, which needs to be stabilized by ensuring more dependable and time-bound contributions by financiers (donors as well as government) or possibly by establishing an 'endowment fund'.

Yet another major challenge is capacity development among all fund management actors. Stakeholder representatives should be drawn from established FOs and trained in their roles and responsibilities, which requires the allocation of adequate financial resources. The Tanzanian ZARF experience also demonstrated that strengthening capacity, particularly of FOs, is crucial for the identification and clear articulation of their demands, and is, in fact, a condition for a strong and inclusive demand-driven innovation system and the start of an interactive learning process. ZARF's multi-stakeholder management teams also require capacity development for financial resource allocation, budgeting and M&E, auditing, value-for-money assessments, communication with stakeholders and downward accountability through participatory M&E.

National policy-makers need to support local efforts to make ZARFs sustainable by helping them to establish procedures for working with low transaction costs, providing for specific district innovation development budget lines and establishing local financing mechanisms, such as district taxes. It is evident that institutionalization of the 'matching fund principle' (e.g. by donors) is often a strong incentive for local fundraising. An important positive outcome of the district-based agricultural innovation funds set up in Tanzania has been the participatory planning approach, including identification of selection criteria and the joint establishment of priorities by village and farmer groups, including village workshops to verify village-level information. Effective and efficient fund operation requires improving the capacity of district staff in planning and in financial and contract management (including development of terms of reference and processing and awarding of contracts). The poor response from researchers and extension staff to district calls for R4D proposals is partly due to the researchers' conventional inward-looking and supply-driven attitudes, inadequate socio-economic research capacity, and the inability of the extension services to facilitate farmers, farmer groups and FOs to articulate their priorities. Major logistical constraints include interpretation of procurement procedures and the time and costs involved in the participatory planning process.

In Benin, the competitive zonal funding mechanisms, which are linked to the national CGS, are part of the overall research planning and management cycle, including peer reviews, multi-stakeholder assessment of R4D proposals, monitoring of implementation, accounting for the funds received and evaluation of the results. The multi-stakeholder meetings have contributed to greater R4D relevance and transparency concerning costs and benefits, enhanced communication among stakeholders, and better understanding of priorities and resource allocation decisions by research management. Separate R4D workshops contributed to enhanced research quality and a stronger performance orientation; researchers also benefited through improved review skills, synergy and focus. These improvements stimulated other donor-funded R4D programmes to have their research proposals and results reviewed through the same multi-stakeholder mechanisms. However, agricultural extension remains the weakest link in the AIS, underlining the need for a more pluralistic and demand-driven agricultural advisory system with adequate resources. Training of FOs in priority setting and participatory planning and implementation of research, as well as client empowerment through cost sharing, are crucial. A comprehensive R4D funding system that provides a balance between strategic, applied and adaptive research, giving attention to priority research topics, is required. It should provide better donor coordination, with national ownership demonstrated through increased financial commitments.

The privatization of TaCRI in Tanzania has resulted in a clear shift towards stakeholder-driven adaptive coffee research and pre-extension services, based on participatory planning and budgeting. The resulting research programmes are more relevant and output oriented; they also achieve a better balance between the currently available research resources and the timing of anticipated practical results. The continuing need to produce public goods in the form of R4D products (particularly for smallholder coffee growers), the need to cope with emerging long-term sector-strategic issues such as food safety and quality (related to new requirements, particularly of the European Union), as well as concerns regarding environmental sustainability and socio-economic well-being of producers provide a strong justification for continued involvement (also financially) of the public sector in coffee R4D. Enhanced coffee production is expected to lead to increased cess levies for research support; but public intervention should continue to ensure special tax arrangements, substantial coffee-sector infrastructure investment and continued small-holder focus. TaCRI needs to further strengthen interaction with FOs through its representation in coffee research management and involve-ment of farmer groups in adaptive coffee research.

The AIC, which represents stakeholders in Benin's cotton sector, has developed a special partnership with public agricultural extension. The financial resources provided through cotton levies are used to recruit and employ extension agents (on a contractual basis), who provide services to cotton-producing farming communities and households. The involvement of village-level FOs has led to enhanced monitoring of extension agents' performance. The partnership has also contributed to a clear separation between the funding and implementation functions of the cotton R4D system. The contracting of service provision with the decentralized entities followed the 'subsidiarity principle': the specifics of extension services to be provided are agreed at village level, technical support is provided from the district level, and management and supervision are organized at provincial level. Commitment by both the government and the FOs is needed to ensure accessible, equitable services on a demand-driven and performance-related basis. Reinforcement of the M&E capacity of the FOs is crucial for the system to work.

CONCLUSIONS

Local R4D funding schemes have contributed significantly to the overall goal of financial diversification for stimulation of agricultural innovation, with a greater contribution by research clients and other stakeholders. However, real and substantial empowerment of farmers and their organizations in controlling the financial resources for adaptive research and pre-extension is still a long haul. This also applies to the private sector in general, although progress has been made, particularly with the commodity-based funds for innovation development. Downward accountability has improved; but real client control of funds has stagnated, partly because of the traditional 'top-down' attitudes of researchers. Farmer representation on the management teams of the CGSs remains weak. In addition, some stakeholders, particularly district governments, shy away from supporting local funds (where they lose direct control) in favour of independent 'contracts' for specific research and/or extension services. This is threatening broad local ownership of such competitive funds, although they still are vehicles for multi-stakeholder control of financial resources (provided mostly by the treasury and donors). Although the new mechanisms at local level work well, more effective mechanisms remain to be developed in order to ensure that stakeholders really own the local funds and that poor farmers, including women, have a real voice in resource allocation through their representatives.

Decentralized and de-concentrated local innovation development funds were found to be more successful in technology generation and also had advantages over other funding mechanisms as a result of the competitive element. This enhanced the quality of research, the sense of ownership by farmers and other stakeholders, and the control over resources by clients. However, some major concerns not yet satisfactorily addressed include the following:

- viable mechanisms for client representation;
- the priority focus and pro-poor status of available funds; and
- the level of cost-sharing and co-financing by truly local stakeholders, which is an indicator for ownership.

CGSs and commodity-based funds for innovation development are insufficiently integrated within an overall national system in which financing from different public and private sources is available for balanced funding of both strategic and adaptive research, as well as funding for pre-extension. The need to make funds available at the local level for enhanced stakeholder participation and R4D impact has trade-offs in terms of effectiveness and up-scaling options, and relatively high transaction costs, as well as limited competition because of insufficient numbers of qualified service providers, which entails a risk of competition between capacities to access funds rather than competition for quality services (to be) provided.

The main opportunities for strengthening local stakeholder-driven funding mechanisms for agricultural innovation can be found in the intensified involvement in fund management by FOs and private-sector actors. To achieve this, PPPs need to be developed that are successful in generating a climate of trust between public- and private-sector actors.

A comprehensive analysis of the roles of all stakeholders in the local agricultural innovation system often results in a clearer identification of the real and most urgent needs for technological, organizational and institutional change. One of the institutional innovations required is the establishment of more effective stakeholder-driven funding mechanisms for agricultural R4D through a participatory process. Capacity develop-ment of the key stakeholders, particularly the farmer organizations, in managing the funding mechanisms and in monitoring the effectiveness of the agricultural research and advisory services provided, is a factor that could contribute significantly to strengthening the entire AIS. Only then will funding make a real difference in local agricultural innovation systems that aim to contribute to pro-poor development.

ACKNOWLEDGEMENTS

A diverse group of stakeholders was involved in the funding mechanisms examined during the review upon which this chapter is based. Several FOs were involved in preparing case studies and in follow-up workshops in which the selected cases were analysed and discussed. The authors wish to specifically acknowledge the authors of the case studies: B. Kapange, S. Lugeye and V. Akulumuka from Tanzania and M. Adomou, C. A. Sogbohossou and R. Fassassi from Benin. However, the authors of the case studies could not have completed their work without the active collaboration of others. The study was financed by the Ministry of Foreign Affairs and Development Cooperation in The Netherlands, but was co-financed and cost-shared in Tanzania by the Department of Research and Training and some of their Zonal Agricultural Research and Development Institutes (ZARDIs) and TaCRI. In Benin, INRAB and the Ministry of Agriculture's Department for Agricultural Extension and Professional Training supported the study.

REFERENCES

Akulumuka, V. and Lugeye, S. (2006) 'District agricultural research funds in Tanzania', in W. Heemskerk and B. Wennink (eds), *Stakeholder-Driven Funding Mechanisms for Agricultural Innovation*, Royal Tropical Institute (KIT), Amsterdam, The Netherlands, pp66–71

Blackie, M., Chema, S., Kinyawa, P., Lyimo, S., Moshi, A. and Ngatunga, E. (2003) Review of the Client-Oriented Research Management Approach (CORMA) and the Zonal Agricultural Research Funds (ZARFs) in Tanzania, Report to the Ministry of Agriculture and Food Security, United Republic of Tanzania, by Smallholder Agricultural Development, Norwich, UK

Carney, D. (1998) *Changing Public and Private Roles in Agricultural Service Provision*, Overseas Development Institute, London

Chema, S., Gilbert, E. and Roseboom, J. (2003) A Review of Key Issues and Recent Experiences in Reforming Agricultural Research in Africa, ISNAR Research Report 24, International Service for National Agricultural Research, The Hague, The Netherlands

Echeverría, R. G. (1998) 'Will competitive funding improve the performance of agricultural research?', Discussion Paper 98–16, ISNAR, The Hague, The Netherlands

Engel, P. G. H. (1997) *The Social Organization of Innovation: A Focus on Stakeholder Interaction*, KIT, Amsterdam, The Netherlands

FARA (Forum for African Agricultural Research) (2006) *Framework for African Agricultural Productivity*, FARA, Accra, Ghana

Gill, G. J. and Carney, D. (1999) *Competitive Agricultural Technology Funds in Developing Countries*, Overseas Development Institute, London

Gotoechan-Hodounou, H., Adomou, M. and Wennink, B. (2006) 'Competitive funds for zonal research programmes in Benin', in W. Heemskerk and B. Wennink (eds) *Stakeholder-Driven Funding Mechanisms for Agricultural Innovation*, KIT, Amsterdam, The Netherlands, pp72–82

Hall, A. Bockett, G. Taylor, S., Sivamohan, M. V. K. and Clark, N. (2001) 'Why research partnerships really matter: innovation theory, institutional arrangements and implications for developing new technology for the poor', *World Development*, vol 29, no 5, pp783–797

Hartwich, F., Janssen W. and Tola, J. (2003) *Partnerships for Agricultural Innovation: Lessons from Partnership Building Between the Public and the Private Sector in Latin America*, ISNAR, The Hague, The Netherlands

Heemskerk, W. and Wennink, B. (2006) Stakeholder-Driven Funding Mechanisms for Agricultural Innovation: Case Studies from Sub-Saharan Africa, Bulletin 373, KIT, Amsterdam, The Netherlands

Kapange, B. and Lema, S. (2006) 'The privatized Tanzania Coffee Research Institute, TaCRI', in W. Heemskerk and B. Wennink (eds) *Stakeholder-Driven Funding Mechanisms for Agricultural Innovation*, KIT, Amsterdam, The Netherlands, pp83–88

Lema, N. and Kapange, B. (2006a) 'The National Agricultural Research Fund in Tanzania', in W. Heemskerk and B. Wennink (eds) *Stakeholder-Driven Funding Mechanisms for Agricultural Innovation*, KIT, Amsterdam, The Netherlands, pp53–58

Lema, N. and Kapange, B. (2006b) 'Zonal agricultural research funds in Tanzania', in W. Heemskerk and B. Wennink (eds) *Stakeholder-Driven Funding Mechanisms for Agricultural Innovation*, KIT, Amsterdam, The Netherlands, pp58–66

Lema, N., Schouten, C. and Schrader, T. (eds) (2003) Managing Research for Agricultural Development: Proceedings of the National Workshop on Client-Oriented Research, 27–28 May 2003, Moshi, Tanzania, Division of Research and Development, Ministry of Agriculture and Food Security, Dar es Salaam, Tanzania

Matthess, A. and Arodokoun, D. Y. (2005) 'Competitive funding as lever for development-oriented agricultural research in Benin, West Africa, Sector Project Knowledge Systems in Rural Areas, Cases: Funding of services', www.gtz.de/de/dokumente/en-funding-cases-2005.pdf

NEPAD (New Partnership for Africa's Development) (2002) *Comprehensive African Agriculture Development Programme*, FAO, Rome, Italy

Rivera, W., Alex, G., Hanson, J. and Birner, R. (2005) 'Enabling agriculture: The evolution and promise of agricultural knowledge framework', International Farming Systems Association Conference, 31 October–3 November, FAO, Rome, Italy

Sogbohossou C. A., Fassassi, R. and Wennink, B. (2006) 'Public and private funding of agricultural extension in Benin', in W. Heemskerk and B. Wennink (eds) *Stakeholder-Driven Funding Mechanisms for Agricultural Innovation*, KIT, Amsterdam, The Netherlands, pp89–98

Spielman, D. J. and von Grebmer, K. (2004) *Public–Private Partnerships in Agricultural Research: An Analysis of Challenges Facing Industry and the Consultative Group on International Agricultural Research*, IFPRI, Washington, DC

Tabor, S. R., Janssen, W. and Bruneau, H. (eds) (1998) *Financing Agricultural Research: A Source Book*, ISNAR, The Hague, The Netherlands

Wennink, B., Heemskerk, W., Vassall, A., Nicolai, M. and Schrader, T. (2003) 'Public private mix for service delivery: Identification of current developments and key discussion issues', Position Paper, KIT, Amsterdam, The Netherlands

Tracking Outcomes of Social and Institutional Innovations in Natural Resource Management[1]

Pascal C. Sanginga, Annet Abenakyo, Rick Kamugisha, Adrienne Martin and Robert Muzira

INTRODUCTION

Much of the literature on NRM refers to social institutions as mediating factors that govern the relationship between a community and the natural resources upon which it depends. The relationship between social capital and natural capital is emphasized (Ostrom, 2000b; Bowles and Gintis, 2002; Pretty, 2003). Social capital is usually defined as the features of social organization (social networks, social interactions, norms, trust, reciprocity and cooperation) that facilitate coordination and cooperation and that enable people to act collectively for mutual benefits (Woolcock and Narayan, 2000). It encompasses the nature and strength of existing relationships between members, the ability of members to organize themselves for mutually beneficial collective action, and the skills and abilities that community members can contribute to the development process (Uphoff and Wijayaratna, 2000).

Fukuyama (1995) and Bowles and Gintis (2002) regard social capital as an instantiated set of informal values or norms, obligations and expectations shared among members of a community that permits cooperation with one another. In this formulation, bylaws are an important dimension of social capital. Bylaws are negotiated rules, social norms, regulations, agreed behaviours and sanctions that exist within communities to prevent and manage conflicts in a way that places community interests above those of individuals (Coleman, 1988; Bowles and Gintis, 2002). In legal and policy terms, bylaws are a body of local laws and customs of a village, town, city or lower-level local government council and provide the local guidelines for implementing sectoral policies and rectifying their inefficiencies.

Recent research has shown the importance of social capital in creating a wide-ranging set of socio-economic outcomes, successful policy interventions and sustainable NRM (Uphoff and Wijayaratna, 2000; Woolcock and Narayan, 2000; Durlauf, 2002; Grootaert and Narayan, 2004). Important conclusions from this body of work are that social capital is often the missing link and that its reinforcement and continued deployment maintain both the existence of particular institutions and the process of institutional innovation (Bridger and Luloff, 2001). The emphasis on strengthening social capital is based on the evidence that communities are more efficient than state structures in managing natural resources (Grootaert and Narayan, 2004).

However, Ostrom (2000a) and Gillinson (2004) caution that some authors have exaggerated claims for the universal efficacy of social capital. The literature on social capital increasingly questions the general presumption that strong social capital has only positive effects (Durlauf, 2002). Cleaver (2005) questions whether social capital can be readily created and used and whether increased social capital necessarily benefits women and the poor. He argues that strengthening social capital may structurally reproduce the exclusion of the poor. Fine (2002), Mayoux (2001) and Coleman (1988) have cautioned that social capital can be used for harmful ends and may result in socially undesirable outcomes. Many scholars have called for a much better understanding of how social capital is constituted and transformed over time, and with what outcomes, so that it is not carried off as a fad (Ostrom, 2000a). A radical critique regards the term social capital as a catch-all phrase 'to mean more or less anything' and therefore not analytically useful (Fine, 2002). Durlauf (2002) questions the empirical evidence to support a significant explanatory role of social capital.

Here, we use empirical evidence from a tracking study to investigate the outcomes and potential impacts of a five-year participatory learning and action research (PLAR) project in Uganda that sought to strengthen the capacity of local communities to formulate and implement bylaws for sustainable NRM. This study, made one year after the end of the PLAR project, documented the specific outcomes, potential impacts and conditions for sustainability of strengthened social capital for improving policies and decision-making in NRM. The main hypothesis was that strengthening social capital will translate into improvements in some of the five capital assets (social, human, natural, financial and physical).

THE CONTEXT

With recent decentralization efforts and the mainstreaming of participatory approaches in policy and development in Uganda, considerable attention is given to devolving decision-making to lower levels. At the base of the local government structure is the local council (LC1) in a village of about 100 households, consisting of a nine-member elected council executive committee. Above the village are the parish (LC2), sub-county (LC3), county (LC4) and district (LC5) councils. The sub-county (LC3) is a critical level in the decentralization system as it has political and administrative powers to develop bylaws and implement development plans. The district (LC5), which is the highest level of local government and the most effective level to link with the central government, has political and administrative powers to enact bylaws, consolidate development plans and allocate budgets.

Although decentralization is regarded as a solution to problems of environmental governance and institutional effectiveness, none of the local councils had formulated bylaws by 2000 when the project began. The inadequacy of human and social capital at the different levels of local government is a key constraint to policy formulation and implementation (Ribot, 2002). Many rural communities are 'powerless spectators': they lack political assets, power, skills, resources, institutions and networks to participate effectively in policy processes (Fabricius et al, 2007). An important outcome of NRM research and development is therefore building the adaptive capacity of community-based organizations to develop institutions for effective participatory governance of natural resources.

Drawing from the body of work and experience that showed the importance of social capital, we facilitated a five-year (2000 to 2004) PLAR project aimed at strengthening social capital for improved policies and decision-making in NRM (see Sanginga et al, 2005). The project was based on the premise that social capital is an important asset drawn on by people in pursuit of their livelihood objectives, for managing their natural resources, and for participation in policy formulation and implementation. The project's strategy was to build on existing social capital and to strengthen it through facilitating participatory social-learning and policy-dialogue processes aimed at empowering rural communities to review and formulate their own bylaws and to develop local institutions for enforcing and monitoring bylaw implementation.

The project was implemented in Kabale District in the highlands of southwestern Uganda. The PLAR framework (see Figure 14.1) included four iterative and complementary processes:

1   community social-learning processes;
2   horizontal linkages between communities;
3   policy dialogues at different levels of decentralization; and
4   stakeholder feedback and learning events.

At the community level, the project facilitated the emergence and functioning of village policy task forces (PTFs) that were meant to create and facilitate platforms for social learning and policy dialogue within and between communities, local government and external organizations. The PTFs were also responsible for monitoring bylaw implementation. The formation of PTFs followed an inclusive and participatory process for selecting 8 to 12 farmers, with at least 40 per cent women. While the village was the level for formulating and implementing bylaws, we recognized that strengthening of community-level processes cannot stand on its own. A key aspect of PLAR was therefore to facilitate dialogue

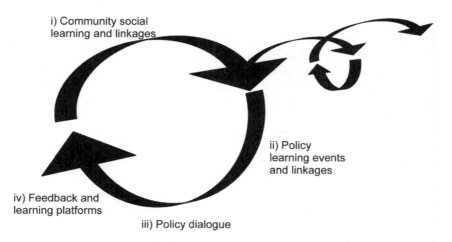

i) Community social
learning and linkages

ii) Policy
learning events
and linkages

iv) Feedback and
learning platforms

iii) Policy dialogue

Figure 14.1 *Operational framework for participatory learning and
action research*

Source: Sanginga (2005)

between these villages and higher structures of local government. We
facilitated formation of bylaw committees or sub-county PTFs made up of
village PTF representatives, sub-county councillors and local government
technical staff to ensure coherence, harmonization, stakeholder consensus,
formalization and legalization of the informal bylaws for general
application in all villages of the sub-county.

Through the PLAR process, the communities developed bylaws for
controlling soil erosion, tree planting, animal grazing, bushfire control
and wetland management. Each bylaw has specific regulations and
enforcement mechanisms (see Sanginga et al, in press). For example, the
tree-planting bylaw requires that anyone who fells a tree must plant two
and ensure their protection. It also requires that only agroforestry trees
can be planted on terrace boundaries and to demarcate plots. The bush-
burning bylaw stipulates that no one may set fire to a bush or part of
it without authorization. If a fire breaks out, all able-bodied community
members will help to extinguish it.

TRACKING OUTCOMES, NOT ASSESSING IMPACTS

In their reviews of participatory research, Okali et al (1994) and Oakley
and Clayton (2000) found that even those projects that claim to be
participatory still emphasize monitoring and evaluating technical outputs
and impacts, with relatively little attention to process and outcomes.
Impacts are long-term, lasting or significant changes in people's lives

brought about by a series of actions and are not the result of a single project. Outcomes are short- and medium-term end-of-project results that usually involve changes in the behaviours, relationships, activities and actions of stakeholders that can be logically linked to, but are not necessarily caused by, a project (Earl et al, 2001). Tracking outcomes is essentially process documentation that helps to assess how the final impacts are reached by looking at intermediate results or changes in how people or organizations behave. Our study focused on tracking outcomes and not assessing impacts because, as Ostrom (2000b) notes, social innovations are not as easy to find, see and measure as is physical and natural capital, which is usually tangible and obvious to external observers.

An important aspect of the tracking study was facilitating a participatory process to develop a set of indicators for documenting change. The community indicators focused on awareness and compliance with community bylaws; participation in mutually beneficial collective action; connectedness and networking; adoption of NRM technologies; conflict management; changes in gender dynamics; inclusion and equity; and performance and sustainability. We conducted eight focus group discussions (FGDs) with male and female farmers, as well as with members of PTFs at village and sub-county level. The FGDs were facilitated using After Action Review (AAR), a participatory technique that helps to structure collective reflection, analysis and learning by talking, thinking, sharing and capturing the lessons learned about a completed activity before they are forgotten (CIDA, 2003). AAR recognizes the explicit interests, different perspectives and judgements of different stakeholders, and provides opportunities for collective learning and reflexivity. Reflective learning practices (Cunliffe, 2004) draw significantly from participatory monitoring and evaluation systems (Estrella et al, 2000) and particularly from empowerment evaluation (Fetterman and Wandersman, 2007). AAR was facilitated using the following questions:

- What was supposed to happen? Why?
- What actually happened? Why?
- What went well? Why?
- What could have gone better? Why?
- What lessons can we learn?

To complement AAR and to obtain quantitative and individual insights, structured interviews were conducted with a stratified subsample of 46 households systematically drawn from a list of 145 households that took part in the baseline survey when the PLAR project started. We also conducted semi-structured interviews with 72 key informants including village leaders (21 per cent), clan elders (17 per cent), farmer-group leaders (27 per cent) and other opinion leaders (18 per cent). The tracking study took five months (May to October 2005) and started one year after completion of the PLAR project.

Human capital outcomes of bylaws

We assessed the extent to which farmers were aware of and complied with the community bylaws. We found that the majority of men and women in the four communities not only had more detailed knowledge of community bylaws and their specific regulations, but also their perceptions of effectiveness of the new bylaws had improved dramatically through the PLAR process. There was significant improvement in the extent of compliance with community bylaws over time (see Table 14.1). The PTF helped to facilitate the flow of information not only on bylaws but also on technologies and other NRM aspects. This role of the PTF as a knowledge-builder led to increased knowledge, skills and access to information and technologies for improving NRM. Several factors account for these improvements, including strong leadership of the village PTF in communities and groups, much sensitization regarding bylaws, regular monitoring and feedback, consistent support of bylaw implementation by non-governmental organizations (NGOs) and the sub-county, and high levels of social capital.

Table 14.1 *Assessment of changes in different dimensions of social capital (percentage of farmers)*

Dimensions of social capital	Improved significantly	Improved slightly	Unchanged	Deteriorated or never happens
Compliance with bylaws	44.8	41.4	3.4	10.3
Participation in community activities	17.2	75.9	6.9	–
Financial contribution	10.3	41.4	20.7	27.6
Cooperation among people	6.9	75.9	10.3	6.9
Altruism (spirit of helping others)	3.4	20.7	10.3	65.5

Social capital outcomes of bylaws

Baseline studies conducted at the start of the project showed that the four communities featured a high density of local organizations and diverse membership within them (Sanginga et al, 2005). However, the communities had weak bridging and linking social capital. 'Bridging social capital' refers to the structural relationships and networks that cross social groupings, involving coordination or collaboration and information-sharing with other groups within and across communities. Simply put, it is the network of horizontal linkages within and outside the communities. 'Linking social capital' refers to the ability of groups to engage with external agencies, either to draw on useful resources or to influence policies, thus linking poor people and those in positions of influence (Pretty, 2003).

The tracking study analysed the extent to which the PLAR process strengthened both bridging and linking social capital. We found that there had been considerable improvement in the horizontal linkages between the PTFs and farmer groups across the four pilot communities and other villages. There was increasing coordination or collaboration with these groups for sensitization, organizing collective action, organizing exchange visits between communities and groups, and, in some cases, mediating conflicts between groups. In three of the four communities, the PTF was embedded in decentralized local government structures at the village level, with most of its members doubling also as local councillors and members of the executive committees of other farmer groups. In these cases, the PTFs played complementary roles to local leadership and existing groups within the communities. In the fourth community, the respondents saw the PTF as a structure parallel to the local government and poorly integrated within existing farmer groups. This undermined the PTF's functioning and its ability to enforce bylaws and to arbitrate and mediate conflicts. It also affected participation in collective action and community meetings.

Gendered outcomes of bylaws

Uphoff and Wijayaratna (2000) stress that mutually beneficial collective action (MBCA) is the most specific outcome of social and institutional innovations. Feminist studies have pointed out the silence of participatory processes on gender and have criticized social capital studies for being gender blind (Mayoux, 2001; Molyneux, 2002). Cornwall (2003) observed that community-driven development, participatory planning and other fine-sounding initiatives that make claims of participation can turn out to be driven by particular gendered interests, leaving the least powerful without voice or much in the way of choice. Similarly, Akerkar (2001) concluded that many participatory projects lack awareness of gender differences: 'Gender was often hidden in participatory research in seemingly inclusive terms: the people, the community, the farmers.' Yet, in Africa, women are central to the forms of social capital that development organizations and governments are keen to mobilize in community development programmes.

One year after project completion, the four communities had organized up to 25 MBCA events related to implementing the bylaws (see Table 14.2). Each of the four PTFs organized an average of five meetings related to community bylaws, ranging from seven in Muguli B to three in Habugarama. The average number of participants varied from 33 to 41, reaching over 100 farmers (almost the entire village) for some events

Table 14.2 *Level of participation in mutually beneficial collective action one year after end of project (standard deviation in parentheses)*

Types of activities	Average number of events	Average number of participants	Average number of women	Maximum number of participants
Making trenches	4.7 (4.7)	25 (17)	11 (7)	100
Planting trees	2.6 (3.7)	20 (20)	10 (9)	70
Managing tree nurseries	4.7 (5.1)	32 (22)	17 (12)	70
Community meetings	5.2 (3.4)	53 (42)	48 (40)	150

organized by the PTF. The most common forms of collective action were making trenches to control soil erosion, planting trees and managing community agroforestry nurseries. The level of participation in MBCA events was consistently high and increased over time.

The linear trend analysis of women's participation (see Figure 14.2) showed an increase in the number of women ($R^2 = 0.83$) attending the different community meetings from below 20 to more than 60 women. The relatively high participation of women agrees with earlier analysis of the patterns and dynamics of participation in farmer organizations in Africa (Sanginga et al, 2006). However, it is interesting to note that, contrary to earlier findings of decreasing participation of men in group activities, this study showed that men's participation was sustained over time. Participation in community meetings on bylaw implementation was relatively regular, with an average of 53 men and 48 women, reaching a maximum of 150 farmers in some villages.

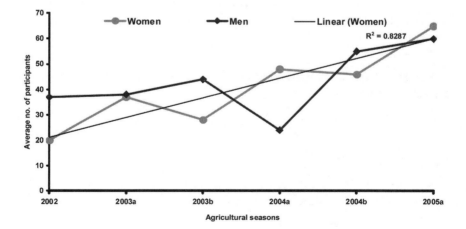

Figure 14.2 *Gender patterns of participation in community bylaw meetings in pilot communities over time*

PLAR had increased women's confidence and changed perceptions of their status within the communities. Most male and female farmers interviewed (96 per cent) indicated that women's participation in decision-making and community leadership positions had improved over the previous three years. On average, women represented 34 to 50 per cent of the membership in village bylaw committees and PTFs. Individual interviews and FGDs revealed that men's respect for and consideration of women had considerably improved (94 per cent and 86 per cent of men and women, respectively). Both men (86 per cent) and women (88 per cent) thought the project had significantly enhanced women's self-esteem and confidence.

Natural capital outcomes of bylaws

The PLAR project was based on the premise that strengthening community capacity to formulate and implement bylaws is an important precondition for adopting NRM innovations and resolving NRM conflicts. Study results showed that the number of NRM technologies practised by farmers and their willingness to purchase and plant more trees had increased significantly. We found that about 43 per cent of households had established new terraces in the recent past, 36 per cent had made further trenches and 28 per cent had used agroforestry technologies to stabilize these trenches (see Table 14.3). There was a clear willingness to use and purchase agroforestry technologies.

There were significant differences in adoption behaviour between communities, as well as significant gender differences within and among communities (see Figure 14.3). For example, Muguli and Karambo communities had the highest number of new trenches: 200 and 169, respectively. On average, male farmers in Muguli B established about 12 trenches each, compared to only three in the case of female farmers. The high involvement

Table 14.3 *New soil conservation measures established one year after end of project (percentage of farmers)*

Soil conservation measures	Female-headed households	Male-headed households	All households
Constructing new terraces	38.6	45.3	42.1
Digging trenches	32.9	38.7	35.9
Stabilizing with agroforestry technologies	25.7	30.7	28.3
Planting grass strips	8.6	9.3	9.0
Using trash lines	5.7	6.7	6.2

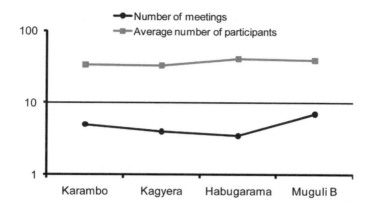

Figure 14.3 *Number of policy task force meetings and average number of participants in meetings*

of men in this village was because the PTF, embedded within the local village structure, was effective in mobilizing men for MBCA.

The probability of adopting agroforestry technologies and constructing new terraces to control soil erosion increased significantly for farmers and villages who complied with and implemented the bylaws on tree planting and on soil and water conservation (Sanginga et al, 2007a). This was consistent with findings that community bylaws played an important role in scaling up agroforestry technologies in eastern Zambia (Ajayi and Kwesiga, 2003). In addition, participatory bylaw formulation and implementation increased the ability of local communities to manage conflicts, minimize their destructive effects and transform conflict situations into opportunities for collaboration for MBCA (Sanginga et al, 2007b). Many cases of NRM conflict (animal grazing, terrace destruction, boundary disputes, tree felling) were resolved through arbitration and negotiation facilitated by PTF members.

PROBABILITY AND CONDITIONS FOR SUSTAINABILITY OF BYLAWS

As a key indicator of sustainability, we assessed the extent to which the PTFs continued to function one year after project completion. We found that:

- The PTFs continued to exist and function in all four communities one year after the end of the project.
- In three of the four communities, the PTFs had a strong and recognized leadership, embedded in other social structures and existing groups

within the communities. They were seen as complementary to the decentralized local government structures, rather than parallel. In contrast, the PTF in Habugarama village, which showed less potential for sustainability, was seen as parallel to the decentralized local government structures. This created conflicts and confusion, and marred the performance of the PTF.

- The four communities had developed their collective visions and community action plans for NRM. This increased their ability to visualize their future, to develop long-terms plans and to learn and reorganize to achieve their vision, and increased their social resilience and sustainability (Marshall and Marshall, 2007). Pretty (2003) and Rudd (2000) note that community visioning serves as a motivating factor that leads to concrete action and collective decision-making and increases the probability of sustainability.

- The PTFs proved to be critical in building support and mobilizing the political, social, human and technical resources needed to sustain participation of local communities in policy dialogue and action. They also supported MBCA and other important dimensions of social capital, such as exchange of information and knowledge, resource mobilization, cooperation and networking. They increasingly became vehicles through which farmers pursued wider concerns, initiated new activities and took the lead in catalysing development within their communities.

- Two of the four PTFs were successful in mobilizing financial resources and achieved recognition from government programmes to serve as resources for other communities on bylaw reform and other community-based NRM initiatives.

- The sub-county expressed genuine interest and willingness to upscale the process beyond the pilot communities to the whole sub-county, and many other villages expressed interest in forming village PTFs for bylaw formulation and implementation.

THE LIMITS OF COMMUNITY BYLAWS

Although the study showed that the outcomes of bylaws have been largely positive, it also revealed some important downsides. Some categories of farmers had difficulty in complying with some of the bylaws: older men and women, widows and orphans who had limited family labour or who lacked money to hire labour or buy the farm implements needed to make conservation structures. There were cases of increased conflict between livestock owners and cultivators, which sometimes led to divisions within communities, as illustrated in Box 14.1.

It was also found that reinforcement of the bylaws did not always ensure fairness. Many of the MBCA events had high social costs for local

---

Box 14.1   Example of conflict resulting from community bylaws

There are two factions that have now emerged in this village as a result of the controlled-grazing bylaw. One group – *Nyang'obutungi* for the rich – dislikes the system of free grazing and does not allow other farmers to graze in their plots. These farmers have their own big farms in which they graze their animals. It is this group that is pushing for strict enforcement of the controlled-grazing bylaw because they have plenty of grazing land. The second faction – *Nkund'obutungi* – is for the poor, who have small and few plots and are forced to confine their animals or be exposed to the bylaw process. They don't have land or people to keep their animals. The *Nyang'obutungi* group passed a bylaw against grazing on their plots that affected the poor who belong to the *Nkund'obutungi* group. In turn, the *Nkund'obutungi* group organized themselves into a strong group for the poor who have limited land or no farms, but own livestock and agreed to always graze in each other's land. This conflict led to the failure of the controlled-grazing bylaw, and implementation was left to the rich, while the poor decided that the poor should continue to graze on poor people's land. We don't even have a mechanism for deciding on this as a community.

Source: focus group discussion in Habugarama village

---

communities, especially for women, poor and elderly farmers, and other vulnerable groups, who ended up taking the burden of paying fines and suffered other forms of social exclusion and coercion. Table 14.4 shows that enforcement of community bylaws seemed to occur at the expense of altruism. This decline reflects some of the negative aspects of participatory processes, which exclude some categories of people, particularly those holding less social and economic power. Some farmers were genuinely unable to participate because of their advanced age and ill health. These were elderly women and men who did not have the labour and other resources required to take part in meetings and collective action. It was also revealed that owners of smallstock, especially women with small farms, had problems with the controlled-grazing bylaw. Strict reinforcement of this bylaw forced the poor to sell their livestock, thereby perpetuating the poverty trap. Older people expressed some distrust of the youth, who began to dominate village committees and farmer organizations and had more links with external organizations.

CONCLUSIONS

This chapter sought to provide an evidence base for the outcomes, potential impacts, performance and sustainability of social and institutional innovations in NRM. A major finding was that the key outcome of

Table 14.4 *Percentage of farmers reporting negative effects of bylaw enforcement*

Negative changes reported	Male respondents	Female respondents	Total
Conflicts between graziers and cultivators	54.5	60.0	58.1
Conflicts with local leaders	18.2	5.0	9.7
Conflicts within homes	9.1	10.0	9.6
Committing old and weak to implement the bylaws	9.1	5.0	6.5
Reduced grazing land	–	10.0	6.5
Much time spent implementing bylaws	–	5.0	3.2
Trees attract grazing animals that destroy crops	9.1	–	3.2
Loss of implements	–	5.0	3.2
Total	100.0	100.0	100.0

strengthening social capital is the creation of more social capital. This is not tautological, considering the different dimensions, types and mechanisms for activating social capital. For example, strengthening bonding social capital (trust, solidarity and cohesion) alone may not result in wide-ranging collective action since such cooperation and reciprocity are confined to group members only. Bonding social capital is limited in impact since its strength is founded on exclusivity. Therefore, other dimensions of social capital need to be strengthened to produce collective norms and rules, and institutions or bylaws that facilitate cooperation beyond the small group. For collective action to take place, the village PTF plays a significant role in facilitating and monitoring the effective implementation of community bylaws.

Results of this study suggest that it is possible to strengthen the cognitive, structural, bridging and linking dimensions of social capital. These include increased awareness and knowledge of bylaws, changes in behaviours and attitudes, and compliance with collective norms that place community interests above those of individuals. Community bylaws gave individuals confidence to invest in collective action, knowing that others would also do so. This created some level of trust that lubricated cooperation and social obligation (Rudd, 2000; Pretty 2003). The PTFs increased the ability of farmer groups to engage in social learning and policy dialogue with external agencies, either to draw on useful resources or to influence policies. This suggests improvement in social capital that links poor people and those in positions of influence. These findings are in line with studies that provide considerable evidence on the effects of institutions in boosting levels of social capital (e.g. Bridger and Luloff, 2001). They also lend credence to studies that point to the role of diverse forms of social capital in enhancing human capital (Coleman, 1988; Uphoff and Wijayaratna, 2000).

In addition to gains in human and social capital, enforcement of bylaws was also an important driver of adoption of agroforestry technologies and a mechanism for dealing with conflict over the use and management of natural resources. The processes of bylaw formulation and implementation proved to be robust over time and led to increased confidence. They continued operating well after the PLAR project ended. However, the study also illustrates the difficulties of addressing the negative dimensions of social capital and lends credence to the emerging literature on the dark side and limits of social capital (Mayoux, 2001; Durlauf, 2002; Fine, 2002). Enforcement of bylaws did not always ensure fairness, especially for women and the elderly with less human, financial, social and political capital. Some authors (e.g. Cleaver, 2005) argue that participatory and community processes may reproduce the exclusion of the poor, who often engage in social and institutional life on adverse terms; they are less able to negotiate their rights and shape social relationships to their advantage. This, in turn, undermines the effectiveness of such innovations to achieve more equitable livelihood impacts and often results in policy resistance and defensive routines (Sterman, 2006) and the tendency for interventions to be defeated by the system's response to the intervention. Policy resistance breeds cynicism about people's ability to improve the world. Defensive routines can hinder learning because powerful stakeholders tend to suppress dissent and seal themselves off from those with different views or, possibly, disconfirming evidence.

This tracking study is an important step towards developing a more robust framework for monitoring and evaluating the tangible and non-tangible benefits of PLAR. Although it is still too early to draw conclusions, the results suggest that social capital can be not only productive, but also persistent. With appropriate catalysis, it can become an important factor of societal production that helps people to meet their livelihood needs better, with whatever other resources are available. The greater challenge is to strengthen social capital in a way that empowers, rather than reinforcing the belief that people are helpless victims of forces that they neither influence nor comprehend (Sterman, 2006).

NOTE

1   This chapter is a modified version of an article submitted to *Society and Natural Resources*. The research was conducted while the first author was with the International Centre for Tropical Agriculture (CIAT) in Uganda.

REFERENCES

Ajayi, O. C. and Kwesiga, F. (2003) 'Implications of local policies and institutions on the adoption of improved fallows in eastern Zambia', *Agroforestry Systems*, vol 59, pp327–336

Akerkar, S. (2001) *Gender and Participation: Overview Report*, Institute of Development Studies, Brighton, UK

Bowles, S. and Gintis, H. (2002) 'Social capital and community governance', *Economic Journal*, vol 112, pp412–426

Bridger, J. C. and Luloff, A. E. (2001) 'Building the sustainable community: Is social capital the answer?', *Sociological Inquiry*, vol 71, no 4, pp458–472

CIDA (Canadian International Development Agency) (2003) *Knowledge Sharing: Methods, Meetings and Tools*, CIDA, Ottawa, Canada

Cleaver, F. (2005) 'The inequality of social capital and the reproduction of chronic poverty', *World Development*, vol 33, no 66, pp893–906

Coleman, J. (1988) 'Social capital in the creation of human capital', *American Journal of Sociology*, vol 94, pp95–120

Cornwall, A. (2003) 'Whose voices? Whose choices? Reflections on gender and participatory development', *World Development*, vol 31, no 8, pp1325–1342

Cunliffe, A. (2004) 'On becoming a critically reflexive practitioner', *Journal of Management Education*, vol 28, no 4, pp407–426

Durlauf, S. N. (2002) 'On the empirics of social capital', *Economic Journal*, vol 112, pp459–479

Earl, S., Carden, F. and Smutylo, T. (2001) *Outcome Mapping: Building Learning and Reflection into Development Programmes*, International Development Research Centre, Ottawa, Canada

Estrella, M., Blauert, J., Campilan, D., Gaventa, J., Gonsalves, J., Guijt, I., Johnson, D. and Ricafort, R. (2000) *Learning from Change: Issues and Experiences in Participatory Monitoring and Evaluation*, Intermediate Technology Publications, London

Fabricius, C., Folke, C., Cundill, G. and Schultz, L. (2007) 'Powerless spectators, coping actors, and adaptive co-managers: a synthesis of the role of communities in ecosystem management', *Ecology and Society*, vol 12, no 1, p29, www.ecologyandsociety.org/vol12/iss1/art29/

Fetterman, D. and Wandersman, A. (2007) 'Empowerment evaluation: yesterday, today, and tomorrow', *American Journal of Evaluation*, vol 28, no 2, pp179–198

Fine, B. (2002) 'It ain't social, it ain't capital and it ain't Africa', *Studia Africana*, vol 13, pp18–33

Fukuyama, F. (1995) *Trust: The Social Virtues and the Creation of Prosperity*, Free Press, New York, NY

Gillinson, S. (2004) 'Why cooperate? A multidisciplinary study of collective action', Working Paper 34, Overseas Development Institute, London

Grootaert, C. and Narayan, D. (2004) 'Local institutions, poverty and household in Bolivia', *World Development*, vol 32, no 7, pp1179–1198

Marshall, N. A. and Marshall, P. A. (2007) 'Conceptualizing and operationalizing social resilience within commercial fisheries in northern Australia', *Ecology and Society*, vol 12, no 1, p1, www.ecologyandsociety.org/vol12/iss1/art1/

Mayoux, L. (2001) 'Tackling the down side: Social capital, women's empowerment and micro-finance in Cameroon', *Development and Change*, vol 32, no 2, pp435–464

Molyneux, M. (2002) 'Gender and the silences of social capital: Lessons from Latin America', *Development and Change*, vol 33, no 2, pp167–188

Oakley, P. and Clayton, A. (2000) 'The monitoring and evaluation of empowerment: A resource document', Occasional Papers Series No 26, International NGO Training and Research Centre (INTRAC), Oxford, www.intrac.org, accessed on 15 May 2006

Okali, C., Sumberg, J. and Farrington, J. (1994) *Farmer Participatory Research: Rhetoric and Reality*, Intermediate Technology Publications, London

Ostrom, E. (2000a) 'Social capital: a fad or a fundamental concept?' in P. Dasgupta and I. Serageldin (eds) *Social Capital: A Multifaceted Perspective*, World Bank, Washington, DC, pp172–214

Ostrom, E. (2000b) 'Collective action and the evolution of social norms', *The Journal of Economic Perspectives*, vol 14, no 3, pp137–158

Pretty, J. (2003) 'Social capital and the collective management of resources', *Science*, vol 32, pp1912–1914

Ribot, J. (2002) *Democratic Decentralisation of Natural Resources: Institutionalizing Popular Participation*, World Resources Institute, Washington, DC

Rudd, M. A. (2000) 'Live long and prosper: collective action, social capital and social vision', *Ecological Economics*, vol 34, no 234, pp131–144

Sanginga, P., Kakuru, A., Kamugisha, R., Place, F., Martin, A. and Stroud, A. (2005) 'Bridging research and policy for improving natural resource management: Lessons and challenges in the highlands of southwestern Uganda', in M. Stocking, H. Helleman and R. White (eds) *Renewable Natural Resource Management for Mountain Communities*, ICIMOD, Kathmandu, Nepal, pp247–266

Sanginga, P., Tumwine, J. and Lilja, N. (2006) 'Patterns of participation in farmers research groups', *Agriculture and Human Values*, vol 23, no 4, pp501–512

Sanginga, P., Kamugisha, R. and Martin, A. (2007a) 'Conflict management, social capital and adoption of agroforestry technologies: Empirical findings from the highlands of southwestern Uganda', *Agroforestry Systems*, vol 69, no 1, pp67–76

Sanginga, P., Kamugisha, R. and Martin, A. (2007b) 'The dynamics of social capital and conflict management in multiple resource regimes: a case of the south-western highlands of Uganda', *Ecology and Society*, vol 12, no 1, p6, www. ecologyandsociety.org/vol12/iss1/art6/

Sanginga, P., Abenakyo, A., Kamugisha, R., Martin, A. and Muzira, R. (in press) 'Strengthening social capital for fostering pro-poor governance of natural resources: participatory learning and action for bylaws reforms in Uganda', *Society and Natural Resources*

Sterman, J. D. (2006) 'Learning from evidence in a complex world', *American Journal of Public Health*, vol 96, pp505–514

Uphoff, N. and Wijayaratna, C. M. (2000) 'Demonstrated benefits of social capital: the productivity of farmers' organisations in Gal Oya, Sri Lanka', *World Development*, vol 28, no 11, pp1875–1840

Woolcock, M. and Narayan, D (2000) 'Social capital: implications for development theory, research and policy', *The World Bank Observer*, vol 15, no 2, pp225–249

IV

Local Innovation Processes

CHAPTER 15

Recognizing and Enhancing Processes of Local Innovation

Ann Waters-Bayer, Laurens van Veldhuizen,
Mariana Wongtschowski and Chesha Wettasinha

INTRODUCTION

Agricultural development is driven by innovation at all levels. At the farmers' level, the term 'innovation' is often used in literature and in practice to refer to farmers' adoption of introduced technologies, in line with Rogers's (1962) theory on diffusion of innovations. Until recently, little attention was given to the new technologies, management practices and institutions that local people have developed themselves – to 'local innovation'. This refers to the *dynamics* of indigenous knowledge (IK) – the knowledge that has developed over time within a group, incorporating both learning from the experience of earlier generations and other knowledge that has been gained from whatever source and fully internalized within local ways of thinking and doing. Local innovation in agriculture and natural resource management (NRM) is the process through which individuals or groups in a locality develop and apply new and better ways of managing the available resources – building on and expanding the boundaries of their IK. The process of local innovation leads to technical, socio-economic and institutional innovations (with an 's').

Farmers – a term used here also to denote other natural resource users, such as pastoralists, forest users and fisherfolk – have been doing most of the experimentation, innovation and adaptation in agriculture and NRM since time immemorial. Before formal research and extension services existed, farmers' own experimentation allowed them to adapt to new situations and, thus, to survive. Sometimes because of sheer necessity, sometimes out of curiosity, sometimes by accident or through serendipity, farmers have found their own ways of improving their farming (e.g. Johnson, 1972; Biggs, 1980; Richards, 1985; Chambers et al, 1989; Kotschi et al, 1989; Reijntjes et al, 1992). Although local innovation has always been happening, it has seldom been recognized even by people who specialize in documenting IK, many of whom regard IK as a treasure that must be documented for posterity – before it is lost – rather than seeing the dynamics in the knowledge of local people.

Already several decades ago, agricultural researchers did recognize that farmers' knowledge – particularly of local conditions – could be valuable for formal research. This realization led to various forms of farming systems research (FSR) or farmer participatory research (FPR), usually involving on-farm trials in which scientists asked farmers to test and possibly adapt the scientists' ideas. Successful technologies were then 'extended' to other farmers. The scientists who developed the technology packages seldom realized how farmers were experimenting informally with package components. For example, when extension promoted new cereal varieties in a package of seed, fertilizer and instructions, many smallholders planted local varieties using the fertilizer intended for the new seed, and some carried out small informal trials to explore (e.g., the best timing and amount of fertilizer to apply to the local varieties) (see, for example, Hansen, 1986). It can probably be said that even after the advent of formal research and extension, most of the original ideas and successful local adaptations of introduced ideas have been developed by farmers without direct support from research. Yet it is often the less creative 'model' farmers who merely demonstrate introduced technologies who are called the 'innovators'.

There is, however, a growing recognition that innovation is not a linear process from formal science through extension to farmer adopters, and that scientists are not the sole and are seldom the most important generators of knowledge (see, for example, Bebbington, 1989; Biggs, 1990; Schreiber, 2002). It is becoming more widely accepted that innovation is a social process involving a multitude of different actors, and that innovation processes can be enhanced by creating more possibilities for actors to interact (Röling, 1996; Engel, 1997; Douthwaite, 2002; World Bank, 2006). This involves many social and psychological processes and requires many personal and institutional changes.

Here we describe how actors in the Prolinnova (Promoting Local Innovation in Ecologically Oriented Agriculture and Natural Resource Management) programme[1] in several countries in Africa and Asia have found practical ways of enhancing the innovation systems in which they are involved. Through joint reflection and analysis of their experiences, facilitated by staff from local non-governmental organizations (NGOs), they are building their own capacities to engage more effectively in innovation processes. The partners in Prolinnova regard themselves as an international community of practice, learning and advocacy. They have formed nested multi-stakeholder platforms at sub-national, national and international level, involving a defined group of Prolinnova members engaged in electronic and face-to-face exchange, and a much larger electronic learning platform open to all interested individuals and institutions.

The initial premise – and the growing experience – in Prolinnova is that a very effective entry point into engaging in participatory research and development is to identify local innovation. Recognizing local creativity and initiative can lead to changes in behaviour and attitudes of all actors in the innovation system and can stimulate institutional change to enhance innovation processes. Here we explain why and how this approach is taken, the gradual changes observed as a result and the challenges faced.

WHY DO LOCAL INNOVATION PROCESSES NEED TO BE ENHANCED?

In order to improve the livelihoods of small-scale farmers in Africa, it is important to enhance local innovation processes for the following reasons:

- *Diversity requires site-specific practices.* Farmers in Africa live and work under a wide range of ecological, climatic, economic and socio-cultural conditions, and the range of farming systems is similarly diverse, not just across regions or countries, but also within districts and even localities. Each farming system has its own dynamics, strengths, challenges and opportunities. It is not possible for scientists to generate the infinite variety of innovations and adaptations required. In the face of this farming diversity, it is wasted effort for them to develop 'perfected' technology to be applied in a blanket-like manner. Local adaptation and locally specific development of options must be key elements in any agricultural research and development (ARD) strategy to alleviate poverty in Africa (IAC, 2004). If scientists accept this, they need not expend so much effort on perfecting technologies and can give more attention to enhancing farmers' efforts to experiment and adapt technologies to fit local realities.
- *Rapidly changing conditions require local capacities to adapt quickly.* No innovation is permanent. A solution to any one problem does not remain valid from now until eternity. Conditions for farmers – including smallholders in resource-poor areas of Africa – are constantly changing. This is especially so for those who try to link with markets, but also for everyone affected by the emergence of new pests and diseases (not only in plants and animals, but also in humans, such as HIV/AIDS), changes in laws and regulations such as in land administration, and climate change. The key to sustainability in farming lies in farmers' capacities to adapt. Recognizing local innovation and then linking local innovators with other actors in the wider innovation system is a way of strengthening farmers' capacities to adapt more quickly to changing conditions.

If many different actors have the opportunity to bring in their different ideas and skills, innovation processes can be accelerated (see, for example, Douthwaite, 2002). If this type of interaction is happening in many different places at the same time, local innovation processes will be widespread. It is to this that Prolinnova aspires. However, good collaboration will develop only if all actors feel that their capacities and potential contributions are valued by the others. The other actors' recognition of farmers' innovativeness stimulates farmers' interest in collaborating in innovation systems.

APPLYING THE THEORY IN PRACTICE

How can diverse actors at local level enter into equitable and effective partnerships for innovation? In many parts of the world, efforts are under way to build multi-stakeholder partnerships in ARD by taking the entry point of recognizing local innovation. Here we describe mainly the experience of the Prolinnova programme, which builds on the experience of the earlier projects Promoting Farmer Innovation (Critchley et al, 1999) and Indigenous Soil and Water Conservation in Africa (Reij and Waters-Bayer, 2001). In Africa, there are established Prolinnova programmes in Ethiopia, Ghana, Niger, South Africa, Sudan, Tanzania and Uganda, and new programmes are being initiated in Burkina Faso, Kenya, Mali, Mozambique and Senegal. Similar work is under way in several countries of Asia, Latin America and the Pacific.

In each country, a national NGO brings together different groups of stakeholders wanting to promote participatory ARD, taking local innovations as starting points. The country programmes share common values and concepts, but are autonomous. Each designs its own plan of action. The essence of this work consists of:

- identifying and giving recognition to innovations developed by local people;
- participatory innovation development (PID): entering into partnerships at field level that combine different types of knowledge, ideas and skills, focused on joint exploration or experimentation that is farmer led and starts from the local innovations identified; and
- joining forces of the different stakeholders involved to bring about policy and institutional change in order to open up more space for PID processes.

Capacity-building activities accompany and strengthen all of these, and mainly take the form of learning through action and reflection. The learning takes place within each country programme, facilitated by the national NGO, and between country programmes at annual face-to-face meetings, facilitated by an international support team[2] and with a strong 'open-space' character. This all forms part of the participatory monitoring and evaluation system within the international programme.

WHY START WITH IDENTIFYING LOCAL INNOVATIONS?

There are four main reasons for making the identification of local innovations the first step in enhancing agricultural innovation systems:

1   First and foremost, it changes the way in which potential partners in an agricultural innovation system regard each other, serves as a tool for learning to understand what farmers are already trying to do, builds mutual respect, and, thus, lays a basis for partnership on a more equal footing.
2   It provides a point of departure for joint exploration and learning (i.e. PID) firmly embedded in local realities.
3   It provides concrete examples for raising wider awareness within formal ARD institutions and for stimulating institutional and policy change.
4   This activity can be fairly quickly and simply introduced into the ongoing work of people involved in ARD. No earthshaking paradigm shift is needed to start this – but it can lead to big changes.

Changing images of others and self

The main reason to start by identifying local innovativeness is a psychological one. In many cases, IK and local innovation are not valued by scientists, and sometimes not even by the farmers themselves. Despite the intellectual discussions about innovation systems, the practice in most African countries still follows the linear model of technology transfer. Researchers, extensionists and farmers see 'innovations' as things coming from outside ('modern' farming) and see farmers as mere receivers of the new technologies and accompanying instructions.

When formally educated agricultural professionals discover farmers' own innovations and informal experiments, they are confronted with the creativity of so-called resource-poor farmers. They begin to see farmers in a different light – as people with something valuable to offer – and see IK and local innovation as being complementary to their own knowledge and skills. Encouraging these professionals to recognize and reflect on farmers' creativity leads them to re-examine their own identity and roles, and changes the way in which they behave towards farmers (De Leener, 2001a, b). Scientists' realization that formal research is not the only source of knowledge and innovation need not demoralize them: on the contrary, it can generate their excitement at the unexpected ideas and energies of the farmers (Kibwana et al, 2001; Tchawa, 2001). Thus, identifying local innovations is a means of changing the attitudes of extensionists and scientists, and of helping them to recognize how they can complement and strengthen the creativity of farmers.

At the same time, the farmers gain in self-esteem. They begin to see themselves not as poor people who need help to solve their problems, but rather as people rich in knowledge, ideas and ingenuity in surviving under difficult conditions – as people to be admired. The recognition that formally educated agricultural professionals give to local innovation generates pride in local knowledge and creativity. Buoyed up with the self-confidence that outside professionals recognize them as researchers in their own right, the farmers are more likely to regard their admirers as potential partners in development. For example, as Kibwana (2001) noted in Tanzania, for farmer innovators and experimenters, 'the most gratifying part of the experience was that they had been treated, at long last, as partners and as equal to the "educated elite".'

Thus, for all actors, identifying local innovativeness changes their images of others and of themselves. It sets the stage and creates enthusiasm for generating new knowledge through equal partnership.

Entering participatory innovation development

The intention in Prolinnova is not to focus exclusively on farmer innovators as independent, isolated individuals, but rather to understand and enhance their links within an innovation system of diverse individuals (e.g. other farmers, traders, craftspeople), institutions and organizations both inside and outside the farming community. All of these actors can play different roles: each of them can be – at different times – a source of new ideas, a channel for communication, a partner in exploration or implementation, or a user of the outputs of an innovation process. By better understanding the complex innovation system in which they are involved, the actors can pinpoint linkages that need to be made or strengthened and information gaps that need to be filled.

Identifying local innovations can bring together holders of local and scientific knowledge in PID around a concrete activity already initiated by the local people. Here again, psychology plays a key role. PID does not start with analysing problems and dwelling on farmers' weaknesses and failures. Instead, it takes a positive approach that starts from local strengths and opportunities that local people can already see. Entering into joint research based on questions that farmers are seeking to answer builds up a spirit of collaboration and a readiness to explore, in addition, options for improvement based on ideas from outside.

PID aims primarily to strengthen the capacities of farmers, extensionists and scientists to collaborate in developing site-appropriate improvements. It may include research by individual farmers or groups of farmers supported by extensionists and/or scientists, as well as work by scientists on research stations or in laboratories to provide experimenting farmers with answers to the questions they raise (Hien and Ouedraogo, 2001; Tchawa, 2001). ARD thus becomes a 'social learning process' (Röling and Jiggins, 1998) in which farmers play the central role, while formally educated professionals strengthen the dynamics that are already under way.

The greatest enthusiasm for recognizing local innovation and venturing into PID with farmers has been observed among the field-based development workers – particularly the 'frontline' extension staff – who see this as a more satisfying approach than trying to convince farmers to accept locally untested technologies (Berhanu and Mitiku, 2001). Where their managers allow them to work in this way, extension workers can encourage farmers to try out and improve new ways of managing agricultural and natural resources (Hocdé and Chacón, 2000). Thus, PID becomes an approach to extension, often without direct involvement of research scientists (Veldhuizen et al, 2005).

Development workers can encourage farmer-led research and development in several ways by (Veldhuizen et al, 1997):

- creating opportunities for farmers to share their innovations as these provide ideas for other farmers to try out;
- offering alternatives to compare with current practices or local innovations;
- improving farmers' experimental design: stimulating farmers to examine how they do their informal experimentation and helping them to explore more systematic methods;
- filling local knowledge gaps: increasing farmers' awareness of resource management principles and providing information on phenomena that they cannot observe on their own in order to help them interpret the results of their experimentation; and
- facilitating mutual learning: creating opportunities for farmers to analyse local and external ideas for improving agriculture and NRM, and to assess the results of farmer-led PID (e.g. through farmer learning groups or exchange visits).

Raising awareness and stimulating institutional change

The personal change described above – 'to make the flip', as Chambers (1991) expressed it – is the first step towards institutional change (i.e. changes in the way that people in organizations think and behave and organize themselves for interaction with others). When scientists and extensionists and their managers examine how the structures and procedures in their institutions help or hinder efforts to support local innovation processes, they begin to see what needs to be changed. In the national multi-stakeholder Prolinnova platforms, people from government institutions and NGOs find space for learning together and for devising strategies for policy influence and institutional change.

A particular concern of Prolinnova partners is that this approach to promoting local innovation becomes integrated within institutions of higher learning so that the next generations of scientists, extensionists and educationists regard and use it as an accepted 'mainstream' approach.

Incorporating into ongoing activities of research and extension

Rather than operating as a separate 'project', each Prolinnova country programme is a multi-stakeholder initiative that seeks to incorporate a farmer-led participatory innovation approach into ongoing ARD work. In order to do so, they have undertaken the following activities.[3]

A core team of keen like-minded people from government organizations of research, extension and education and from local NGOs made an inventory of in-country experiences related to promoting local innovation and PID. In a national workshop, all major stakeholders jointly analysed these experiences and considered whether and how they wanted to collaborate in order to scale them up.

In different regions of the country, members of the core team arranged brief (one- or two-day) workshops involving extensionists, scientists and university staff to introduce the concepts of local innovation and PID in an innovation systems perspective. They drew out the participants' own experiences and observations about this and included local examples of farmer- or community-led innovation. The participants were then given follow-up assignments to identify and document local innovation, informal experimentation or participatory research processes in their working areas.

The participants completed these assignments during their regular work. The extensionists often documented cases that they had previously observed but had never mentioned because they were only supposed to extend technologies coming from research, not to inform researchers about technologies being developed by farmers.

In a follow-up workshop, the original participants brought farmer innovators to explain what they had developed or were trying out. All workshop participants reviewed the local innovations and selected those to be explored further in farmer-led joint research. This workshop was usually combined with further training in PID.

At different sites, small research groups composed of one or more local innovators and other nearby interested farmers, extensionists and – wherever possible – one or two scientists from a nearby research centre or university planned and implemented farmer-led joint research (see Box 15.1).

Such PID processes are under way in several countries. Partners in these processes are reflecting jointly on their experiences and identifying what factors help and hinder the experimentation and innovation processes and what can be done to improve them. In this way, the process of institutional change begins from below.

Box 15.1 *Participatory innovation development in beekeeping in Tigray, Ethiopia*

The Northern Typical Highlands team of Prolinnova-Ethiopia brought together farmer innovators at an Innovative Farmers Workshop held in Axum in central Tigray in April 2005. Here, the farmers explained their innovations to each other and to formal researchers and technical experts. The workshop participants selected beehive modification and queen-rearing innovations by a woman beekeeper, Gidey Aregay, and a male beekeeper, Gebrehiwot Mehari, to be explored further in joint research.

Each of these two innovators served as a nucleus in her/his village, working together with three to four local farmers with similar interests. They looked into:

- the optimal ratio of mud, dung and other materials for constructing beehives with a view to durability, regulation of temperature and insulation against noise;
- estimating colony size and assessing the quality and quantity of honey production; and
- understanding the seasonal aspects in the life cycle of the queen in order to improve the queen-rearing business.

Each group met every second weekend to assess what was happening in their experiments and to plan next steps. They met without facilitation by outsiders. Sometimes, other local farmers joined to observe and comment. Occasionally, the local development agents and district-level subject matter specialist joined the meetings and helped to document the farmer-led research.

Source: Hailu and Abera (2006)

At the same time, at national or regional (provincial) level, the multi-stakeholder learning platforms (members of which are stimulating and advising the above-mentioned local-level processes) try to bring about institutional change at higher levels so that PID processes can be accommodated – or even encouraged. These Prolinnova platforms raise awareness among research managers, development administrators and policy-makers. They facilitate exposure to, and discussion of, local innovation and PID. They organize events such as farmer innovation markets. They bring policy-makers to visit innovative farmers and bring innovative farmers to workshops, conferences and agricultural exhibitions where the farmers can show and explain what they are doing. They publicize the innovations and PID processes in catalogues, posters, photographs, video films, radio, etc., and, in some cases, help farmers document their own innovations (see Wettasinha et al, 2006).

Five country programmes (Cambodia, Ethiopia, Nepal, South Africa and Uganda) are piloting alternative funding mechanisms to promote local innovation. The most powerful way for farmers to exert influence on ARD is through controlling funds. Prolinnova is therefore exploring ways of giving local people access to and control over resources for experimentation and innovation in the Farmer Access to Innovation Resources (FAIR) action research project funded by the French government. Using local innovation support funds, smallholder farmers and community-based organizations can 'hire in' research support to fit local agendas and needs (Waters-Bayer et al, 2005; Krone et al, 2006). This piloting includes exploring ways of institutionalizing such funding mechanisms without external support.

Thus, the seemingly simple activity of identifying local innovations marks the start along what becomes a long and far-reaching path. It is an activity carried out within the existing ARD institutions, facilitated in such a way that it leads to a complex process of reflection and change.

When the researchers and development agents start to bring examples of what they think are local innovations, and when farmers start showing what they regard as innovations – then everyone becomes involved in discussions about what is traditional and what is innovative; what is an invention and what is an innovation; is it something that is new here or new everywhere in the world; can an innovation here be a tradition there; where do the ideas for local innovation come from; what is indigenous and what is exogenous; does it make a difference in the end where the idea comes from if local people can make something useful out of it? This discussion is necessary to help the actors see each other's perspective and approach a common understanding of innovation systems and their potentials. Struggling to define 'local innovation' is part of the process of becoming more deeply aware of it. Each country programme within Prolinnova has come up with a somewhat different definition of local innovation – and that definition changes as the discussion and learning continue (Wettasinha et al, 2006).

WHAT IS BEING SHARED AND SCALED UP?

A question that many people pose about promoting local innovation is: to what extent can the local innovations be scaled up? The NGOs that conceived Prolinnova in a workshop in Rambouillet, France, in 1999 (Rambouillet Group, 2000) were originally thinking along these lines, and much of the discussion was about using a database and various media to store and disseminate locally developed technologies. But then we realized that this puts too much emphasis on the innovations, rather than on the process of social interaction to enhance innovation.

The aim in identifying local innovations and further developing them in PID is *not* primarily to disseminate them in a transfer of technology extension mode. Local innovations are site specific. Results from farmer-led research and innovation in one locality can seldom be copied exactly ('adopted') somewhere else. In the diverse conditions of smallholder farming in Africa, the spread of a local innovation beyond the locality would not be a good indicator of success. However, sharing new ideas that have been discovered and developed in the course of PID can stimulate farmers' experimentation and innovation elsewhere. It can provide other farmers with options that they could try out and adapt for their own circumstances.

In the Prolinnova programme, identification of local innovations is meant to provide entry points for engaging in farmer-led participatory research as a *learning ground* for changes in stakeholders' attitudes and behaviour, in institutions and in policies – and, above all, in order to empower farmers in decision-making about ARD.

Box 15.2 *Vision, mission and goal formulated by Prolinnova partners*

Vision: a world in which farmers play decisive roles in agricultural research and development for sustainable livelihoods.

Mission: to foster a culture of mutual learning and synergy in local innovation processes in agriculture and NRM.

Goal: to develop and institutionalize partnerships and methodologies that promote processes of local innovation for the environmentally sound use of natural resources.

Source: Prolinnova (2005)

From the process of promoting local innovation, the major outcomes that are suitable for wider dissemination are therefore not the specific innovations, but rather:

- field-tested methods of discovering and stimulating local innovation processes (e.g. Wettasinha et al, 2006);
- lessons from experience in supporting personal and institutional change so that the formal ARD sector can support local innovation (e.g. Wettasinha et al, 2003); and
- lessons about building partnerships at local level and higher institutional levels, forging alliances and engaging in policy dialogue to create enabling conditions for enhanced processes of local innovation (e.g. Critchley et al, 2006).

Analyses of and information about these processes allow others to find out what has been applied in real-life situations and to adapt the methods and tools for application in their own settings. The specific local innovations are often of only local relevance, whereas the principles and processes of building partnerships and learning to support farmer-led ARD are of global relevance.

As an international community of practice, Prolinnova provides a platform where these experiences can be shared. The partners in the different countries describe, analyse and exchange views on how they are giving recognition to local innovation, engaging different stakeholders in PID and stimulating institutional change, including the development of educational and training curricula and modules. Opportunities for mutual learning are created (e.g. through electronic discussion groups, international workshops, joint publications, and supporting South–South mentoring – particularly between existing and emerging country programmes).

CHALLENGES AND CONCLUSIONS

The challenges are many in trying to stimulate actors in formal ARD to recognize and enhance local innovation processes. Bringing about change in attitudes and behaviour is a long and slow process, particularly in research organizations. It is difficult to break habits: even scientists who recognize local innovation tend to dominate as soon as they enter into on-farm research. A great deal of reflection and self-critique are still needed before participatory research can become truly farmer led.

Some development agents lack confidence to embark on a PID approach because they fear sanctions for not meeting their superiors' expectations in transferring predetermined technologies. The middle level of extension management, in particular, finds it easier to monitor field staff according to the number of farmers whom they convince to adopt an introduced technology, rather than the degree to which they have strengthened farmers' capacities to experiment and innovate. There is still a great need for the managers to rethink how extension is done and how development agents are rewarded for their work.

Because scientists are normally assessed according to other criteria than helping farmers develop what works on the ground, not many of them have been eager to engage in PID (see, for example, Ejigu and Waters-Bayer, 2005). In many countries, public funds for research and extension services are decreasing as privatization expands. There are fewer researchers and development agents available to engage in PID with smallholders. On the other hand, many research institutions are now under greater pressure to do work relevant for smallholders in order to meet the Millennium Development Goal of reducing poverty and hunger. This could be an opportunity as scientists may now be more willing to link up with farmers and other local actors engaged in PID.

We have described here how Prolinnova is trying to transform the theories of agricultural innovation systems into practical action at a local level in a way that leads to institutional innovation, above all to a change in culture, procedures and policies in formal ARD. We see promoting local innovation not primarily as an approach to research, but rather as an approach to development – not only of technologies and rural communities, but also of organizations. Recognizing local creativity serves as a point of entry into building partnerships for farmer-led joint research, which, in turn, triggers internal reflection and institutional change at higher levels. In this way, some space – however small – can be created to allow multi-stakeholder learning processes and, thus, innovation to happen from the grassroots upwards.

NOTES

1 Prolinnova is a Global Partnership Programme under the umbrella of the Global Forum on Agriculture Research (GFAR).
2 The International Support Team currently (2008) comprises advisers and trainers from ETC EcoCulture and the Centre for International Cooperation of the Vrije Universiteit Amsterdam, The Netherlands; the International Institute of Rural Reconstruction (IIRR), The Philippines; Innovations, Environnement Développement Afrique (IED Afrique), Senegal; and the FAIR sub-project coordinator, South Africa.
3 Reports on these activities – particularly on the national and provincial PID workshops – can be found on the Prolinnova website (www.prolinnova.net) on the country programme webpages. They are also documented in two recent booklets (Critchley et al, 2006; Wettasinha et al, 2006).

REFERENCES

Bebbington, A. (1989) 'Institutional options and multiple sources of agricultural innovation: Evidence from an Ecuadorian case study', Agricultural Research and Development Network Paper 11, Overseas Development Institute, London
Berhanu, H. and Mitiku H. (2001) 'Liberating local creativity: Building on the "best farming practices" extension approach from Tigray's struggle for liberation', in C. Reij and A. Waters-Bayer (eds), Farmer Innovation in Africa, Earthscan, London, pp310–324
Biggs, S. D. (1980) 'Informal R&D', Ceres, vol 76, pp23–26
Biggs, S. D. (1990) 'A multiple source of innovation model of agricultural research and technology promotion', World Development, vol 18, no 11, pp1481–1499
Chambers, R. (1991) 'To make the flip: Strategies for working with undervalued-resource agriculture', in ILEIA (ed) Participatory Technology Development in

Sustainable Agriculture: An Introduction, Information Centre for Low-External-Input and Sustainable Agriculture, Leusden, The Netherlands, pp5–9

Chambers, R., Pacey, A. and Thrupp, L. A. (eds) (1989) *Farmer First: Farmer Innovation and Agricultural Research*, Intermediate Technology Publications, London

Critchley, W., Cooke, R., Jallow, T., Lafleur, S., Laman, M., Njoroge, J., Nyagah, V. and Saint-Firmin, E. (eds) (1999) *Promoting Farmer Innovation: Harnessing Local Environmental Knowledge in East Africa*, RELMA/UNDP, Nairobi, Kenya

Critchley, W., Verburg, M. and van Veldhuizen, L. (2006) *Facilitating Multistakeholder Partnerships*, IIRR, Silang/Prolinnova International Secretariat, Leusden, The Netherlands

De Leener, P. (2001a) 'From technology-based to people-oriented synergies', IFAD Technical Workshop on Methodologies, Organization and Management of Global Partnership Programmes, 9–10 October, Rome, Italy

De Leener, P. (2001b) 'Towards an evolving conceptual framework to effectively capture the complexity of collaborative research', IFAD Technical Workshop on Methodologies, Organization and Management of Global Partnership Programmes, 9–10 October, Rome, Italy

Douthwaite, B. (2002) *Enabling Innovation: A Practical Guide to Understanding and Fostering Technological Change*, Zed Books, London

Ejigu J. and Waters-Bayer, A. (2005) *Unlocking Farmers' Potential: Institutionalising Farmer Participatory Research and Extension in Southern Ethiopia*, FARM-Africa, London

Engel, P. (1997) *The Social Organization of Innovation*, Royal Tropical Institute, Amsterdam, The Netherlands

Hailu, A. and Abera, G. (2006) 'PID in beekeeping in Tigray, Ethiopia', in C. Wettasinha, M. Wongtschowski and A. Waters-Bayer (eds) *Recognising Local Innovation*, IIRR, Silang/Prolinnova International Secretariat, Leusden, The Netherlands, p45

Hansen, A. (1986) 'Farming systems research in Phalombe, Malawi: The limited utility of high-yielding varieties', in J. R. Jones and B. J. Wallace (eds) *Social Sciences and Farming Systems Research*, Westview, Boulder, CO, pp145–169

Hien, F. and Ouedraogo, A. (2001) 'Joint analysis of the sustainability of a local SWC technique in Burkina Faso', in C. Reij and A. Waters-Bayer (eds) *Farmer Innovation in Africa*, Earthscan, London, pp257–266

Hocdé, H. and Chacón, M. (2000) '"This is my own innovation": The history of Limpo grass', *ILEIA Newsletter*, vol 16, no 2, pp31–32

IAC (InterAcademy Council) (2004) *Realizing the Promise and Potential of African Agriculture: Science and Technology Strategies for Improving Agricultural Productivity and Food Security in Africa*, InterAcademy Council, Amsterdam, The Netherlands

Johnson, A. W. (1972) 'Individuality and experimentation in traditional agriculture', *Human Ecology*, vol 1, no 2, pp149–159

Kibwana, O. T. (2001) 'Forging partnership between farmers, extension and research in Tanzania', in C. Reij and A. Waters-Bayer (eds) *Farmer Innovation in Africa*, Earthscan, London, pp49–57

Kibwana, O. T., Mitiku H., van Veldhuizen, L. and Waters-Bayer, A. (2001) 'Clapping with two hands: bringing together local and outside knowledge for innovation in land husbandry in Tanzania and Ethiopia', *Journal of Agricultural Education and Extension*, vol 7, no 3, pp133–142

Kotschi, J., Waters-Bayer, A., Adelhelm, R. and Hoesle, U. (1989) *Ecofarming in Agricultural Development*, Margraf, Weikersheim, Germany

Krone, A., Amanuel, A., van Veldhuizen, L., Waters-Bayer, A. and Wongtschowski, M. (2006) 'Reflections on Prolinnova's FAIR project: Local Innovation Support Fund prospects and challenges in promoting innovation', Innovation Africa Symposium, 20–23 November, Kampala, Uganda

Prolinnova (2005) Report on Prolinnova Country Programme Coordinators Meeting, 5–7 June 2005, Entebbe, Uganda, Prolinnova International Secretariat, Leusden, The Netherlands

Rambouillet Group (2000) 'New mechanisms for reinforcing partnerships in agro-ecology/natural resource management research and development', Global Forum on Agricultural Research, 21–23 May, Dresden, Germany, Document GFAR/00/18–04

Reij, C. and Waters-Bayer, A. (eds) (2001) *Farmer Innovation in Africa: A Source of Inspiration for Agricultural Development*, Earthscan, London

Reijntjes, C., Haverkort, B. and Waters-Bayer, A. (1992) *Farming for the Future: An Introduction to Low-External-Input and Sustainable Agriculture*, Macmillan, London

Richards, P. (1985) *Indigenous Agricultural Revolution: Ecology and Food Production in West Africa*, Hutchinson, London

Rogers, E. M. (1962) *Diffusion of Innovations*, Free Press, New York, NY

Röling, N. (1996) 'Towards an interactive agricultural science', *European Journal of Agricultural Education and Extension*, vol 2, no 4, pp35–48

Röling, N. and Jiggins, J. (1998) 'The ecological knowledge system', in N. Röling and M. A. E. Wagemakers (eds) *Facilitating Sustainable Agriculture: Participatory Learning and Adaptive Management in Times of Environmental Uncertainty*, Cambridge University Press, Cambridge, pp283–311

Schreiber, C. (2002) 'Sources of innovation in dairy production in Kenya', Briefing Paper 658, International Service for National Agricultural Research, The Hague, The Netherlands

Tchawa, P. (2001) 'Participatory technology development on soil fertility improvement in Cameroon', in C. Reij and A. Waters-Bayer (eds) *Farmer Innovation in Africa*, Earthscan, London, pp221–233

Veldhuizen, L. van, Waters-Bayer, A. and Zeeuw, H. (1997) *Developing Technology with Farmers: A Trainer's Guide for Participatory Learning*, Zed Books, London

Veldhuizen, L. van, Waters-Bayer, A. and Wettasinha, C. (2005) 'Participatory technology development where there is no researcher', in J. Gonsalves, T. Becker, A. Braun, D. Campilan, H. de Chavez, E. Fajber, M. Kapiriri, J. Rivaca-Caminade and R. Vernooy (eds) *Sourcebook on Participatory Research and Development for Sustainable Agriculture and Natural Resource Management*, vol 1, CIP-UPWARD, Los Baños, The Philippines, pp165–171

Waters-Bayer, A., van Veldhuizen, L., Wongtschowski, M. and Killough, S. (2005) 'Innovation support funds for farmer-led research and development', IK Notes 85, World Bank, Washington, DC

Wettasinha, C., van Veldhuizen, L. and Waters-Bayer, A. (eds) (2003) *Advancing PTD: Case Studies on Integration into Agricultural Research, Extension and Education*, IIRR, Silang, The Philippines/CTA, Wageningen, The Netherlands

Wettasinha, C., Wongtschowski, M. and Waters-Bayer, A. (2006) *Recognizing Local Innovation*, IIRR, Silang/Prolinnova International Secretariat, Leusden, The Netherlands

World Bank (2006) *Enhancing Agricultural Innovation: How to Go Beyond the Strengthening of Agricultural Research*, World Bank, Washington, DC

Building Institutions for Endogenous Development: Using Local Knowledge as a Bridge

Jeanne T. Gradé, John R. S. Tabuti and Patrick Van Damme

INTRODUCTION

Pastoral communities live in the midst of the natural resources upon which they depend directly for their livelihood, with a very narrow margin of survival. They use water and grass to feed livestock; trees for medicine, food, and firewood; and wildlife to supplement their diet. When pastoralists are nomadic, environmental stress (even if extreme) is generally short lived because people and their livestock move elsewhere, allowing resources to recover. The constant challenge of coping with nature creates a depth of alignment in a community's customs and habits of daily life.

This chapter addresses change agent efforts in the Karamoja semi-nomadic pastoral area in northern Uganda to relieve environmental stresses by creating infrastructure to revive indigenous approaches to natural resource management (NRM), with a focus on medicinal plants.

Karamoja's cluster extends into Sudan, Kenya and Ethiopia (see Figure 16.1). The semi-arid environment, remoteness from major urban centres, poor infrastructure and poor access to services have forced the Karamojong to rely heavily on the traditional elders' system and indigenous knowledge (IK) for survival livelihoods.

When assessing the current situation of natural resources in their area, elders cited several problems that were not present, or were less severe, when they were young. These include less rain, fewer water catchment ponds and more contamination of the limited available water; diminished plant life, including fewer savannah trees; dramatically fewer wild animals; erosion; and lower soil fertility. They felt that the traditional IK system and communal efforts to manage resources had weakened. For example, local leaders used to guide teams to dig ponds for catching rainwater and to protect ponds to keep animals out. The communities had been proud of their ponds. Later, when the United Nations World Food Programme gave money to dig or de-silt ponds, pastoralists perceived these ponds as being

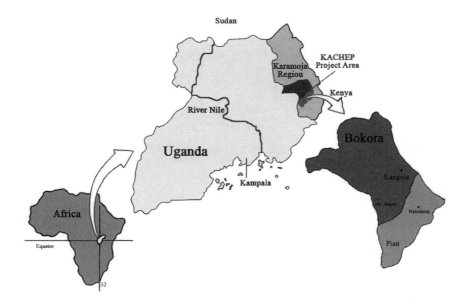

Figure 16.1 *Map of study area*

Source: Drawn by John O. Gradé

owned by outsiders. Local people sometimes even refused to clean their ponds unless paid. In an effort to help 'modernize' the culture, outsiders did not recognize traditional leadership structures and did not deal with Karamojong leaders directly. Communities were divided between looking to their traditional leaders or taking money from external players; and outside money trumped tradition.

Elderly pastoralists recalled a clan of women, the Ngiyepan, who protected trees in various ways, such as by enforcing conservation-related taboos through song, dance, drama and stories. As a group, they were persuasive verbally and sometimes even physically. It is unclear why they are no longer active, although people still remember some Ngiyepan songs.

During interviews, Karamojong were hard-pressed to recall any positive innovations that had come from outside during their lifetime or that of their parents. They could remember that there were more trees and more peace, that they did not have many cows but the food was enough, that colonists had built dams in the valleys but only one is still functional, that many boreholes were made but there was more water and food before that. They also recalled the strength of culture and respect they had for their parents and elders.

They perceived two more recent external innovations – automatic weapons and enforced schooling – as having caused much damage to their culture and, thus, to their way of managing natural resources. AK-47s (Kalashnikov rifles) turbo-charged an existing self-destructive element in the culture, while both 'innovations' encouraged pastoralists to become more sedentary. Fear and insecurity caused people to band together in larger groups where schools were commonly built. This increased pressure on already scarce resources in certain areas, while vegetation flourished in uninhabited areas within a day's journey from the settlements. The school system also led people to look outside their own 'culture box' for new answers to the same problems. Elders saw tradition and schooling as mutually exclusive options.

In endogenous development, people seek a balance between modern and traditional practices. It is a collaborative process for local institutional development that involves a careful blending of internal with some external processes (e.g. 'modern' schooling with the traditional system).

One potential bridge to this process is the institution of traditional healers. Karamojong pastoralists are proud of their knowledge of how to use local plants. One elder explained that, even though Karamoja does not have 'powerful hospitals, the limited access to "modern medicines" has allowed us to greatly utilize our ancestor's medicine'. Because the Karamojong depend upon cattle for both subsistence and cultural pride, one of the most important forms of IK is ethnoveterinary knowledge (EVK): the local knowledge, skills, practices and beliefs about the care of livestock (McCorkle, 1986). This includes various different treatments: store bought, homemade, prescribed and/or prepared by traditional healers, mainly from plant extracts.

This EVK needs to be conserved because it is threatened by the pull toward modernity; yet modern medicines are almost entirely out of the pastoralists' reach. Therefore, our goal was to help to conserve and revive this knowledge by helping the pastoralists set up their own local organizations – for example, non-governmental organizations (NGOs) and community-based organizations (CBOs) – and encourage documentation and practice of EVK, as well as NRM, through agroforestry and plant conservation.

FROM STUDYING ETHNOVETERINARY KNOWLEDGE TO FORMING AN NGO

In 1998, a project was launched by the Bokora Livestock Initiative (BoLI), a cooperation by three NGOs working with livestock keepers in Bokora County of Moroto District: the Lutheran World Federation, the Church of Uganda's Livestock Extension Programme and Christian International Peace Service. These organizations agreed to harmonize their veterinary services to fill gaps in Bokora's limited veterinary infrastructure. One joint BoLI activity was 'training of trainers' workshops for community animal health workers (CAHWs) facilitated by the international agency, Christian

Veterinary Mission (CVM). BoLI mandated a study of local EVK in order to integrate this with introduced veterinary practices within the CAHW training (Gradé and Shean, 1998).

To strengthen EVK infrastructure, a participatory action research (PAR) approach and an ethnographic framework were used. PAR is research that involves all relevant parties in actively examining together current action (which they experience as problematic) in order to change and improve it. PAR is not just research which is hoped will be followed by action. It is action which is researched, changed and re-researched, within the research process by participants (Wadsworth, 1998). Specific methods used included direct observation, semi-structured interviews, scoring and ranking, participatory field trials, exchange visits and 'free-listing' (Martin, 1996). The methods were continually readjusted in response to participant input in identifying problems and solutions in joint experimentation as all learned together how to strengthen EVK.

During the study of EVK, the first author began training community members in documenting IK using a modelling framework created by CVM (Shean, pers. comm., February 1998). Livestock diseases, their prevention and treatments were documented through group discussions involving community members and BoLI staff. Staff were trained in ethnoveterinary surveys covering formal and local names of diseases, species of animal treated, name of treatment, description of medication, method of treatment (preparation, administration and dosage), pharmacological rationale and efficacy. BoLI extension workers identified pastoral communities and traditional healers to be contacted.

From these group discussions, active community members with obvious knowledge and commitment (traditional healers) were selected by BoLI staff. The first author confirmed these selections in one-on-one interviews. These community members – both men and women – were then involved in focus group discussions, together with BoLI staff, to identify priority medicines to promote and to brainstorm on how to form a network of livestock healers. The initial plan was only to make an EVK database and to hand it over to BoLI for the CAHW training programme. However, during the PAR process, the healers initiated monthly meetings and recruited new members. The first author, impressed by the depth and breadth of the healers' knowledge, also became more emotionally involved with the pastoralists' concerns and eagerness to cooperate in seeking solutions.

After finalizing the BoLI document, two young Karamojong men, who had secondary-level education and were freshly trained in documenting EVK, joined the first author to go to the neighbouring county, Pian (see Figure 16.1) to compare their EVK with that of Bokora. The same PAR process was followed, bolstered by the lessons learned in Bokora. In addition to sharing and comparing EVK in Bokora and Pian, we invited livestock healers and NGOs to discuss ways of disseminating the most confidently used EVK within their communities.

With renewed Karamojong interest in EVK and its dissemination, the first author's work in Karamoja continued. What was to be a six-week programme continued without a distinct end, just with a desire to be part of the healers' PAR cycle. When the supervisory international NGO World Concern Africa opted to leave Uganda in 2002, the staff – which included pastoral community members – decided to form an indigenous NGO to continue EVK development.

FROM ETHNOVETERINARY KNOWLEDGE TO AGROFORESTRY

A natural extension of EVK in a damaged environment is medicinal plant agroforestry. The healers' associations selected particular plants for domestication and multiplication based on several factors, primarily their confidence in the plants' medicinal efficacy. Confidence levels were established through ranking, scoring and defining 'best bets' (Martin 1996). They also assessed whether the disease treated was common in Karamoja and whether, therefore, the plant would have potential economic benefit for disease control. They gave high priority to multi-purpose plants. For example, one plant selected provides medicine for three diseases (one as a best bet – that is, highest-ranking plant), fodder for livestock, food for people during hunger periods, and highly valued wood for construction and making charcoal. The species were then evaluated for their economic value for the local market. Low threat of bio-piracy was another factor used in selection (e.g. if synthetic medicines were available for the disease for which a plant is used, that plant was considered 'safer' or less likely to be exploited). All of the above factors were used to rank a long 'free list' of plants.

An additional key activity was to develop medical product micro-enterprises using local EVK. This involved multiplying the species in production orchards.

THREE KEY OUTPUTS OF THE PARTICIPATORY ACTION RESEARCH

Institutions built

Within the course of ethnographic action research, four organizations with a common mission to preserve, promote and protect EVK were formed and registered at national level: Bokora Traditional Livestock Healers Association (BTLHA), Pian Traditional Livestock Healers Association (PTLHA), Karamoja Ethnoveterinary Information Network (KEVIN) and Karamoja Christian Ethnoveterinary Programme (KACHEP). The first two are CBOs; the third is a consortium of government, NGOs and CBOs; and the fourth is an NGO.

A group of 12 male Bokora healers, who formed the BTLHA, first gathered in mid 1998 as a focus group when EVK in Bokora was being catalogued. The Pian group (PTHLA) first gathered in the kraals (mobile cattle camps) in February 2000, when Bokora and Pian EVK was being compared. Since 2001, both associations have been setting their own schedules for meetings, at least quarterly, and both are registered in Uganda. Both were created with the aim that Karamoja would utilize their natural resources and EVK for sustainable development and poverty reduction.

The BTLHA has grown from the original 12 men to 50 subscribed members. After Pian healers first met, the ten core healers continued to meet with EVK project staff and other KEVIN members at rotating locations, either near one of their *manyattas* or near the kraals, depending upon the season. Pian membership grew to 22, then dropped slightly, but then expanded to 44 over the last two years. Initially, members were only elderly men; but, as the association grew, younger men and women became interested and were invited to join. By 2007, over 92 association workshops had been held in Karamoja. Membership of BTLHA and PTLHA is open to livestock healers living in Bokora and Pian, respectively. Other individuals and organizations who share the associations' mission may also subscribe. The associations have elections for executive members, who, in turn, run their meetings, frequently inviting an external member to teach on a specific topic.

At association workshops, members from different communities take turns teaching and learning. They discuss cases they have treated (both failures and successes) in order to share new information and gain advice from other members. They are then able to pass on this information to their family and neighbours. The livestock healers promote the best practices with their direct contacts at household level and continue to experiment with EVK. Other activities of the associations include agroforestry and micro-enterprise, adding value through medicine extraction, packaging and distribution.

The network (KEVIN) originated from a three-day EVK-sharing workshop held in Pian in July 2000, which brought together regional stakeholders who shared case studies and best practices for livestock husbandry, disease prevention and disease cure. A unique feature of this workshop was that participants paid for it, making it 'locally owned'. At most gatherings of this type in Uganda, the organizer not only pays for transportation, lodging, food and training materials, but also provides *per diems* (also called allowances or 'motivation'). This practice was introduced during the colonial era to 'encourage' attendance and is reinforced by NGOs and government agencies to this day throughout Africa. KEVIN's formation created a forum for continuous sharing of ways to preserve, promote and protect EVK in Karamoja.

KEVIN members operate in four of the five Karamojong districts (a different group, the Labwor, who currently focus less on cattle, inhabit the fifth district). Membership is open to all government agencies and NGOs in Karamoja that are involved in any aspect of livestock management. Members include district veterinary officers and veterinary officers from Moroto and Nakapiripirit, as well as ten local NGOs and CBOs. Four member NGOs have agroforestry schemes, two are involved in EVK research and development, and four incorporate EVK in their CAHW training.

KEVIN is a conduit for disseminating EVK through extension workers who originate from, live in and work throughout Karamoja. The network empowers each stakeholder organization to use the collective information to integrate IK within agriculture and livestock training at community and household level and to encourage the adoption of a variety of innovations developed by the Karamoja healers.

KACHEP was registered in Uganda as a local NGO in June 2004, having grown out of the collaborative EVK project funded through CVM. The project was initially managed by the first author; but KACHEP is now run by a core staff of her former assistants, all of whom are local people. It seeks to preserve, promote and protect EVK in Karamoja through research and development, as well as building the capacity of the livestock healers' associations. It is the key liaison agency that identifies EVK users and innovators and links them with interested organizations.

Plants conserved

One component of EVK preservation is documentation and conservation of medicinal plants. As mentioned earlier, there used to be a clan of women, Ngiyepan, who protected trees. According to Nalem Rose, a Pian healer: 'When these women were active, we had plenty of rain and the tall [tree] shrines were well cared for'. Tree planting is not otherwise a part of Karamojong culture. The healers' associations, however, now promote agroforestry and protection of medicinal plant species. Their agroforestry scheme has focused on domestication of 32 tree species, 24 of which are indigenous (see Table 16.1). The species were selected by livestock healers and other community members based on their confidence that the plant treats endemic livestock diseases effectively and on the importance of these diseases to the local economy. They identified internal and external parasites as key problems. Purchased medicines are not regularly available in these remote and resource-poor areas; therefore, plants with pesticidal qualities were given high priority, alongside plants used for treating wounds, snake bites and retained placentas.

Exotic fruit trees – custard apple, guava, papaya and pomegranate – were chosen because they are drought resistant and have been cultivated in Karamoja for at least 50 years at Christian missions. In addition to its edible fruits, papaya is also used for medicinal purposes. Four of the medicinal plants domesticated by the Karamojong also provide valued edible fruits. Three medicinal trees – neem (from India), fish bean or *Tephrosia vogelii* (from Zambia) and *Moringa oleifera* (from Arabia and India) – are not indigenous, but have long been domesticated in Karamoja.

More than 70 healers' communities are involved in efforts to conserve medicinal trees through agroforestry schemes. Additionally, they teach family members and neighbours about conservation and sustainable harvesting techniques. Many make thick fences from thornbush branches to protect crops from wild animals and raiders. As a result of the efforts of the healers' associations, 50 communities now have living fences around their homes and/or medicinal gardens. These fencing plants are medicinal, fruit-producing and/or protective. Live fences reduce the cutting of thorn-bush and help to protect against sun and wind. In addition, 40 com-munities have established 0.5ha to 1ha woodlots, each with 60 to 200 trees of 15 to 25 different species of slow-growing indigenous medicinal trees, and 45 communities have backyard medicinal gardens with at least 12 indigenous and two exotic medicinal species. At least 12 communities have prepared nursery beds of medicinal, fruit and general-purpose tree seedlings. According to KACHEP's 2005 field report, more than 100,000 medicinal, fruit and live-fencing trees are growing around the 70 local healer communities in Pian and Bokora.

Knowledge shared

Four primary schools have created EVK clubs and established medicinal plant demonstration gardens on the school grounds. These clubs and gardens have encouraged preservation and promotion of IK in surrounding communities.

At least once a year since 2000, the BTHLA and PTHLA come together for a joint healers' workshop. Two exchange visits have taken place: 28 Karamojong went to southwest Uganda in the Ankole pastoralists' cattle corridor, and 12 healers and project staff visited Samburu and Turkana healers in Kenya. The Kenyan healers later attended a Karamojong joint healers' workshop organized by KACHEP. In the words of Dengel Lino, a livestock healer from Bokora: 'We used to share food and knowledge only with our family, but now I feel comfortable sharing with other healers from Pian and Kenya. It has helped me with my cattle.' The focus on sharing knowledge extends to association members sharing with other healers in their community, NGOs and neighbours. Peace has been an unintended consequence of these meetings, but is a critical component of development.

The healers' associations decided to focus on agroforestry schemes and the prevention of endemic diseases, mainly of cattle. Significant endemic diseases originate from internal and external parasites. For example, ana-plasmosis, East Coast fever, babesiosis, and heartwater are all common and serious tick-borne diseases in the area. Unfortunately, allopathic

Table 16.1 *Plant species in agroforestry schemes in Karamoja*

Local name	Botanical name
Eyelel	Acacia drepanolobium Harms ex B.Y. Sjöstedt
Eminit	Acacia gerradii Benth.
Ewalongor	Acacia sieberiana DC
Ekadokodoi	Acacia senegal Willd
Eyelel	Acacia seyal Delile
Ekwakwa	Albizia amara (Roxb.) Boiv
Ekapangiteng	Albizia anthelmintica Brongn
Ecucukwa	Aloe spp
Custard apple	Annona spp
Neem	Azadirachta indica A. Juss
Ekorete	Balanites aegyptiaca (L.) Del.
Ekadolia	Capparis tomentosa Lam.
Papaya	Carica papaya L.
Ekadeli	Commiphora abyssinica (O. Berg) Engl.
Kei apple	Dovyalis caffra Warb
Jeriman	Euphorbia bongensis Kotschy and Peyr
Ekalie	Grewia mollis Juss
Epongae	Grewia villosa Willd
Ekere	Harrisonia abyssinica Oliv
Eligoi	Kleinia odora DC
Moringa	Moringa oleifera Lam
Ebuto	Neorautanenia mitis (A. Rich) Verdc
Edapal	Opuntia cochenillifera DC
Epapai	Piliostigma thonningii (Schumach) Milne-Redh
Guava	Psidium guajava L.
Pomegranate	Punica granatum L.
Abukut	Sanseveria spp
Elamoru	Steganotaenia araliacea Hochst
Lokile	Synadenium grantii Hook. f.
Epederu	Tamarindus indica L.
Fish bean	Tephrosia vogelii Hook. f.
Abwach	Warburgia ugandensis Sprague

Source: Field Report, KACHEP, 2005

medicines are rarely locally available, commonly mishandled and often ineffective against tick-borne diseases, even if administered properly. Therefore, the focus is on prevention rather than cure.

It was once common practice to remove ticks by hand; but this was all but abandoned after the colonial government constructed cattle dips with modern acaracides. A few problems resulted from this well-meaning introduced technology: limited resources to buy drugs led to increased strain on already inadequate finances; when acaracides are used properly, tick load is heavily reduced, leading to decreased resistance to tick-related diseases; and there have been some accidental poisonings of people.

Healers advocate keeping tickload at minimal levels, and recognizing and treating tick-borne diseases early, before the blood parasites infiltrate the entire circulatory system. Karamojong keep tickloads low by reverting to removing ticks from animals by hand daily, or with regular use of plant-based dips or, more rarely, commercial products. Pian had been using one plant to treat against ticks, Bokora another and Turkana a third. Since all three plants are found in each area, they now have greatly increased the availability of effective botanical medicines, allowing for more regular treatments. Before the healers' sharing network, it would have been virtually unheard of for these groups to exchange knowledge with one another. It is rare even today; but with adopters in the ranks of the networks, they spread the other group's knowledge to their own neighbours, and all the pastoralists benefit.

Impact of sharing knowledge

Many tangible benefits, including self-sufficiency, have been realized through knowledge sharing. In the words of 28-year-old Pian healer Augustino: 'Our cows' milk yield has increased and people are eating a more balanced diet from the cows' milk, our new fruits and Moringa leaves ever since our *manyatta* put up a backyard pharmacy.'

There have also been less tangible benefits. Regular meetings between healers from groups that are often at war have helped to improve, at least to some extent, relationships between groups. For example, in 2001, Loduk Joachim, a Pian traditional healer, escaped being shot at close range when an opposing Bokora warrior recognized him as a healer who had earlier taught him about a remedy that cured his prize bull.

An additional intangible, but vitally important, benefit of the promotion of EVK has been increased respect from both within and outside the culture for the knowledge of the traditional healers. A medical student said his teachers used to mock the slow students by telling them: 'Don't be like the Karamojong and get left behind!' A non-Karamojong teacher based in Bokora said: 'I never thought the Karamojong knew so much. Now I use their EVK for my poultry and have taught my family about some of their treatments.'

CONCLUSIONS

Increased conservation

With the growth of a viable EVK network in Karamoja, local attention to nature conservation is increasing. Communities have planted their own trees after handpicking the best seeds. At least four workshops take place each year to share knowledge about conservation and harvesting techniques. Twenty-four indigenous tree species have been domesticated and over 100,000 trees planted. Thousands of seedlings are growing in members' nurseries to be planted during the wet season. Also outside the network, there is increased interest in indigenous tree species such as gum arabic, shea butter and amarula. This shows progress toward the objective of the four EVK organizations: preservation of medicinal trees.

Increased sharing

Whereas in the past this would never even have been considered, today, inroads have been made towards open discussion between those formally educated and those not, between Pian and Bokora schoolchildren and their parents, and even with communities outside of Karamoja and Uganda. Over 92 workshops have been held. This contributes to fulfilling the objective of protecting and promoting EVK.

Increased interest in EVK

Membership in the healers' associations has grown from 12 in 1998 to 94 in 2006. School children are also keen to learn about EVK. All livestock NGOs in Karamoja are members of KEVIN. The high interest was evident in 2000, when individuals and organizations not only attended but also paid for Karamoja's first EVK workshop. Sharing within and between healers' associations has brought about institutional change. In the past, knowledge about healing animals was shared only with close friends and neighbours. Today, the blanket of hospitality is spread more widely. Sharing of EVK has increased the number of people using it, which means that more medicinal plants are grown, protected and used.

Increased trust and security

Knowledge sharing is multiplied at monthly and annual gatherings, when healers from as many as five tribes share case stories and learn from each other. This sharing leads to greater respect among local people, including the youth, for their culture and for one another. Gathering to share knowledge necessarily involves sharing food, water, firewood and other resources. In the Karamoja culture, after two people have shared a meal, they are like kin and cannot harm one another. This leads to decreased fighting, raiding and ambushing. More peace leads to more sharing, and the virtuous cycle continues. Therefore, encouraging EVK and increasing medicinal plant availability benefit not just the livestock, but also the people who depend upon them. The sharing has encouraged dialogue between antagonistic groups, within families, clans and tribes, and even across borders. Strengthening local institutions that address NRM may thus produce peace as a by-product. Further analysis could increase understanding of how bringing people together to share EVK can lead to increased trust and security.

The success may be due partly to the fact that it has been an endogenous movement (from within) rather than exogenous (initiated or led from outside). Since indigenous people have led the processes from day one, they have developed the skills and local capacity to continue without help from outside.

ACKNOWLEDGEMENTS

Sincere thanks are due to the traditional livestock healers and other keepers of knowledge in Karamoja.

REFERENCES

Gradé, J. T. and Shean, V. S. (1998) Investigations of Karamojong Ethnoveterinary Knowledge in Bokora County, Moroto District, Internal Trade Secrets Report for BoLI, Lutheran World Federation, Moroto, Uganda

Martin, G. J. (1996) Ethnobotany: A Methods Manual, Chapman and Hall, London

McCorkle, C. M. (1986) 'An introduction to ethnoveterinary research and development', Journal of Ethnobiology, vol 6, no 1, pp129–149

Wadsworth, Y. (1998) 'What is participatory action research?', Action Research International, www.scu.edu.au/schools/gcm/ar/ari/p-ywadsworth98.html, accessed 2 August 2007

CHAPTER 17

Village Information and Communication Centres in Rwanda

Silvia Andrea Pérez, Amare Tegbaru, Speciose Kantengwa and Andrew Farrow

BACKGROUND

Rwanda, like many developing nations, relies heavily on agriculture for both domestic consumption and exports, and more than 85 per cent of the population depend upon agriculture for their livelihoods. Starting in 2003, the Agricultural Technology Development and Transfer (ATDT) project initiated 30 village information and communication centres (VICs) in Rwanda as a mechanism for sharing information to improve livelihoods and natural resource management (NRM). This project was managed by the International Centre for Tropical Agriculture (CIAT) and the Institute of Agronomic Sciences of Rwanda (ISAR) with support from the US Agency for International Development (USAID). The findings of a rapid assessment of these VICs are presented here.

The model of VICs studied here has its origin in the late 1990s as part of an integrated pest management (IPM) project with bean growers in Hai District, northern Tanzania. It was a pilot in technology development and dissemination, involving farmer groups who creatively developed different dissemination mechanisms for printed materials on agricultural technologies produced by researchers (CIAT, 2004). Farmers in Hai District used to visit the research station looking for information; but they encountered many difficulties and had to walk long distances on bad roads. Consequently, the project decided to take the information closer to them and established a village information and communication centre.

In Rwanda, farmers are commonly organized in a nested hierarchy of associations. Every association has crop and livestock production as their main activity. Such associations host and manage the VICs, ensuring institutional support and sharing the costs of running them.

The ATDT project selected the associations based on three criteria:

1   willingness to host an information centre;
2   strength of the association in terms of organizational structure; and
3   the physical capacity to host a VIC (having space to put in shelves and allow activities that attract people, such as selling agricultural inputs).

All of the associations visited in this study (see Table 17.1) have extensive experience in all aspects and are therefore suited to promote VICs.

Table 17.1 *Membership of the seven farmer associations hosting village information and communication centres visited*

Association/co-operative	VIC in district	Total membership	Women	Men
Impabaruta	Kamonyi	2524	1484	1040
Impakomu	Muhanga	2530	1596	934
Abahujumugambi	Bugesera	4353	2256	2097
Tuswanyianzara	Ngoma and Kirehe/ Old Kibungo	82	47	35
Cooperative Urunana	Gatsibo	1230	930	300
Abajyinama	Nyabihu	2235	1232	1003
Indangamirwa	Nyamagabe	861	396	465
Total/potential VIC users		13,785	7941	5874

Initially, all VICs in Rwanda were equipped with reading and training materials on extension, crops, livestock and nutrition. Farmer associations, community extension officers and other trainers could use this material to share information and knowledge with members of the community. Each VIC is set up as a public area with some basic furniture and some information material, such as books, posters, extension leaflets, booklets or other publications. Farmers can easily access this material and read and share the information with co-farmers and other service providers.

The VICs in Rwanda were intended to facilitate a multidirectional flow of information and communication among farmers and other service providers. This included content also produced by the users of VICs, thus sharing their lessons and experiences in the process of learning. It was envisaged that information and knowledge would be shared not only within a VIC, but in the field as well. Each VIC has a coordinator or facilitator who is responsible to a given community and preferably resides within the community. The coordinator's job is similar to that of a village librarian, who interprets and facilitates the information-sharing process with the user community. She or he assists in managing the information (classification of books by themes), maintains records of publications and user visits, maintains the building and promotes the information. The co-ordinator is expected to monitor the information and training needs of users, and to communicate those needs to those who can deal with them.

Two main levels for scaling up VICs were envisaged:

1 within the associations that host VICs; and
2 within higher-level organizations such as ISAR, CIAT and other NGOs.

CONCEPTUAL FRAMEWORK FOR VILLAGE INFORMATION AND COMMUNICATION CENTRES

VICs support the flow of technology, the sharing of information and know-ledge, communication, learning, interaction and negotiation between different actors/organizations of an agricultural innovation system. These actors participate in an innovation process where they turn an idea into a product or service and are empowered to use and share these results. Innovation is seen as a process of network-building, social learning and negotiation (Leeuwis, 2004).

Management of information and knowledge is a very important issue for all organizations, but often poorly understood. Andrews and Herschel (1998) argue that the concepts of 'information' and 'knowledge' are often considered synonymous; the authors challenge this notion and contend that knowledge is a step beyond information:

> Information does not always lead to understanding. In contrast to inform-ation, knowledge goes beyond the facts, connecting and explaining them. Knowledge further refines information and seeks to reconcile seemingly disparate findings. It is knowledge, not information, that can best contribute to empowerment. (Andrews and Herschel, 1998)

Farmers can access a variety of information in VICs and this, in itself, leads to empowerment (World Bank, 2002). Apart from access, the pos-sibility of sharing their own information and knowledge within the VICs can promote empowerment of farmers by allowing them greater free-dom, autonomy and self-control over their work, and responsibility and involvement in decision-making.

Farmers feel confident in applying information and getting new know-ledge about agricultural technologies and other subjects that help to improve their livelihoods. They apply this knowledge in their daily work in order to produce better results, and, in the process of solving individual and common problems, generate new knowledge through socialization with farmers and others. In this respect, empowerment enables them to utilize the knowledge and information acquired (Andrews and Herschel, 1998).

VIC users share the information that they find in the centres (explicit or codified knowledge, transmittable in systematic language) and also share knowledge that they have as individuals (tacit or individual's knowledge acquired by experience: a personal quality), creating between them new knowledge through a communication process (Polanyi, 1966). At the level of the associations, this knowledge is 'organizationally' amplified by their members and crystallized as part of the knowledge network of the farmer association and the process of institutionalization.

The process of sharing information and knowledge using informational materials of the VICs is gradually translated, through interaction and a process of trial and error, into different aspects of tacit knowledge. The interaction between tacit and explicit knowledge tends to become larger in scale and spreads faster as more actors in and around the organization become involved. Thus, creation of organizational knowledge can be viewed as an upward spiralling process, starting at the individual level, moving up to the collective (group) level, and then to the organizational level, sometimes reaching out to the inter-organizational level (Nonaka, 1994).

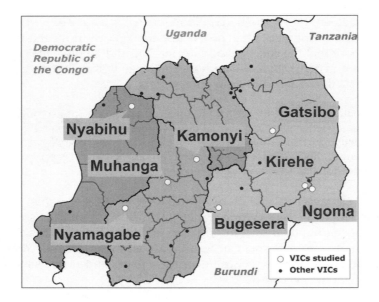

Figure 17.1 *Sample of village information and communication centres visited in Rwanda*

Source: Map drawn by Andrew Farrow (31 August 2006) based on NIMA (1997) for country boundaries, NUR (2006) and Speciose Kantengwa (pers. comm., 31 August 2006) for locations of VICs in Rwanda.

DATA COLLECTION AND ANALYSIS

Data for the assessment were collected in eight of the 30 VICs (see Figure 17.1) set up in Rwanda. These are identified by the district in which they are found. The criteria for selection include representation of the three agro-ecological zones of the country, geographic location (north, south, central, east and west Rwanda) and demographic variations. Each farmer association hosts one VIC, with the exception of Tuswanyianzara, which hosts two.

The total of potential users of the eight VICs in the associations is 13,785, of whom 7941 are women and 5874 men (see Table 17.1). This represents a target group to which the eight VICs are offering their services. It was observed that, while the VIC in Bugesera covers a potential user community of 4353 members of the Abahujumugambi Farmers' Association, there are two VICs located in Ngoma and Kirehe, covering a potential user community of just 82 members of the Tuswanyianzara Association.

The data collection methods employed in this rapid assessment are largely qualitative. The main instrument was focus group discussions guided by checklists, supported by key informant interviews with selected farmers (women and men). In addition to these instruments, a semi-structured questionnaire was used to collect quantitative data on issues such as VIC activities, usefulness of the information, association membership, etc., disaggregating according to gender. In one of the VICs (Ngoma), no focal group was convened because the VIC had been closed. It was possible, however, to have interviews with the manager of the VIC and two members of the Tuswanyianzara Association.

Krueger and Casey (2000), Pini (2002) and Shortall (2002) recommend focus group discussions as a participatory and dynamic approach for identifying important issues associated with a particular theme or situation. Focus groups are often used before a more structured survey in order to understand a range of issues as diverse as empowerment, gender relations, HIV/AIDS (Pool et al, 2001) and urbanization (Bah et al, 2003). In this case, the focus groups constituted users of VICs, members of user communities, some members of management teams of farmer associations that host the VICs, and the individual in charge of a VIC. Efforts were made to ensure full participation of women and members of different groups in the community. The discussions covered:

- uses of VICs;
- type of information that users require;
- type of printed materials found in VICs;
- usefulness of the materials; and
- records that are being kept (VIC profile).

Users of VICs were also asked about the main economic activity in their respective communities, the groups or associations that exist, the decision-making processes, levels of participation, and profiles of communities in terms of gender and equity (communities and association profiles).

In order to understand farmers' own perception of the organization and its institutional arrangements, users of VICs were asked to draw on paper the structures and positions of their respective groups. The idea was to have their own description of organizational structures and to gain information about the decision-making processes and flow of information and communication. Andrews and Herschel (1998) state that discussion about such drawings gives more information on how the direction of communication flow in an organization depends upon the structure of the organization; however, changes in the direction of communication flow, intentional or otherwise, can alter the shape of the organizational structure.

Farmers and user groups predominantly speak *Kinyarwanda* (with its own local dialect as spoken by Rwandese). English and French are the official languages, but are spoken largely by people in key professional sectors and students in training institutions. High levels of illiteracy further limited the use of some data collection instruments.

Data from the focus group discussions, questionnaires and key inform-ant interviews were first used to describe and characterize each VIC. This was followed by an analysis of the strengths, weaknesses, opportunities, and threats (SWOT) of the VIC concept and its implementation in Rwanda. In addition, photography was used as a means of enriching the data and triangulation.

FINDINGS

In general, VICs serve as meeting and contact points for disseminating information on subjects broader than agriculture. Based on the demand of the rural communities, there are information materials on health, food security, education, gender, development, strategic planning of interventions, institution-building, leadership and management of associations. The study revealed that more than 6600 farmers had visited one particular VIC in 2005 to obtain information on new technologies and training materials.

VICs are community spaces where farmers can find many resources and services for their work under one roof. These include information and knowledge services; sales outlets for agricultural inputs; collective marketing of products to gain better prices; cooperative banking; training facilities; community meeting places, etc.

Discussions in the focus groups revealed information about gender and equity in the VICs. In Rwanda, women outnumber men (4,249,105 female and 3,879,448 male). This ratio is also reflected in the membership of farmer associations in the study sample (see Table 17.1). Women are a power to reckon with in the farmer associations by dint of their number, especially when decisions are made democratically. Women are also a powerful labour force. Most associations follow a principle of gender balance, which facilitates the process of empowerment.

In Bugesera VIC, women are expanding their agricultural production capacity through the use of new agricultural technologies, which they access through the VIC and share amongst themselves. They are not just producing for sale in the market; they are also producing for consumption by their families, thus ensuring food security. They are increasing their income and are becoming more independent of men economically and more confident about their own capacities.

Men said that they like VICs because they can read books about different crops and check out information on crop diseases. Women said that they learned about new techniques in agriculture, gained information about seeds and field extension, and found out about the use of various drugs for livestock. They also learned about fish farming and maize production and treatment. In addition, they improved their nutrition through initiatives such as the kitchen garden in Gatsibo District.

Some associations that are hosting VICs are also investing resources in their management and development. The Impakomu Association, for example, is investing in human resources and has hired an agronomist to coordinate the activities of the VIC. Abahujumugambi has invested in physical infrastructure and now has a dedicated room for the VIC.

Strengths

Users said that VICs are places where their communities have access to useful information and share knowledge that transforms their livelihoods, increases their income, provides food security and improves nutritional standards. 'Information is useful only if it is available, if the users have access to it, in the appropriate form and language – i.e. if it is communicated, if it circulates among the various users with appropriate facilities, if it is exchanged' (Mundy and Sultan, 2001). They have found that the information materials at VICs are understandable, educative and translated into the local language.

All users consider the VIC as a very important tool to build new knowledge. They mentioned gaining new knowledge and improving their skills. The information at VICs is shared and used by farmer associations. Members of some associations share the information in formal training using materials from the VIC. Some farmers share what they have learned with their families, friends and neighbours; agricultural teachers do the same with their students. This supportive environment of sharing knowledge is called a 'community of practice' (Wenger, 1998; Hildreth et al, 2000). Wenger says that communities of practice are an organization's most resourceful and dynamic knowledge source and the centre of its ability to know and learn. Learning is wished for to bring new knowledge to the organization, allowing people to create new and better results – in other words, to innovate.

VICs are hosted by organizations that have a horizontal structure, knowledge of community needs, and a sense of community and solidarity. Members of organizations are involved in participatory processes of decision-making about community problems.

VICs are public (community) places where farmers can find many resources for their work and community services (training, banking facilities, etc.) in one place close by. This saves them time and money that otherwise would have been spent on going far to obtain the same.

Weaknesses

A major weakness identified in this study was the lack of a clear definition of a VIC. The principles underlying the VIC are undefined, yet are necessary for the process of institutionalization by farmer associations. The VICs started out without defining the basic resources for their implementation. Apart from one or two exceptions mentioned earlier, VICs, in general, have limited financial resources for building capacity and development. This could affect their sustainability.

The flow of information and communication with regard to the organizational structure was consistent in all of the VICs studied (see Figure 17.2). In general, the flow is multidirectional; however, this becomes unidirectional at the level of the research organizations. The ATDT project is facilitating a process of information sharing, but does not appear to have built in a capacity to receive and respond to feedback from users of VICs. The incorporation of feedback from VICs into the activities and decision-making processes of research and development (R&D) organizations could be a good indicator of scaling up of VIC principles within higher-level organizations.

The VICs have not developed training to improve skills for their management – in particular, information management.

Opportunities

VICs can be an important source of feedback and very useful for the work of organizations that focus on livelihood improvement and NRM. VICs can consolidate a strategy that uses an integrated system of information and communication to value the community's local knowledge and capacity, and to promote its own ways or means of sharing information. Communities can design an integrated information and communication system complementing existing materials in VICs and utilizing other available media. Communities can systematically use VICs to share their information and experiences with other communities and R&D organizations.

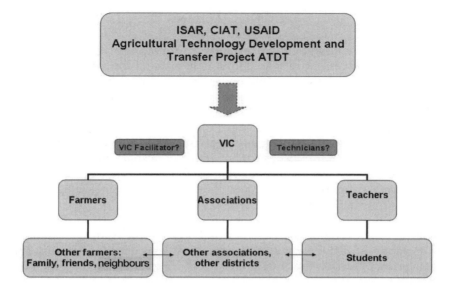

Figure 17.2 *Flows of information and communication between village information and communication centre users and agricultural research and development organizations*

VICs can reinforce cooperation, participation, communication, sharing of knowledge and involvement of different members within organizations and communities. VICs can be a tool for social development of marginal and poor communities. Some associations have started a process where VICs convert available printed material into formal training resources.

VIC users were encouraged to share their experiences (bad and good) acquired through the VIC, using different media available in the community to reach more people, to share knowledge and to convert tacit knowledge to explicit knowledge (Polanyi, 1966).

Threats

Possibly the biggest threat to the realization of the opportunities envisaged above is the limited human and financial resources (facilitators in VICs and technicians visiting VICs, etc.). This is particularly relevant given the uncertain/unstable financial environment of the ATDT project.

Other threats include changes in the administration at various political levels. This was seen clearly in the former district of Kibungo, where a change in boundaries has resulted in a VIC being unable to offer services to some communities.

CONCLUSIONS

Encouraging communities to invest in an information facility and to share knowledge, targeted towards helping them meet their own vision of improved and sustainable livelihoods, can eventually lead to empowerment of rural and marginalized communities. VICs should therefore be understood as organized public spaces where community members can access information, share knowledge and obtain services.

Free access to and democratization of information are key principles in setting up and operating VICs. Organization of farmer groups in a farmer association is the basic requirement for establishing a centre. This ensures institutional support for hosting and managing the VIC, including sharing the costs for running it.

From the initial stage of establishing the VICs, it should be clear that the centres are demand-driven and address the information needs and priorities of the rural and marginalized communities (women and men, youth, the sick, the elderly and the disabled). In other words, VICs should be organized in a manner that facilitates a broad ownership base since the objective is to benefit as many farmers as possible on a range of issues related to agriculture and rural development. Efforts should also be made to ensure that VICs promote multidirectional flows of information and communication between different stakeholders, R&D partners, farmer associations and other community groups, including sharing of 'expert' as well as traditional knowledge of communities. The long-term sustainability of VICs lies in community empowerment, which also demands a strategic vision developed in partnership with farmers and investment in human resources, in the physical condition of the VICs and in training. The more information that is available and internalized by users, the more it dramatically influences 'the way organizations are structured, the ways people lead and attempt to share power with others, and the very nature of organization and organizational communication' (Andrews and Herschel, 1998).

There is a variety of focal areas (agriculture, education, health, poverty alleviation, etc.) and contexts and, accordingly, a variety of mechanisms for sharing information and communication, such as information kiosks, village libraries, community telecentres, village phones, community radio, internet radio, local area networks, mobile phones, etc. Choices should be made, taking into account the levels of literacy of the communities involved and other important aspects, such as levels of participation and flows of information and communication. VICs are just one of the mechanisms, and do not exclude other mechanisms or media. On the contrary, most of them can be complementary and included within an integrated system of information and communication.

The inclusion of ICTs in different initiatives is an important step in the democratization and participation of marginalized communities in the information society. However, there is still a long way to go until rural communities which have low levels of literacy and differing local languages can start participating actively and share knowledge with other communities, using the internet on their own. This is without considering accessibility problems, capacity-building, high operational costs, connectivity problems and technical dependency, which can affect the sustainability of these initiatives.

REFERENCES

Andrews, P. and Herschel, R. (1998) *Organizational Communication*, Houghton Mifflin, Boston, MA

Bah, M., Cissè, S., Diyamett, B., Diallo, G., Lerise, F., Okali, D., Okpara, E., Olawoye, J. and Tacoli, C. (2003) 'Changing rural–urban linkages in Mali, Nigeria and Tanzania', *Environment and Urbanization*, vol 15, no 1, pp13–24

CIAT (International Centre for Tropical Agriculture) (2004) 'Village Information Centres', *Highlights: CIAT in Africa*, vol 16, CIAT, Cali, Colombia

Hildreth, P., Kimble, C. and Wright, P. (2000) 'Communities of practice in the distributed international environment', *Journal of Knowledge Management*, vol 4, no 1, pp22–37

Krueger, R. A. and Casey, M. A. (2000) *Focus Groups: A Practical Guide for Applied Research*, Sage, Thousand Oaks, CA

Leeuwis, C. (2004) *Communication for Rural Innovation: Rethinking Agricultural Extension*, Wageningen Academic Publishers, Wageningen, The Netherlands

Mundy, P. and Sultan, J. (2001) *Information and Communication Technologies for Rural Development and Food Security: Lessons from Field Experiences in Developing Countries*, FAO, Rome, Italy

NIMA (National Imagery and Mapping Agency) (1997) *VMAP_1V10: Vector Map Level 0 (Digital Chart of the World)*, NIMA, Fairfax, VA

Nonaka, I. (1994) 'A dynamic theory of organizational knowledge creation', *Organization Science*, vol 5, no 1, pp14–37

NUR (National University of Rwanda) (2006) 'Secteurs, Districts and Provinces of Rwanda', digital dataset, accessed 29 May 2006, Center for GIS (Geographic Information Systems), National University of Rwanda, Butare, Rwanda

Pini, B. (2002) 'Focus groups, feminist research and farm women: Opportunities for empowerment in rural social research', Journal of Rural Studies, vol 18, no 3, pp339–351

Polanyi, M. (1966) *The Tacit Dimension*, Routledge and Kegan Paul, London

Pool, R., Nyanzi, S. and Whitworth, J. A. (2001) 'Attitude to voluntary counselling and testing for HIV among pregnant women in rural south-west Uganda', *AIDS Care*, vol 13, pp605–615

Shortall, S. (2002) 'Gendered agricultural and rural restructuring: A case study of Northern Ireland', *Sociologia Ruralis*, vol 42, no 2, pp160–175

Wenger, E. (1998) *Communities of Practice: Learning, Meaning, and Identity*, Cambridge University Press, New York, NY

World Bank (2002) 'A framework for empowerment: Summary', based on *Empowerment and Poverty Reduction: A Sourcebook*, Poverty Reduction Group, World Bank, Washington, DC, http://siteresources.worldbank.org/INTEMPOWERMENT/Resources/486312-1095094954594/draft.pdf

Farmer Field Schools for Rural Empowerment and Life-Long Learning in Integrated Nutrient Management: Experiences in Eastern and Central Kenya

André De Jager, Davies Onduru, Louis Gachimibi, Fred Muchena, Gituii Njeru Gachini and Christy Van Beek

INTRODUCTION

In Africa, maintaining and improving soil fertility are major factors to attain food security, reduce poverty and address environmental degradation (Sanchez et al, 1997). Formal agricultural research has generated fundamental insights into various aspects of soil fertility management and has developed improved technologies. However, farmers' application of the results has been below expectations. This is largely because the prevailing extension approaches did not stimulate farmers to assess technologies critically, make adaptations for their specific conditions, learn how to develop the technologies further and assist in farmer organization. Within the diverse and variable environments of rain-fed farming in Africa, farmers already have a wide body of knowledge in addressing soil fertility problems. Research and development programmes should build on these experiences and further develop farmers' expertise. They should strengthen farmers' decision-making and action-taking capabilities that are informed by principles and methods and are aided by instruments and tools developed through links with science.

To address shortcomings in extension in integrated pest management (IPM), the United Nations Food and Agriculture Organization (FAO) developed the farmer field school (FFS) approach in Asia during the early 1990s. In IPM, insects were the entry point for a different approach to innovation in small-scale irrigated rice production. Likewise, integrated nutrient management (INM) offers an entry point for a different approach to innovation in rain-fed farming in Africa. INM aims at the 'best' combination of available nutrient management technologies that are biophysically relevant, economically attractive and socially acceptable (Smaling et al, 1996). FFSs addressing INM combine a technical focus on a locally feasible and sustainable mix of nutrient management strategies with a developmental focus on stimulating farmers' creativity in capturing local opportunities to make farming more profitable.

In this chapter, we assess whether FFSs focused on long-term farmer organization, experimentation and learning are an appropriate approach to effect innovation in soil fertility management in East Africa. We include the results of a four-year pilot project in eastern and central Kenya.

FARMER FIELD SCHOOLS: AN EVOLVING APPROACH

The FFS approach was first developed to address a major threat to food security in Asia: rice yield losses caused by the brown planthopper (Pontius et al, 2002). FFS is a learner-centred approach: through observation, experimentation and evaluation leading to understanding, farmers are equipped to address challenges and introduce appropriate changes in managing their farms. Farmers are the main actors. Outsiders – extensionists, researchers and non-governmental organizations (NGOs) – serve as facilitators or sources of information. Experiences in implementing FFSs in IPM have been extensively documented (e.g. Kenmore, 1991; Van de Fliert, 1993; Davis, 2006). Over the years, the FFS approach has been extended to include other issues in agriculture and rural development, such as natural resource management, animal husbandry, conservation agriculture, HIV/AIDS, food security and nutrition (FAO, 1998; Minjauw et al, 2002; UPWARD, 2003). More recently, it has been seen as an appropriate vehicle for empowering rural actors in life-long learning processes, strengthening local institutions and networks, and stimulating social processes and collective action, leading to improvement in rural livelihoods (Hounkonnou et al, 2004).

The FFS approach and adult-learning processes triggered a paradigm shift in agricultural knowledge systems, building on two earlier major shifts: from a commodity to a (farming) systems approach and from the linear research–extension–farmer model to dynamic models of innovation systems combining multiple sources of knowledge. FFSs built on these approaches by embracing the dimensions of collective action and strengthening farmer organization.

Debates about options for large-scale implementation emerged after three studies funded by the World Bank assessed the impact of IPM–FFS programmes in The Philippines and Indonesia (Quizon et al, 2001; Feder et al, 2004; Van den Berg, 2004). They concluded that FFSs are fiscally unsustainable because up-scaling costs are high, no long-term effects on pesticide expenditures and rice yield were observed, and knowledge was not diffused to neighbouring farmers. Others (e.g. NARC, 2004; Braun et al, 2006) criticized that broader impacts such as adult education, social organization and farmer empowerment were not considered in these studies. The 'one-size-fits-all' approach of FFSs was critically reviewed and a call made for a more flexible methodology adapted to local situations (Davis, 2006). There is evidence of positive impacts of FFSs on rural communities as long as the impact is measured not only in terms of technology transfer (e.g. Tripp et al, 2004). A comprehensive impact assessment methodology must still be developed to cover the broader development impacts of FFSs on empowerment, education, farmer organization, farmer–research linkages and social cohesion. Comparative cost–benefit analyses of the FFS approach and other research and extension models are also still lacking.

In Kenya, the FFS approach was first implemented in 1995 by the Ministry of Agriculture (MoA) and the FAO. By 2003, about 1000 FFSs and 250 facilitators were active and 34,000 farm households had taken part in FFS activities focused on arable crops, horticulture, livestock and soil management (FAO et al, 2003). In Kenya, Uganda and Tanzania, the FAO initiated various FFS programmes for improving soil fertility. Experiences presented by implementing agencies and policy-makers at a regional conference on FFS experiences in soil fertility management, held in Uganda in 2006, revealed a large variation in approaches, intensity and quality of the learning process and impacts.

The IPM–FFS approach had to be modified to deal effectively with the more complex issues of rain-fed agriculture in Africa. Based on literature review and discussions with stakeholders, we made the following modifications in the pilot programme for FFSs in soil fertility management:

- The simple but rather rigid structure of the activities in the IPM–FFS was replaced by a more flexible set of activities depending upon the priorities of the FFS members.
- Instead of being for only one growing season, permanent FFSs were to be established to address long-term challenges of improving soil fertility and facilitating farmer organization.
- As much as possible, use was made of already existing community groups to which farm household members belonged.
- In addition to central-plot experimentation, on-farm experimentation was stimulated in order to capture diversity in farm systems and to allow for individual adaptation of technologies.
- No initial grants were provided since these jeopardize the sustainability and up-scaling of FFSs; instead, commercial activities to cover costs were stimulated.
- Systematic monitoring and impact assessments were included in the activity plan.
- District- and national-level policy-makers were involved in the process to facilitate future up-scaling of the approach.

METHODOLOGY

Mbeere and Kiambu districts in eastern and central Kenya, respectively, were selected to implement the modified FFS approach. These districts had experience with FFSs, faced decline in soil fertility and represented major but contrasting agro-ecological zones and farming systems. In each district, we selected a representative catchment, organized community workshops to explore farmers' interest and willingness to participate, and identified existing groups or readiness to form new groups. Four pilot FFSs were formed: Kamugi (30 farmers, 50 per cent women) and Munyaka (31 farmers, 74 per cent women) in Mbeere, and Kibichoi (30 members, 40 per cent women) and Ngaita (26 members, 56 per cent women) in Kiambu. The FFSs in Kamugi and Munyaka were based on existing community groups. Baseline surveys at the four sites recorded current farming practices, revealed how farmers manage soil fertility and captured their farm management dynamics. All FFS members then joined a participatory diagnostic activity using a nutrient monitoring approach (Van Beek et al, 2004) that produces soil nutrient balances per cropping season instead of the more commonly used annual balances. Diagnosis of farm management activities covered one cropping season (March to August 2002). Results of the diagnostic activity were discussed at FFS level, and individual farm households received a diagnostic report on soil fertility management and economic performance indicators.

These start-up activities were followed by a five-season curriculum consisting of experimental design, central-plot and individual farmer experiments, agro-ecosystems analysis (AESA) (Gallagher, 2003), special topics and group dynamic activities. The FFSs met every two weeks. Experimental design was an integrated process in which farmers, scientists and extensionists shared views and decided on options and methods for experimentation. All FFSs started experiments on a central learning plot. An experiment typically consisted of a pair-wise design with two to four treatments, including a control, on plots of 20 square metres to 50 square metres. The FFSs formulated hypotheses such as: 'If we apply DAP [diammonium phosphate, an inorganic fertilizer; 18-46-0] when planting maize variety Cargil 4141, grain yields will increase because DAP improves crop nutrient status; if rains are adequate, good-quality seeds are planted and planting is done early in the season.' Agreements were made on implementation, meetings, observations, group regulations, etc. Some farmers also carried out experiments on their own farms and reported their experiences during FFS meetings.

The FFS members monitored and evaluated the experiments using an AESA format and various pictorial and scoring tools. They agreed on indicators to observe (e.g. yield, pests and diseases, leaf colour, plant health, soil moisture, weed incidence, plant vigour and labour inputs). Based on the first season's experimental results, a new cycle of experimental design was initiated before the next season. The FFS members, jointly with the facilitators and resource persons, also determined the curriculum for special topics during the season (see Table 18.1). A graduation ceremony marked the end of the facilitated FFS period and was the starting point for continuation of farmer-led FFSs, with only limited periodic backstopping from facilitators. A one-day policy workshop was organized in each district to share results of the FFS approach with stakeholders and district-level policy-makers, leading to an action plan to facilitate implementation of the FFS approach.

One year after graduation, the facilitators assessed the contributions of the FFSs to general livelihood improvement in the target areas and particularly to the adoption of sustainable soil fertility management practices. The impact assessment included a longitudinal (comparison before and after joining the FFS) and latitudinal (comparison between FFS members and non-members) analysis. It focused on knowledge and skills, changed practices, farm-level impacts and livelihood impacts.

Table 18.1 *Summary curriculum of special topics in farmer field schools*

Area	Curriculum topics
Integrated nutrient management (INM)	Soil properties and functions; soil nutrient supply and deficiencies; mineral fertilizer use; green manure and *Tithonia*; cover crops; water harvesting; composting; manure management; soil organic matter management; biological sources of fertility (legumes, *Rhizobium*); soil and water conservation practices; agroforestry; soil physical fertility; mulching
Production aspects of specific crops	Cowpea; soybean; sweet potato; climbing bean; watermelon; grafted fruit trees; beans; vegetables; Irish potato; cassava; kitchen gardening; crop storage; drip irrigation; natural crop protection and pest management
Livestock management (general, feeding, housing, health, breeding)	Dairy goats; cattle; dairy cattle; beekeeping and honey processing; calf rearing; poultry; pigs; rabbits; feeding and feed preservation; Napier grass
General farm management	Farm planning; record-keeping; tree nursery management; organic farming and use of local farm resources; tillage practices
Home economics	Cookery; human nutrition; fireless cookers; cake baking; juice/jam making; soap making; milking salve; yoghurt preparation
Others	HIV/AIDS; leadership and team-building

Using a semi-structured questionnaire, the facilitators discussed matters with individual FFS members on their farms, as well as with all members during an FFS meeting. A sample of non-members was selected: half from within the village where the FFS activities were conducted and half from neighbouring villages. Non-members were purposively sampled to ensure that FFS members and non-members were comparable in terms of production resources (land and livestock holdings). In total, 80 FFS members and 31 non-members were interviewed.

RESULTS

Experiments and their results

Most experiments on the central FFS plots were on food crops, testing various combinations of organic and inorganic nutrient sources. In Kiambu, livestock experiments were also conducted, focused on feed production, feeding regimes and manure management. The experiments and results in Kibichoi and Munyaka are presented in Tables 18.2 and 18.3, respectively.

In Kiambu, yields and financial returns were increased by:

- applying DAP or triple superphosphate (TSP) combined with manure and/or *Tithonia* on maize;
- deep digging; and
- applying DAP and *Rhizobium* on local beans.

The *Tumbukiza* (Kiswahili for 'placing in a hole') system of planting Napier grass (eight canes per hole versus one cane in the traditional system, and over six times as much farmyard manure applied in the planting hole), combined with an improved variety, increased the yields and financial returns and reduced the nitrogen (N) mining of Napier compared with the traditional system. An experiment with storing manure in a pit lined and covered with polythene was conducted on seven farms. After 11 weeks of storage, the concentrations of potassium (K), magnesium (Mg), copper (Cu), manganese (Mn) and zinc (Zn) were, on average, higher than when storage began; the percentage increase ranged from 1.5 to 53.5 per cent. At the end of the storage period, an N loss of 1.7 per cent was recorded, much lower than 40 per cent N losses reported by Lekasi et al (2001) during storage and/or composting of uncovered manure heaps.

In Mbeere, combinations of manure and DAP with and without *Tithonia* showed positive impacts on yields and financial returns in maize, while combined TSP and *Rhizobium* application showed similar positive impacts in beans. Farmers' rankings of preferred technologies generally correlated with yield levels rather than financial returns. At both sites, DAP or TSP application combined with organic manure resulted in value–cost ratios exceeding two, indicating short-term financial benefits.

Table 18.2 *Summary of experimental results on a central plot in Kibichoi, Kiambu District, 2002–2005 (standard deviation in parentheses)*

Treatments and crops	Yield (kg ha⁻¹)	Gross margin (US$ ha⁻¹)	Benefit: cost ratio	VCR[1]	N balance[2] (kg ha⁻¹)	Mean FFS score[3]
Maize 2002 SR[4] + 2003 LR[5] (n = 8)						
Normal tillage	3958 (1971)	1013 (461)	3.3	–	17	5.2
Double digging	4729 (1161)	928 (210)	2.6	0.4	5	14.8
Maize: 2002 SR + 2003 LR (n = 4)						
FYM[6] (18 tonnes ha⁻¹)	3709 (699)	848 (171)	2.5	–	33	8.2
DAP[7] (120kg ha⁻¹)	5000 (2193)	1249 (441)	5.3	–	-89	3.1
FYM (18 tonnes ha⁻¹) + DAP (120kg ha⁻¹)	4917 (1708)	1030 (306)	2.7	0.2	33	4.7
FYM (18 tonnes ha⁻¹) + CAN[8] (120kg ha⁻¹)	3750 (1664)	755 (317)	2.3	-0.7	66	4.0
Beans: 2003 SR (n = 1)						
FYM (22 tonnes ha⁻¹)	1750 (–)	532 (–)	2.0	–	59	3.0
FYM + Rhizobium[9] (0.27kg ha⁻¹)	1500 (–)	392 (–)	1.7	-54.5	69	4.0
FYM + Rhizobium (0.27kg ha⁻¹) + DAP (105kg ha⁻¹)	2000 (–)	681 (–)	2.2	4.2	50	13.0
Maize: 2004 SR (n = 2)						
Control (zero application)	5227 (321)	1320 (71)	7.5	–	-117	3.3
Tithonia (11 tonnes ha⁻¹)	8182 (857)	1795 (151)	7.1	6.4	-116	4.5
Tithonia (11 tonnes ha⁻¹) + TSP[10] (120kg ha⁻¹)	8788	1856 (191)	6.3	4.7	-128	5.5
Tithonia (11 tonnes ha⁻¹) + TSP + CAN (120kg ha⁻¹)	(1071)[11]	–	–	–	–	6.7
Napier grass: 2004 LR + 2004 SR + 2005 LR[12] (n = 2)[13]						

Conventional tillage + local variety[1]	8925 (2928)	1220 (507)	5.3	–	–174	2.7
Conventional tillage + *Kakamega*[1]	13259 (220)	1561 (176)	6.4	–	–260	5.0
Tumbukiza tillage + local variety[1]	13691 (3557)	1108 (677)	2.0	0.9	–63	2.8
Tumbukiza tillage + *Kakamega*[1]	17308 (326)	1726 (341)	2.5	1.6	–157	9.5
Napier grass: 2005 LR[14] (n = 6)[15] Cattle manure (10 tonnes ha^{-1})	2073 (956)	60 (169)	1.2	–	–10	2.5
Cattle manure (10 tonnes ha^{-1}) + slurry (11.7 tonnes ha^{-1})	2487 (852)	26 (153)	1.1	0.7	–3	17.5

Notes:

1 VCR = (gross value treatment – gross value control)/(variable costs treatment – variable cost control).

2 N balance = partial nitrogen balance (IN1 + IN2 – OUT1 – OUT2).

3 FFS score = 20 points divided over treatments on preference of technology.

4 SR = short rains (October to February).

5 LR = long rains (March to August).

6 FYM = farmyard manure.

7 DAP = diammonium phosphate (an inorganic fertilizer; 18-46-0).

8 CAN = calcium ammonium nitrate (an inorganic fertilizer; 26-0-0).

9 No estimation of additional input in nitrogen balance was made.

10 TSP = triple superphosphate (an inorganic fertilizer; 0-46-0).

11 Due to shading of plot, unreliable yield data.

12 Data based on total of three cuts.

13 Napier grass dry matter yields.

14 Data based on one cut.

15 Napier grass dry matter yields.

Table 18.3 *Summary of experimental results on a central plot in Munyaka, Mbeere District, 2002–2005 (standard deviation in parentheses)*

Treatments and crops	Yield (kg ha⁻¹)		Gross margin (US$ ha⁻¹)		Benefit: cost ratio	VCR[1]	N balance[2] (kg ha⁻¹)	Mean FFS score[3]
Maize: 2002 SR[4] + 2003 LR[5] (n = 2)								
FYM[6] (16 tonnes ha⁻¹)	2530	(1017)	28	(122)	1.1	–	–22	4.0
DAP[7] (216kg ha⁻¹)	2960	(1479)	185	(228)	1.7	–0.4	–22	4.5
FYM (16 tonnes ha⁻¹) + DAP (216kg ha⁻¹)	3741	(918)	114	(121)	1.3	2.2	–2	5.2
FYM (16 tonnes ha⁻¹) + DAP (216kg ha⁻¹) + *Tithonia* (3.6 tonnes ha⁻¹)	4350	(772)	203	(114)	1.4	2.7	1	6.3
Beans: 2003 SR (crop failure)								
Control	–		–		–	–	–	–
TSP[8] (100kg ha⁻¹)	–		–		–	–	–	–
Rhizobium (0.27kg ha⁻¹)	–		–		–	–	–	–
TSP (100kg ha⁻¹) + *Rhizobium* (0.27kg ha⁻¹)	–		–		–	–	–	–
Cowpeas: 2004 LR + 2004 SR (n = 2)								
Control	1100	(115)	131	(28)	1.8	–	–38	4.8
Rhizobium[9] (0.17kg ha⁻¹)	1175	(206)	149	(53)	1.9	9.6	–40	4.8
TSP (104kg ha⁻¹)	1387	(322)	171	(86)	1.9	2.1	–48	4.8
TSP (104kg ha⁻¹) + *Rhizobium*[10] (0.17kg ha⁻¹)	1700	(216)	253	(57)	2.3	4.2	–58	5.6

Notes:

1 VCR = (gross value treatment – gross value control)/(variable costs treatment – variable cost control).

2 N balance = partial nitrogen balance (IN1 + IN2 – OUT1 – OUT2).

3 FFS score = 20 points divided over treatments on preference of technology.

4 SR = short rains (October to February).

5 LR = long rains (March to August).

6 FYM = farmyard manure.

7 DAP = diammonium phosphate (an inorganic fertilizer; 18-46-0).

8 TSP = triple superphosphate (an inorganic fertilizer; 0-46-0).

9 No estimation of additional input in nitrogen balance was made.

10 No estimation of additional input in nitrogen balance was made.

Commercial activities and institutionalization

In the second year, all FFSs initiated commercial activities to generate income to cover FFS costs and to test the viability of doing so as a group. The facilitators helped in making contacts to obtain needed inputs (seeds and materials), in formulating business plans and, where necessary, in arranging short-term loans. The following activities were undertaken: growing watermelon and Irish potato, milk processing and marketing (yoghurt), and keeping improved goats. The groups used the cash to meet various group needs: buying inputs to continue commercial activities, creating cash reserves in the group's bank account and paying for hired labour. One FFS employed a community member to manage milk product sales. The scheme to upgrade dairy goats with improved bucks resulted in additional income for the FFS and improved goat herds in the surrounding villages. In addition to generating cash, the small-scale processing plant provided group members with an opportunity to sell their milk at higher prices, bypassing brokers. Each week, the Kibichoi group processed more than 100 bottles of fresh milk into yoghurt for sale, in addition to selling fresh milk and cakes and running a tea kiosk.

The FFSs were registered with the Department of Social Services to facilitate participation of the groups in other rural development programmes. One year after graduation and withdrawal of regular facilitation, all four FFSs were still operational, holding regular meetings and carrying out experimental and commercial activities.

Impact assessment

Knowledge and skills
Households that took part in FFS activities gained more knowledge on soil fertility management and were aware of more types of management practices to address declining soil fertility than were non-participants in FFSs (see Table 18.4). Prior to joining the FFS, 75 to 90 per cent of all households reported having conducted on-farm trials. During the season of assessment, almost all FFS households were conducting one or more trials on their farms, versus 52 per cent of the non-participants in FFSs (see Table 18.5). FFS households engaged in a wider variety of trials beyond the common testing of crop varieties and planting methods. The considerably lower proportion of FFS members reporting to have conducted crop variety trials before joining the FFS (43 per cent) compared to the non-participants (81 per cent) is difficult to explain. It could be that, at the time of assessment, FFS farmers associated experimentation with soil fertility issues and 'forgot' their regular experimentation with crop varieties. Although the assessment provided little information about the quality of the learning process, observations during FFS meetings gave a positive impression. For instance, most FFS members could explain major soil

Table 18.4 *Technologies perceived by households to address soil fertility decline: Comparison between farmer field school members and non-members (percentage of households mentioning particular technology)*

Technology	FFS (n = 80)	non-FFS (n = 31)
Fertilizers	76	81
Manure	75	87
Terraces/grass strips	50	48
Tithonia	18	0
Compost	35	0
Crop residues	11	3
Crop rotation	9	10
Double digging	29	19
Green manure	4	0
Agroforestry	7	0
Mulching	6	3
Lime	3	10
Average number of technologies/farms	3.5	2.8

Table 18.5 *Households conducting on-farm experiments: Comparison over time for farmer field school members and non-members (percentage of households conducting experiments)*

	Before FFS/last three years		After FFS/currently	
	FFS (n = 80)	non-FFS (n = 31)	FFS (n = 80)	non-FFS (n = 31)
Farm households experimenting (%)	75	90	98	52
Average number of experiments/farm	1.6	1.5	1.7	1.4
Type of experiment (%):				
Rhizobium	0	0	24	0
Manure/fertilizer	0	0	11	0
Fertilizer (+ ridges)	22	27	13	26
Tithonia	0	0	10	0
Manure	7	15	10	0
Crop varieties	43	81	29	61
Planting method	16	12	20	17
Double digging	3	13	10	0
Composting	8	4	5	0
Tumbukiza Napier	0	0	22	0
Vegetables/spices	0	0	10	19

fertility processes such as the role of nitrogen, phosphorus and potassium (NPK) in crop growth and how *Rhizobium* increases N available to crops.

Changed practices
All households reported changes in soil fertility management practices over the previous three years, illustrating the dynamics of smallholder farming in the region (see Table 18.6). However, the FFS households reported considerably more changes and more diversity in types of adopted practices than did the non-FFS households. Some technologies tested in the FFSs were well adopted by members – for example, application of *Rhizobium* (75 per cent of the households) and *Tithonia* (45 per cent) in Mbeere and double digging (65 per cent) and *Tumbukiza* Napier (40 per cent) in Kiambu. Other technologies tested were not adopted by many farmers, such as *Rhizobium* in Kiambu (60 per cent of the households because it was not locally available and/or not economical) and TSP application in Mbeere (55 per cent of the households because of unavailability and high costs). The FFS activities also led to other changes in management practices, such as in livestock husbandry and feeding (40 to 60 per cent of the households), record-keeping (40 per cent of households in Kibichoi) and early planting. All farms reported changes in cash-generating activities, with no difference between FFS members and non-members (see Table 18.7). Vegetable production (kale, watermelon and tomato) was an important new activity at all sites. In Mbeere, fruits, goats and khat (*Catha edulis*) were new activities, as were dairy cattle, goats and poultry in Kiambu. The FFS-supported yoghurt and cake/jam making activities were adopted widely by farmers in Kiambu. The impacts of the FFSs outside the group were limited to the village where the FFS was located. Most households in neighbouring villages knew of the existence of the FFS, but received little information and adopted few technologies originating from the FFS (see Table 18.8).

Farm-level impacts
The majority (>90 per cent) of households reported higher yields and financial returns as a result of adopting new soil fertility management practices. Adoption of new cash-generating activities contributed to increased income and food security. The additional income was used mainly to buy food items (60 to 80 per cent of the households), non-food items (25 to 30 per cent) and for school fees (15 to 20 per cent). Investments in agriculture (inputs and hired labour) were reported by 10 to 20 per cent of the households in Kiambu and 80 per cent of those in Kamugi.

Livelihood impacts
Most households observed a positive trend in livelihood aspects (health, soil fertility, water, cash flow, reserves for catastrophes, networks and relations, the role of women in decision-making, access to markets, food security and diversity in sources of income) over the previous three years

Table 18.6 *Soil fertility management practices adopted since farmer field school participation (or for farmer field school non-members) compared to three years ago (percentage of households mentioning type of management practice)*

Soil fertility management practices	Kibichoi		Ngaita		Munyaka		Kamugi	
	FFS (n = 19)	non-FFS (n = 9)	FFS (n = 19)	non-FFS (n = 6)	FFS (n = 25)	non-FFS (n = 9)	FFS (n = 17)	non-FFS (n = 7)
Rhizobium	11	–	–	–	76	–	71	–
Manure	63	56	26	50	64	56	47	57
Fertilizer	68	44	63	67	48	67	53	57
Tithonia	53	–	–	–	40	–	47	–
Manure/fertilizer	–	11	16	17	24	22	6	–
Crop residues	–	–	11	–	20	–	6	–
Mulching	11	11	32	–	12	–	–	–
Ridges	–	–	–	–	4	33	–	–
Terraces	32	11	16	–	8	33	6	71
Compost	42	22	47	–	8	22	12	–
Double digging	68	11	84	–	4	11	24	29
SWC[1]	–	–	–	–	4	11	12	–
Tumbukiza Napier	11	–	42	–	–	–	–	–
Agroforestry	–	–	16	–	–	–	–	–
Crop rotation	5	–	16	–	4	–	–	–
Planting method	–	–	5	–	–	11	–	–
Average number of practices/farms[2]	3.7	1.9	3.8	1.3	3.3	2.8	2.9	2.3

Notes:
1 SWC = soil and water conservation.
2 Maximum of four practices/farms were recorded.

Table 18.7 *New commercial activities since farmer field school participation (or for farmer field school non-members) compared to three years ago (percentage of households mentioning type of commercial activity)*

Commercial cash-generating activities	Kibichoi		Ngaita		Munyaka		Kamugi	
	FFS (n = 13)	non-FFS (n = 9)	FFS (n = 9)	non-FFS (n = 6)	FFS (n = 25)	non-FFS (n = 9)	FFS (n = 17)	non-FFS (n = 7)
Farms with new activities	100	100	100	100	96	89	82	86
Maize and/or beans	15	11	11	–	50	–	57	–
Fruits (mango/pawpaw/passion)	–	–	–	–	33	38	21	33
Livestock (poultry/goats)	31	44	22	17	21	–	21	–
Butternuts	–	–	–	–	13	13	–	–
Cassava	–	–	–	–	13	–	–	–
Khat	–	–	–	–	–	25	50	50
Tobacco	–	–	–	–	4	50	–	–
Dairy cattle	46	44	67	83	–	–	–	–
Vegetables (kale, melon and tomato)	77	55	44	–	47	63	21	67
Coffee	–	–	22	17	–	–	–	–
Cut flowers	–	–	11	17	–	–	–	–
Bananas	23	11	–	–	–	–	–	–
Average number of activities/farms	2.3	1.9	2.0	1.7	2.0	2.0	2.0	1.7

Table 18.8 *Dissemination of information from the farmer field school members to non-members within villages and to neighbouring villages (percentage of households responding positively to indicated statements)*

	In same village (n = 14)	In neighbouring village (n = 17)
Farmer field school (FFS) is a major source of information	86	42
Aware of existence of FFS	100	66
Technologies adopted from FFS	76	32
Information received from FFS	78	51
Willingness to start/join FFS	87	95

(60 to 75 per cent of respondents were positive on these aspects). The FFS households noted a generally positive contribution of these activities to their livelihoods, with a low score only on health aspects.

Farmer field school methodology
The FFS members (90 to 100 per cent) evaluated positively all of the activities in the FFS, but only about 75 per cent gave a positive rating on commercial activities, in respect of which many farmers would have preferred more attention. In two FFSs, problems with leadership were encountered because of poor financial transparency, and new elections were necessary so that activities could continue smoothly. More attention to time management was needed: duration of meetings and long decision-making processes were noted as negative points. All FFS members expressed a willingness to continue with FFSs; 50 per cent of respondents wanted to focus on developing commercial activities, 20 per cent on group savings and only 15 per cent on research and technology development.

DISCUSSION AND CONCLUSIONS

The learning in FFSs had a positive impact on the members' knowledge, skills and capacities to experiment and innovate. The households were selective in adopting the technologies they tested. Seven of the crop-related technologies tested (manure, fertilizer, composting, double digging, *Tumbukiza* Napier, *Tithonia* and *Rhizobium*) and modified livestock husbandry and feeding practices were adopted by 40 to 70 per cent of the farmers. As a result, households reported higher yields and financial returns, while partial soil nutrient balances showed less nutrient depletion. Since impact was assessed only one year after FFS facilitation ended, no information could be gathered about farmers abandoning newly adopted technologies. Another assessment after two to three years could provide valuable information about the sustainability of technology adoption.

The major adaptations made in the FFS approach – long-term group process, flexible type and frequency of activities, on-farm experimentation in addition to central-plot experimentation, and no initial grants – appeared suitable for addressing soil fertility management in complex smallholder farming systems. One year after facilitation ended, all four FFSs were still operating. Implementation of joint commercial activities was the dominant driving force for sustaining the group process, rather than learning and innovation on soil fertility issues. Although experimentation on the individual farms and the central plot continued, the FFS activities were commercially focused. In field experiments, farmers did not perceive risk of yield loss as a major constraint, while farmers were very risk averse in experiments involving livestock. Aspects of risk should receive more attention in the FFS process.

The potential impacts of FFSs extend beyond participatory learning and innovation in farm management. FFSs can be regarded as a stepping stone to empowering rural people. Striking illustrations are men and women farmers confidently presenting the results of experiments during FFS meetings, FFS members sharing experiences and expressing their needs during meetings with district-level policy-makers, and initiatives in group-based commercial activities. Experiences in this project showed that development activities leading to improved income and livelihoods were taken up by well-functioning community groups. In Africa, where the degree of organization of rural people is low, policy-makers, education specialists and private-sector partners should give high priority to facilitation of farmer organization.

The viability of FFSs engaging in a wide range of activities (innovation and learning, commercial activities, group savings, etc.) requires good management skills within the group and calls for flexible and multi-disciplinary support from service providers. The leadership problems encountered in two FFSs indicate the need for more attention to leadership and group management issues. Facilitation of FFSs is provided mainly by agricultural experts from extension or NGOs, supported by researchers. A wider array of FFS activities also calls for other types of support, such as in marketing, processing, cooperatives and micro-finance management.

Although learning, experimentation and observation are endogenous processes in many farm households in East Africa, the role of outsiders such as extensionists and researchers is essential to provide the necessary impulses for a dynamic process of innovation that meets the demands of smallholders in a quickly changing environment. Long-term relationships with research and service providers are necessary for an effective farmer-led innovation process.

Links to markets and inclusion of commercial activities are essential for the long-term sustainability of FFSs. In the life-long learning process envisaged in this approach, the FFSs need to generate cash to cover the costs of group activities and service providers. The synergy achieved in the FFS approach through strengthening farmer organization, linking farmers to markets, empowering rural people and stimulating experimental learning is an example of a sustainable and effective farmer-led process of innovation in smallholder agriculture in East Africa.

The study shows relatively limited diffusion of knowledge to non-FFS households. This raises questions about the role of FFSs in extension strategies. The results of this project suggest that FFSs are cost effective compared to other approaches. Kenya, Tanzania and Uganda have included the FFS approach in their national research and extension strategy. Up-scaling the experiences and the required enabling conditions are priority issues to be addressed by national policy-makers and the international development community.

ACKNOWLEDGEMENTS

We gratefully acknowledge the enthusiastic participation of all household members in the four FFSs and the extension staff of the MoA. We also thank the European Commission and the Dutch Ministry of Agriculture, Nature and Food Quality for financial support.

This study refers to unpublished data contained in numerous internal papers and reports of the Integrated Nutrient Management to Attain Sustainable Productivity Increases in East African Farming Systems (INMASP) project, which operated from December 2001 to March 2006 in Ethiopia, Kenya and Uganda. This project was implemented by the Agricultural Economic Research Institute (LEI) of Wageningen University and Research Centre (WUR) in The Hague, The Netherlands, in collaboration with the National Agricultural Research Foundation (NAGREF), Athens, Greece; ETC East Africa, Nairobi, Kenya; the Kenya Agricultural Research Institute (KARI), Nairobi, Kenya; Debub University, Awassa, Ethiopia; SOS-Sahel, Addis Ababa, Ethiopia; Environmental Alert, Kampala, Uganda; Makerere University, Kampala, Uganda; and the Animal Sciences Group of WUR in Lelystad, The Netherlands. The project was co-financed by the European Union and the Department of Knowledge (DLO) of The Netherlands Ministry of Agriculture, Nature and Food Security.

REFERENCES

Braun, A., Jiggins, J., Röling, N., van den Berg, H. and Snijders, P. (2006) A Global Survey and Review of Farmer Field School Experiences, Report prepared for International Livestock Research Institute, Endelea, Wageningen, The Netherlands

Davis, K. (2006) 'Farmer field schools: A boon or bust for extension in Africa?', *Journal of International Agricultural and Extension Education*, vol 13, no 1, pp91–97

FAO (United Nations Food and Agriculture Organization) (1998) *Farmers Field School on Integrated Nutrient Management: Facilitator's Manual*, FAO, Bangkok, Thailand

FAO (United Nations Food and Agriculture Organization), ILRI (International Livestock Research Institute) and KARI (Kenya Agricultural Research Institute) (2003) Farmer Field Schools: The Kenyan Experience, Report of the Farmer Field School Stakeholders' Forum, 27 March 2003, FAO, ILRI and KARI, Nairobi, Kenya

Feder, G., Murgai, R. and Quizon, J. B. (2004) 'The acquisition and diffusion of knowledge: The case of pest management training in farmer field schools, Indonesia', *Journal of Agricultural Economics*, vol 55, no 2, pp221–243

Gallagher, K. D. (2003) 'Fundamental elements of a Farmer Field School', *LEISA Magazine*, vol 19, no 1, pp5–6

Hounkonnou, D., Offei, S. K., Röling, N., Tossou, R., Van Huis, A., Struik, P. C. and Wienk, J. (eds) (2004) 'Diagnostic studies: A research phase in the Convergence of Sciences Programme', *Netherlands Journal of Life Sciences*, vol 52, nos 3/4, pp209–210

Kenmore, P. E. (1991) *Indonesia's Integrated Pest Management: A Model for Asia*, Intercountry Programme for Integrated Pest Control in Rice in South and Southeast Asia, FAO, Rome, Italy

Lekasi, J. K., Tanner, J. C., Kimani, S. K. and Harris, P. J. C. (2001) *Manure Management in the Kenya Highlands: Practices and Potential*, Henry Doubleday Research Association, Kenilworth, UK

Minjauw, B., Muriuki, H. G. and Romney, D. (2002) 'Development of farmer field school methodology for smallholder dairy farmers in Kenya', International Learning Workshop on Farmer Field Schools: Emerging Issues and Challenges, 21–25 October, Yogyakarta, Indonesia

NARC (National Agricultural Research Centre) (2004) *Impacts of the Group FFS Activities on the Organizational Capacities of the Farmers: Evidence from Pakistan*, NARC, Islamabad, Pakistan

Pontius, J., Dilts, R. and Bartlett, A. (2002) *From Farmer Field School to Community IPM: Ten Years of IPM Training in Asia*, FAO, Rome, Italy

Quizon, J. B., Feder, G. and Murgai, R. (2001) 'Fiscal sustainability of agricultural extension: The case of the farmer field school approach', *Journal on International Agricultural and Extension Education*, vol 8, pp13–24

Sanchez, P. A., Shepherd, K. D., Soule, M. J., Place, F. M., Buresh, C. G. and Woomer, P. L. (1997) 'Soil fertility replenishment in Africa: An investment in natural resource capital', in R. J. Buresh, P. A. Sanchez and F. Calhoun (eds) *Replenishing Soil Fertility in Africa*, Soil Science Society of America and ICRAF, Special Publication 51, Madison, WI

Smaling, E. M. A., Fresco, L. O. and De Jager, A. (1996) 'Classifying, monitoring and improving soil nutrient stocks and flows in African agriculture', *Ambio*, vol 25, no 8, pp492–496

Tripp, R., Wijeratne, M. and Piyadasa, V. H. (2004) 'What should we expect from FFS? A Sri Lanka case study', *World Development*, vol 33, no 10, pp1705–1720

UPWARD (Users' Perspectives with Agricultural Research and Development) (2003) *Farmer Field Schools: From IPM to Platforms for Learning and Empowerment*, UPWARD, International Potato Centre, Los Baños, The Philippines

Van Beek, C. L., De Jager, A., Onduru, D. D. and Gachimbi, L. N. (2004) *Agricultural, Economic and Environmental Performance of Four Farmer Field Schools in Kenya*, Integrated Nutrient Management to Attain Sustainable Productivity Increases in East African Farming Systems Project Report 18, Wageningen University and Research Centre, Wageningen, The Netherlands

Van de Fliert, E. (1993) 'Integrated pest management: Farmer field schools generate sustainable practices: A case study in Central Java evaluating IPM training', *Wageningen Agricultural University Papers*, vol 93, no 3, Pudoc, Wageningen, The Netherlands

Van den Berg, H. (2004) IPM–FFS: A Synthesis of 25 Impact Evaluations, Report prepared for Global IPM Facility, Wageningen, The Netherlands

From Strangler to Nourisher: How Novice Rice Farmers Turned Challenges into Opportunities

Geoffrey Kamau and Conny Almekinders

INTRODUCTION

Research and development work on rice in Kenya prior to 1999 was mandated to the National Irrigation Board (NIB), the government agency in charge of irrigation schemes in the country. NIB provided services at a cost charged on the farmers' produce at the end of the season. These services included land preparation, supply of production inputs, infrastructure maintenance, water abstraction regulation, and research and extension services (Nguyo et al, 2002). During 2000, NIB's authority was challenged after a farmers' protest demonstration at the Mwea Rice Irrigation Scheme (MRIS) of Kenya, which led to loss of lives and property (Kabutha and Mutero, 2001; Nguyo et al, 2002). Consequently, NIB stopped providing services to the scheme's farmers, who took over control of their rice cultivation. This also led to the emergence of 'out-of-scheme' rice cultivation by novice farmers in stream and river valley bottoms formerly infested with reeds and papyrus vegetation, which was illegal according to NIB bylaws (GoK, 1967; Wangui, 2000). Growing rice in this niche marked the beginning of the *jua kali*, or 'informal' rice system.

Jua kali rice cultivation started with little or no available technical information apart from the farmers' assumption that their fields were suitable for rice cultivation because water was available in the swamps. This group of novice farmers tried out rice-growing practices acquired from the MRIS farmers through direct contacts, hiring-in casual labourers or offering themselves as labourers in the scheme to gain experience. Such acquired practices from the scheme did not necessarily work in the new river valley fields owing to differences in the two environments. The farmers therefore had to adapt what they learned from the rice scheme to their new niche through trial and error, which eventually resulted in a thriving rice production system. The innovations included seed pre-treatment, fertilization and input acquisition, marketing practices, and farmer organization.

Noteworthy among the innovations was one to control the previously unknown water weed *Azolla* spp that invaded the rice fields in the year 2000 to 2001. Efforts by the farmers to control the weed by manual removal were rendered futile owing to the weed's rapid spread. The floating weeds would suffocate any transplanted rice plants, causing spindly growth and eventual death. This led the farmers to liken the infestation symptoms to those of the human disease AIDS – hence, the name *kaukimwi*, or 'little AIDS'.

This chapter describes the farmers' experiences in dealing with this weed and other production constraints using their own innovative and adaptive capacity, without any extension or research support. This is juxtaposed to a conventional research approach in which researchers attempted chemical control on the weed. The chapter argues for the need by researchers to recognize and utilize farmers' innovative capacity as the initial building blocks to develop relevant technologies, instead of introducing completely new technologies. It advocates the use of researchers' input for 'feeding' local technical innovation by linking farmers with sources of new ideas, as argued by Loevinsohn (1990). Resultant technologies in such situations are more likely to be pertinent and appropriate to farmers' circumstances, which are dynamic in nature.

LITERATURE REVIEW

Rural development in Africa has been constrained because change agents or development agents have been transferring external knowledge, without recognizing the local knowledge and development initiatives of the farmers (Mbithi, 1994; Veldhuizen et al, 1997). This view has its roots in the development-from-above paradigm grounded in neoclassical economic theory (Walter and Taylor, 1981). Based on this paradigm, the transfer-of-technology model assumes that rural change is exclusively technological and all that is required is an emphasis on farmers' technical mastery of the physical environment to ensure success, which is assumed to improve as the level of farming technology advances. Technology was expected to change society, while the technology remained unchanged. This view has been challenged by the realization that technology is normally not applied in a vacuum. In addition, it has been shown that farming activities are social activities and that new technologies often tend to increase social problems rather than solve them, and their rejection is often a dismissal of their social implications (Mbithi, 1994).

In this context, men and women farmer innovators who take their own initiative to change local agriculture should be considered key allies in agricultural development. The 'inventive self-reliance' of small-scale farmers who continuously experiment, adapt and innovate has been well documented by Chambers et al (1989), Richards (1989), Veldhuizen et al

(1997) and others. These findings add support to the development-from-below paradigm based on maximum mobilization of each area's natural, human and institutional resources and the use of appropriate technology rather than the highest technology (Walter and Taylor, 1981). This view renders support to the fact that people operating in a given agro-ecological setting have a good sense of local risks and possibilities. They are able to judge the appropriateness of new ideas through informal experiments, the generation of new ideas and practices, and the adaptation of others' ideas to their own conditions (Prain and Fujisaka, 1998).

Farmers are keen to obtain appropriate information to solve their problems and particularly value information from others working under comparable conditions. This is especially evident in exchange visits between farmers, where they observe and later adapt what they see to fit their local contexts. Extension agents and researchers need to become more aware of local creativity to be able to stimulate it. Policy-makers, on the other hand, need to be exposed to convincing information from both farmers and scientists working with them. Conditions for land husbandry are constantly changing. Therefore, innovation and joint learning must be a continuing process, where innovation in this context includes both technical and institutional change (Leeuwis and Van den Ban, 2002).

The process of innovation involves various stakeholders who generate, adopt and adapt novel ideas, approaches, technologies or ways of organizing, and builds on the capacity of participating stakeholders, including farmers (Biggs and Matsaert, 2004; Kaaria et al, 2004). This creates a sustained collective capacity focused on improving livelihoods and the management of natural resources. The enhanced capacity for innovation enables rural people to develop new technologies, products and markets, and ways of organizing, as well as policies and institutional arrangements, which catalyse and enable innovativeness. The interaction of systems in the farmer's environment is therefore critical for the successful adoption of any innovation, and understanding the system as a whole is important in order to effect changes (Dillon and Hardaker, 1993; Goncalves, 1995). A versatile approach to working with communities hinges on several related elements, such as farmer experimentation, social and human capital formation, access to information and leadership, and entrepreneurship. Extensionists and researchers 'feed' local innovation by providing farmers with links to external knowledge, ideas to explore and options to test (Loevinsohn, 1990). They can also facilitate communication between farmers, who examine local innovations, discuss the advantages and disadvantages, and consider who would like to try them out.

A major handicap to the innovation process is the attitude of extensionists and researchers, who assume rigid roles in the development and extension of new technologies, of which the farmers are assumed to be passive users. This is a view that limits the actors and, hence, curtails the innovativeness that is expected in agricultural production systems characterized by changing circumstances. Enhancing farmer experimentation could help to develop site-appropriate technology more quickly and, in turn, strengthen local capacities to adapt to new conditions (Haverkort et al, 1991; Veldhuizen et al, 1997). This issue is, however, inadequately addressed in many agricultural research organizations, including those in Kenya, in spite of overwhelming evidence of farmers' innovative capacity. This is illustrated by this study conducted in 2003 and 2004 as part of a wider study on researcher–farmer information exchange in the Kenyan public agricultural research system, using two Kenyan Agricultural Research Institute (KARI) research centres as case studies (Kamau, 2007). The *jua kali* rice system was selected as an emerging agricultural innovation system. It provided a chance to study how farmers' innovations arise and how they are shared, enriched and improved as they are exchanged between and among farmers operating in different contexts.

METHODOLOGY

Location of study

The study was conducted in central Kenya and covered six villages spread over four locations of Ndia Division in Kirinyaga District. The study area was divided into three clusters based on land-ownership categories, labour sources and information-flow systems. In East Kagio, farmers hired or owned land for rice cultivation and had frequent interaction with MRIS farmers and casual labourers. Northeast Baricho consisted of farmers who owned land for rice cultivation, obtained their information from outside the cluster, but also used their own experiences. Southwest Baricho consisted of farmers who own land and generated most information from their own experiences. Unlike the other two clusters, the farmers in this third cluster did not use hired labour.

Sample size and study approach

A total of 92 farmers were interviewed in the three clusters through individual farmer interviews, key informant interviews, focus group discussions and observations:

- *Individual farmer interviews:* farmers in the three clusters were interviewed as they worked in their fields. This allowed the detailed examination of various ongoing issues and the identification of farmers to be involved in focus group discussions.
- *Key informant interviews:* these interviews involved key players in the system, such as rice traders, agricultural-input suppliers, middlemen, group leaders, area extension agents and researchers from the NIB rice research substation.

- *Focus group discussions:* these involved discussions with selected farmers, nominated by other farmers, on the basis of their experience in certain aspects of rice cultivation. Using a checklist, discussions were held in groups of six to eight farmers involving a total of 28 farmers.
- *Observations:* in the course of the study, the activities and interactions of the farmers were observed. Actors in the rice system such as brokers, traders and input suppliers were also observed. These observations supplemented the information gathered from the interviews.

FINDINGS AND DISCUSSIONS

Rice varieties grown

The main rice cultivars grown in this area are IR 2793 (commonly called 'small B' by the farmers), Sindano BW 196 (called *bkubwa*, or 'big B') and basmati (called *pishori*), while a few farmers grew mixed varieties (see Figure 19.1). The majority of the farmers started by growing variety small B, which performed well initially but deteriorated after a few seasons on account of decreased soil fertility. At the time of the study, only a few farmers in one cluster were growing it (see Figure 19.2), while the rest had changed to *bkubwa* and *pishori*, which continued performing well on the low-fertility soils. The deterioration was manifested in yellowing of the rice foliage and poor grain setting.

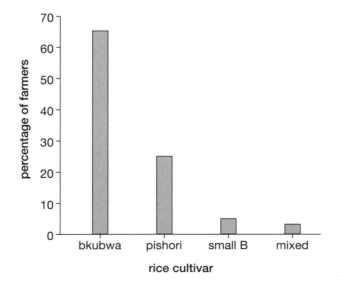

Figure 19.1 *Rice cultivars in* jua kali *rice*

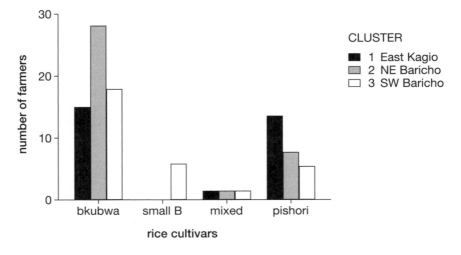

Figure 19.2 *Cultivars by clusters*

Land preparation

Preparation of land involved clearing of reeds and other water plants from the fields, followed by hoe digging to remove the roots. Structures called *kipandes* (pieces) measuring 10m × 10m were constructed by heaping soil clods on the sides of the earmarked portions, forming polders. The embankments served both as water reservoirs and as drainage structures. A corner of this plot would be earmarked for locating the nursery, which later was observed as a yellow patch amidst green rice.

In some cases, the nursery was established outside the polder, with its size dependent upon the extent of land to be transplanted and the quantity of available seed. The soil in the nursery was made into a fine tilth with a mixture of goat or chicken manure and ash added to it. Cattle manure, according to the farmers, was too coarse for the seedlings, while fertilizers would scorch the fragile seedlings.

Pre-germination and nursery planting

Rice seeds were pre-germinated before seeding in nurseries, using one of three methods. The first method was to soak seeds in water, pack them in a sisal bag and bury the bag in a hole on top of which a fire was lit to provide warmth. The bag was turned on the second day to allow for even distribution of warmth to hasten germination. In the second method, seeds were packed in a bag and buried in manure heaps, where warmth would stimulate sprouting. The third method involved covering bag-packed seeds with heaps of rice straw. These three methods were farmer innovations and differed from the method recommended by rice experts of soaking seeds in water for 24 hours, followed by covering them in rice straw for 48 hours. According to the farmers, 'the hole and fire method gives very good germination and is faster than the rice straw method recommended by extension owing to low night temperatures' (farmer interviews, 2004). Pre-germinated seeds would then be broadcast in the nursery and, after a month, the seeds would be ready for transplanting, followed by fertilizer application, pest control and weeding.

Transplanting and other field operations

Before transplanting, the excess water was drained from the rice fields, leaving wet muddy fields behind. Transplanting involved a synchronized operation of digging a hole by hand in the soft wet ground and placing a seedling in it. This was followed by re-flooding of the fields until the appearance of *Azolla* spp in the year 2000. The weed formed dense mats on the water surface and choked the rice plants submerged in the water.

Fertilization involved application of urea and di-ammonium fertilizer to the seedlings a few weeks after transplanting, with each 10m × 10m plot receiving 1kg of fertilizer. This unit of measure contrasted with the researchers' hectare/acre-based unit. Pest control and weeding operations took place at about the same time as fertilization. The major pests were leaf cutters and stem borers. The farmers scouted for the presence of leaf cutters by checking for leaf pieces floating on the water surface. Weeding involved manual removal of the weeds by hired or family labour and heaping them on the sides of the polders. In some cases, the weeds would be buried in the mud 'to rot and provide food for the rice plants', as stated by the farmers.

Prior to the emergence of the *Azolla* weed, the level of water after trans-planting was not of concern to the farmers. The dense mats of weeds that covered the water surface, however, changed the situation. Farmers put great efforts into removing the weed from the water surface. However, the weed would resurface within a short time and suffocate any rice seedlings growing beneath the water surface. The stifling and subsequent death of seedlings, as well as the futile control efforts, were likened to the human scourge AIDS. A fortuitous observation led to a discovery that provided a solution to the *Azolla* challenge. This happened through water-level reduction in one rice-growing cluster because excessive use of water upstream led to insufficient quantities of water in the polders. The water level remained low and the floating weed stayed below the canopy of the rice seedlings. In cases of complete water deficiency, *Azolla* covered the ground under the rice seedlings. The farmers noticed that weeds remaining below the rice plants did not interfere with the rice crop, while all the fields that had weeds covering the ground did not suffer from moisture stress. This led to the discovery that by controlling the water level at transplanting time, the weed could co-exist with the rice seedlings and have no adverse effect on rice yields. A concurrent observation was that, when the *Azolla* was collected and heaped on the side of the field, it decomposed rapidly and led to vigorous growth of plants around such heaps. Following this observation, the farmers started utilizing the *Azolla* compost in combination with rice straw mulch on crops, such as kale and tomato, leading to high yields and reduction in watering frequency from four to two days a week. At weeding time, *Azolla* buried in the rice plots was also found to stimulate a vigorous rice crop.

These farmers' experiments contrasted with researchers' attempts to control the weeds, which resulted from the observations on *Azolla* by a weed scientist. Five herbicides were tested at the nearby Mwea Research Centre. The experimental field was prepared by hand at the research farm in September 2001 and *Azolla* was introduced into the plots. The treatments consisted of five herbicides, a weed-free and a weedy check, as well as a conventional hand-weeded treatment. The treatments were arranged in a randomized complete block design (RCBD) and replicated four times. Data on percentage weed control (PCW) were collected 10, 28, 39 and 68 days after treatment application, while percentage crop damage (PCD) was based on a visual scale of 0 to 100 (where 0 = no control and 100 = complete control of the weed). Rice yields from all the treatments were assessed.

The results from this experiment indicated that the weeds reinvaded the field in high numbers in all the treatments, apart from two of the herbicides, which had residual effects. The conclusion, according to the technical report, was that *Azolla* control using herbicides was possible; but weed reinvasion occurred quickly when herbicides with no residual effect were used. The rapid weed regrowth combined with the complexity of the herbicide technology for resource-poor farmers rendered the researchers' efforts futile. In any case, this technology would not have been accepted by the farmers owing to the prohibitive costs and the effective solutions found through their own observation and experimentation.

Rice harvesting and marketing

The farmers harvested the rice three to four months after transplanting, depending upon the variety. The plants were cut with a sickle and tied into bundles. These bundles were then threshed by dashing them on the ground to separate the grain from the chaff. The paddy was then collected in bags. The grain was sold to middlemen or traders in the nearby shopping centre and, at times, stored by farmers until the prices were more favourable. Some farmers also entered into financial contracts with brokers at the beginning of the season, or during the vegetative stage, when money was advanced to accomplish the rest of the field operations. Such money would be recouped from the rice harvests. This, in effect, meant that the crop was sold while it was still in the field. The broker collected his money's worth of paddy and the farmer would be left with the rest. Of the interviewed farmers, 45 per cent were found to have entered into agreements with brokers in the three previous seasons, 35 per cent had sold their rice to traders in the market and 15 per cent had kept their rice until prices appreciated.

Rice straw as fodder and mulch

In the course of exploring rice varieties, farmers encountered non-booting or flowering types that they used for feeding their dairy cows, resulting in an increase in milk yields and improved condition of the animals. According to 10 per cent of farmers interviewed, rice in the vegetative stages and ratoons that develop on leftover rice stubble provide very good fodder for animals. In one case, a farmer left his ready-for-drying-off animal to graze on the ratooning crop and, in his own words, 'an increase of three bottles of milk left me undecided whether to dry it out or not' (farmer interview, 2004). This led to the discovery that green rice plants are good for dairy cows. Animal nutrition specialists explained that this was due to the abundant proteins and other nutrients at this stage of crop growth (researcher interview, 2004).

Information sources

Farmers gathered information, particularly on nursery establishment, transplanting and threshing, from various sources. Thirty per cent of the interviewed farmers obtained their information from hired labourers, 10 per cent from offering their labour to experienced farmers, and 15 per cent from trying out a practice and then comparing the results with their neighbours (see Figures 19.3 and 19.4). In all cases, however, a substantial amount of information used by the farmers came from their own experience. Farmer-to-farmer exchange was a key mechanism for flow of information through local rice-growing support groups. These groups were formed out of other groups that used to meet early in the morning to discuss issues related to irrigation water. Elderly farmers played key moderating positions in these groups, while young and middle-aged farmers in the groups scouted for new information from outside their immediate areas. In most cases, these farmers were also the first ones to experiment with new ideas on their own farms.

DISCUSSION AND CONCLUSIONS

Several authors hold the view that successful innovation systems are those in which institutions facilitate flows of information in a 'bazaar' approach (Douthwaite, 2002; Biggs and Matsaert, 2004). As illustrated in the current study, the flow of information between the farmers and other actors led to widespread utilization of technical innovations that were not initially available. As documented by Stolzenbach (1994) and Wolley

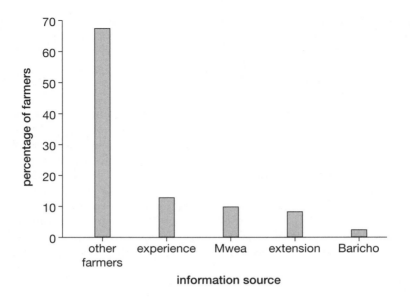

Figure 19.3 *Rice information sources*

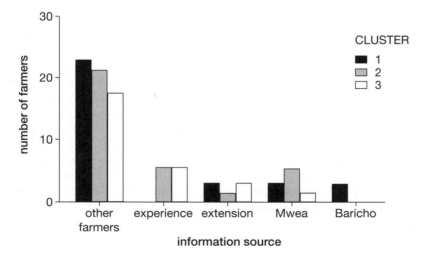

Figure 19.4 *Information source by cluster*

(2002), many decisions that farmers, artisanal fisherfolk and others make are based on years of careful observation and research. The research takes the form described as 'adaptive performance' by Richards (1989), where the research design shifts during experimentation depending upon the farmers' perception of his or her best options. The farmers in this *jua kali* rice system made their observations and then decided to switch varieties because of deteriorating rice performance over time. Another example is the observation on *Azolla,* where farmers discovered the water-level control technique. This agrees with Biggs and Clay (1981) and Reece and Sumberg (2003), who argue that farmers will innovate within the limits of their technical capacity to solve problems of simultaneous adaptation, fine-tuning it in an attempt to fit it into the physical and socio-economic aspects of an environment.

This innovative capacity of farmers seems to receive a lukewarm reception by researchers, as illustrated by an interview with a former rice researcher. He felt that the *jua kali* rice farmers had poor yields because of weak agronomic practices and poor varietal selection. While this view is informed by rice varieties and production in a different environment, it is an indicator of the views and mindset of many researchers that high grain yields are the only objective of any grain variety. As observed and evidenced by the farmers' own information, so-called low-yielding rice varieties have been converted into fodder, which improves their livestock. The 'low yields' are also innovatively used by the farmers in a speculative manner and stored until prices improve. The farmers, in fact, acquired food and cash from areas that were deemed unsuitable for crop growth.

As evidenced by the researchers' approach to the weed infestation, solutions that may work for one situation do not always work in a similar situation, but in a different context. The herbicides tested by the researchers in efforts to control the weed performed dismally, whereas the farmers' practical solution based on observation and related to water management was effective. Vigorous plant growth from using *Azolla* compost was also observed, confirming the findings of Ventura and Watanabe (1993), Galal (1997) and others about the utility of the *Azolla* weed as a source of nitrogen. The farmers' experiences with the water weed confirm the synergy hypothesis advanced by Sumberg and Okali (1997), who advocate for drawing on multiple sources of innovation. This is due to the realization that farmers have an intimate knowledge of their local environment, conditions, problems, priorities and criteria for evaluation, which is often out of the reach of outsiders. Similarly, the results of formal agricultural research are often inaccessible or inappropriate for farmers.

The findings of this study concur with Bebbington et al (1994) that investments have to be made in local farmers' innovation capacity if research organizations are intent on being client-driven. This is further emphasized by Collinson (2001), who states that traditional applied agricultural experimentation promotes allegiance to commodities and disciplines, thus slowing progress in improving the relevance of research output to smallholders. Farmers' experiments can be a basis for collegial relationships between farmers and researchers, where conventional and participatory research play complementary roles (Thiele et al, 2000). Good participatory research should improve the relevance of conventional research, and good participatory research without strong conventional research to back it up will not be effective. However, putting this into practice still remains a challenge in many research organizations. Organizational and managerial problems are often encountered in trying to operationalize end-user participation with an emphasis on the identification and dissemination of 'successful' examples and 'best practice' (Veldhuizen et al, 1997; Sumberg et al, 2003). Many researchers are determined to recognize participatory research, where farmers' innovative capacity is given its due place. They use experiences of innovation, such as the informal rice-growing system documented here, as springboards to successful participatory research.

As farmer and research environments change rapidly, it is important for both farmers and researchers to modify their roles and attitudes towards improved dialogue with each other. Farmers have been known to innovate, either because they have resources and can take risks, or because they do not have resources and are forced to look for new ways of doing things (Chambers et al, 1989). They can be young with some formal education or old without any formal education, and include both men and women. Farmers may innovate if they were used to a certain way of doing things and circumstances have forced them to do things in another way in order to survive. In this study, the innovating farmers were of mixed education levels and resource endowment. It is apparent that this did not prevent them from being innovative. These aspects should be considered by change agents when introducing new innovations into an area. It is also important to consider how farmers use the information generated. In this case, farmers have used the innovation to obtain food and cash and to convert the valley bottoms into productive niches. They have also innovatively acquired nutrients for their crops and fodder for their livestock. The diversity in utilization of the rice crop makes the innovation more appealing and adds value to the farmers' activities.

Through the advent of this innovation, new systems of organizing have also been developed by the farmers in the form of water-user groups turned into rice-support groups, who discuss rice cultivation and other issues. Contrary to the perceptions of extensionists and researchers, who schedule their farmer meetings in the afternoons, the farmers hold early morning meetings, thus allowing themselves sufficient time to get through their other chores during the rest of the day.

Professionals in research organizations who are intent on improving or increasing the impact of their work with farmers have to recognize farmers' propensity for innovation. This capacity of farmers arises out of living through situations that they encounter every day in their fields. Researchers have to understand the how and why of farmers' practices before introducing new practices with the assumption that they will work as well in the given setting as they have elsewhere. Furthermore, researchers need to recognize farmers' socio-organizational and other innovations that may support any new technical innovation introduced to them. Finally, it emerges from this account that the statistically analysable technical data that researchers primarily collect may not be the only data that are important. There is a need to 'make in-roads' through the confines of statistically analysable data in order to understand that farmers' problems may not only be solved by stand-alone technical innovations. These are issues that must be embraced if participatory research efforts being made by many research organizations are to bear any fruit. This may not happen without institutional changes in the management and policies governing the formal research and development sector that encourage a paradigm shift to build new partnerships and to avoid the business-as-usual attitude.

REFERENCES

Bebbington, A. J., Merrill-Sands, D. and Farrington, J. (1994) 'Farmer and community organizations in agricultural and extension: Functions, impacts and questions', Agricultural Administration Research and Extension Network (AgREN) Paper 47, Overseas Development Institute, London

Biggs, S. D. and Clay, E. J. (1981) 'Sources of innovation in agricultural technology', World Development, vol 9, no 4, pp321–326

Biggs, S. and Matsaert, H. (2004) 'Strengthening poverty reduction programmes using an actor-oriented approach: Examples from natural resources innovation systems', AgREN Paper 134, Overseas Development Institute, London

Chambers, R., Pacey, A. and Thrupp, L. (eds) (1989) Farmer First: Farmer Innovation and Agricultural Research, Intermediate Technology Publications, London

Collinson, M. (2001) 'Institutional and professional obstacles to a more effective research process for smallholder agriculture', Agricultural Systems, vol 69, pp27–36

Dillon, J. L. and Hardaker, J. B. (1993) Farm Management Research for Small Farmer Development, Farm Systems Management Series 6, FAO, Rome

Douthwaite, B. (2002) Enabling Innovation: A Practical Guide to Understanding and Fostering Technological Change, Zed Books, London

Galal, Y. G. (1997) 'Estimation of nitrogen fixation in an azolla-rice association using the nitrogen-15 isotope dilution technique', Biology and Fertility of Soils, vol 24, no 1, pp76–80

GoK (Government of Kenya) (1967) Irrigation Act: Chapter 347 of the Laws of Kenya, revised edition, Government of Kenya, Nairobi, Kenya

Goncalves, V. S. P. (1995) 'Livestock production in Guinea-Bissau: Development potentials and constraints', PhD thesis, University of Reading, Reading, UK

Haverkort, B., Kamp, J. and Waters-Bayer, A. (eds) (1991) Joining Farmer's Experiments: Experiences in Participatory Technology Development, Intermediate Technology Publications, London

Kaaria, S., Kirkby, R., Delve, R., Njuki, J, Twinamasiko, E. and Sanginga, P. (2004) 'Enhancing innovation processes and partnerships', Uganda Journal of Agricultural Sciences, vol 9, pp819–837

Kabutha C. and Mutero, C. M. (2001) 'From government to farmer-managed smallholder rice schemes: The unresolved case of the Mwea irrigation scheme', FAO–IWMI Seminar on Private Sector Participation and Irrigation Expansion, October 2001, Accra, Ghana

Kamau, G. M. (2007) 'Researching with farmers: Farmer participatory research in the Kenya Agricultural Research Institute in context', PhD thesis, Wageningen University, Wageningen, The Netherlands

Leeuwis, C. and Van den Ban, A. (2002) *Communication for Innovation in Agriculture and Resource Management: Building on the Tradition of Agricultural Extension*, Blackwell Science, Oxford

Loevinsohn, M. (1990) 'Feeding farmer innovation', *ILEIA Newsletter*, vol 6, no 1, pp14–15

Mbithi, P. (1994) *Rural Sociology and Rural Development: Its Application in Kenya*, Kenya Literature Bureau, Nairobi, Kenya

Nguyo, W., Kaunga, B. and Bezuneh, M. (2002) 'Alleviating poverty and food insecurity: The case of the Mwea Irrigation Scheme in Kenya: Broadening access and strengthening input market systems (BASIS)', University of Wisconsin, Madison, WI, www.ies.wisc.edu/ltc/live/bashorn0209.pdf

Prain, G. D. and Fujisaka, S. (eds) (1998) *Biological and Cultural Diversity: The Role of Indigenous Agricultural Experimentation in Development*, Intermediate Technology Publications, London

Reece, J. D. and Sumberg, J. (2003) 'More clients, less resources: Toward a new conceptual framework for agricultural research in marginal areas', *Technovation*, vol 23, pp409–421

Richards, P. (1989) 'Agriculture as a performance', in R. Chambers, A. Pacey and L. A. Thrupp (eds) *Farmer First: Farmer Innovation and Agricultural Research*, Intermediate Technology Publications, London, pp39–43

Stolzenbach, A. (1994) 'Learning by improvisation: farmers experimentation in Mali', in I. Scoones and J. Thompson (eds) *Beyond Farmer First: Rural Peoples' Knowledge, Agricultural Research and Extension Practice*, Intermediate Technology Publications, London, pp155–159

Sumberg, J. and Okali, C. (1997) *Farmers' Experiments: Creating Local Knowledge*, Lynne Rienner, Boulder, CO

Sumberg, J., Okali, C. and Reece, D. (2003) 'Agricultural research in the face of diversity, local knowledge and the participation imperative: Theoretical considerations', *Agricultural Systems*, vol 76, pp739–753

Thiele, G., Van de Fliert, E. and Campilan, D. (2000) 'What happened to participatory research at the International Potato Center?', *Agriculture and Human Values*, vol 18, pp429–446

Veldhuizen, L. van, Waters-Bayer, A., Ramirez, R., Johnson, D. A. and Thompson, J. (eds) (1997) *Farmers' Research in Practice: Lessons from the Field*, Intermediate Technology Publications, London

Ventura, W. and Watanabe, I. (1993) 'Green manure production of *Azolla microphylla* and *Sesbania rostrata* and their long-term effects on rice yields and soil fertility', *Biology and Fertility of Soils*, vol 15, no 4, pp241–248

Walter, B. S. and Taylor, F. D. R. (1981) *Development from Above or from Below? The Dialectics of Regional Planning in Developing Countries*, John Wiley and Sons, Chichester, UK

Wangui, K. (2000) 'Mwea rice farmers petition Wako: Irrigation Act repressive', *The People Daily*, 24 September

Wolley, C. (2002) 'They scorn us because we are uneducated: Knowledge and power in a Tanzanian marine national park', *Ethnography*, vol 3, no 3, pp265–298

V

Building Capacity
for Joint Innovation

Strengthening Inter-Institutional Capacity for Rural Innovation: Experience from Uganda, Kenya and South Africa

Richard Hawkins, Robert Booth, Colletah Chitsike, Emily Twinamasiko, Moses Tenywa, George Karanja, Thembi Ngcobo and Aart-Jan Verschoor

INTRODUCTION

Since 2004, research and development (R&D) partners in Uganda, Kenya and South Africa have been working with the International Centre for development oriented Research in Agriculture (ICRA) to improve the capacity for rural innovation. The overall strategy followed in all three countries consists of four main components:

1 building inter-institutional steering groups that increase awareness of the need for collective action and oversee joint capacity-strengthening (CS) initiatives;
2 building a national team of facilitators that designs and implements in-country CS programmes;
3 strengthening the skills of current professionals and the ability of their institutions to work together to promote rural innovation; and
4 reviewing and revising the academic teaching programmes that produce future professionals.

A number of lessons have been learned in the implementation of these programmes that relate to individual and institutional involvement. The integration of CS activities within ongoing programmes, work plans and budgets across multiple organizations – at both local and national level – requires considerably more planning and follow-up than normal 'training programmes', and the intensive sensitization of institutional heads, strategic units and new partners or stakeholders (including funding organizations). Focus and loyalty have to be changed from an institutional basis to one of a collective challenge – not an easy change, especially where there is no dissatisfaction among senior managers with the current state.

In-service capacity strengthening for rural innovation needs to respond to opportunities for adding value to selected multi-stakeholder R&D

programmes, rather than being organized as a stand-alone training activity determined by the training needs of selected individuals or institutions. Improving the competencies of future professionals requires inclusion of 'meta-disciplines', as well as addressing curriculum development procedures, improving teaching methods, and exposing students to interdisciplinary interaction and real-world learning contexts. Above all, students need to recognize that agriculture and rural development are human activities; hence, promoting rural innovation requires social skills and a personal ethos, as well as technical ability.

RURAL INNOVATION SYSTEMS

The linear model of technology development in agriculture, where research generates new technologies that are transferred – and hopefully adopted – by end users, is increasingly seen as a poor picture of rural innovation. Rather, an innovation systems model, in which a variety of individuals and organizations interact in a complex relationship and according to their interests and opportunities, is seen as a better representation of reality (e.g. Spielman, 2005; World Bank, 2006).

Rural innovation systems are complex because of the different interests and organizational levels of the various stakeholders involved. These actors will typically include producers and their organizations, purchasers and processors of rural products, input suppliers, and technical, financial and business services. These stakeholders are organized at household, communal, district, national and, in some cases, regional or even global levels. There are usually competing interests for the scarce natural resources of land and water, for maximizing income flows and for control of the system.

Because of the complexity and competing interests, rural development rarely proceeds through the efforts of one stakeholder or institution alone or by a simple technological 'fix'. Rather, it requires the integration of technological, policy and institutional factors to provide commercial, social and institutional solutions that achieve broad and multiple objectives, including poverty alleviation, environmental protection, social and gender equality, and linking resource-poor farmers to increasingly demanding local and international markets. As Röling (see Chapter 2 in this volume) stated in his keynote address to the Innovation Africa Symposium: 'When innovation is the emergent property of interaction, promoting innovation becomes a matter of facilitating the interaction process.'

Publicly funded research and technical services therefore need to act in partnerships within collective innovation systems: jointly learning, defining development needs and opportunities, and generating and diffusing the knowledge required to realize these opportunities. They need to be able to analyse and manage innovation systems that integrate the different actors and organizational levels, and that respond to changing policy and market environments.

In the last few years, a number of R&D approaches have been introduced to promote collective innovation systems. Among these are Integrated Natural Resource Management (INRM) (ICARDA, 2005), the Competitive Agricultural Systems and Enterprises (CASE) approach (AISSA, 2005) and Enabling Rural Innovation (ERI) of the International Centre for Tropical Agriculture (CIAT)-Africa. Each of these approaches tends to have a different focus (natural resource management, linking farmers to markets, etc.), and they are, in any case, actively evolving as approaches. However, more important than any differences is their common emphasis on improving stakeholder linkages and collective action around a commonly agreed focus.

Within the partnerships described in this chapter, we refer to this integrated approach for collective action as agricultural research for development (ARD) (South Africa) or integrated agricultural research for development (IAR4D) (Uganda and Kenya). For simplicity and the purposes of this chapter, we will use these terms of ARD, IAR4D and 'rural innovation' synonymously.

STRENGTHENING CAPACITY FOR RURAL INNOVATION

Competency in process facilitation

Improving rural innovation processes is not achieved by consulting with farmers, 'beneficiaries' or 'stakeholders' to see what they regard as constraints or needs. Nor is it just a set of stepwise activities that can be followed in all circumstances, or even just applying an integrated R&D approach such as INRM or CASE – the value of these approaches notwithstanding.

The tasks of analysing and managing innovation systems require more than the traditional disciplinary-based technical knowledge and skills that are the basis of most academic programmes and professional form-ation. If, as Röling said, promoting innovation is a matter of facilitating the interaction process, then 'innovation professionals' need to be competent in communication, in facilitation and management of functional partnerships, and in teamwork – often referred to as 'soft skills'.

Strengthening capacity for rural innovation is therefore not just a matter of improving professional knowledge of the underlying concepts or even skills in suitable research methods. Rather, it requires a change in attitude and mentality: a different way of looking at the world, of thinking and analysing, of interacting with others. It also requires a different way of organizing and managing institutions to enable them to work more effectively with other stakeholders and provide incentives for activities that lead to change, to rural innovation, rather than to knowledge generation *per se*. In other words, agricultural research needs a human as well as a scientific face.

Learning to think in new ways and changing attitudes is a complex and slow process which involves questioning assumptions that have underpinned actions and careers to date. It requires some uncomfortable confrontation: both with ourselves and with others. It involves a change from working individually to working with others in ever-changing teams. It involves a shift from teaching to learning, which is especially difficult to make. It cannot be achieved by the same methods that instilled the current paradigms and that are still the basis for much professional development at educational institutions and for the organization and management of many of our professional institutions.

ICRA's experience

ICRA was formally established as the International Course for development-oriented Research in Agriculture in 1981 by European members of the Consultative Group on International Agricultural Research (CGIAR).[1] The working group charged to establish the course specified that 'the training programme should be designed to provide a cadre of agricultural scientists able to apply their specialized knowledge to the development problems of agriculture in developing countries' (ICRA, 1979). The 'flavour' of the courses offered by ICRA has reflected evolving R&D approaches over nearly three decades, which have tended to include an increasing array of stakeholders. These approaches have included, for example, farming systems research (FSR) (Collinson, 2000), farmer participatory research (FPR) (e.g. Okali et al, 1994), rapid appraisal of agricultural knowledge systems (RAAKS) (Engel and Salomon, 1997) and the Sustainable Livelihoods Approach (www.livelihoods.org).

The main evolution of ICRA's strategy, however, has involved a shift from a training focus with individual scientists, to interacting more with institutional partnerships in targeted countries in an attempt to build national capacity. Under this new strategy, participation in ICRA's European programmes is by inter-institutional teams, who are then expected to develop and implement national strategic plans for capacity strengthening in IAR4D.

Fundamentals of IAR4D capacity strengthening

ICRA has increasingly based its CS programmes on a number of fundamental characteristics, as follows.

Basis in real world challenges
Stakeholders in rural innovation systems are not neutral observers, passive beneficiaries or even analysts, but actors with roles to play and interests

to defend. Learning about such systems can take place only in a real world environment where these interests are in play. Even a case study approach cannot replicate the complexities encountered or offer opportunities for trying, reflecting upon and, hence, improving communication and facilitation skills. Learning about rural innovation is therefore best organized around a shared R&D challenge or opportunity, or an agreed entry point, which serves as a 'platform' where stakeholders can come together on the basis of mutual interest and clearly defined institutional roles and commitments. This means locating learning within ongoing programmes and projects – with all their organizational and operational constraints.

Multi-stakeholder and interdisciplinary interaction
Working with other disciplines, institutions and diverse stakeholders is easier in theory than in practice. Each discipline or institution has its own culture: forms of communication, norms and patterns of behaviour. An appreciation of the contributions of different stakeholders or disciplines, and a willingness to listen to and learn from diverse viewpoints are fundamental in rural innovation partnerships.

Teamwork
Learning how to be an effective team member and to lead effective teams does not often come naturally to graduates who have spent years studying alone and producing individual theses. In fact, higher education is a process that tends to select and reward individuals who are happiest working alone. Nevertheless, learning in small groups is more effective for most people than is learning alone. The opportunity to participate and contribute is greater in small groups of five to six individuals than in larger groups or a classroom where a person can easily hide (or go to sleep). Ideas are exchanged and consolidated, and one's own knowledge is therefore more effectively constructed in a small group that maximizes social learning.

Action research learning cycles
Everyone has a different learning style. However, for most people – particularly adults – learning from experience is undoubtedly important. Kolb (1984) expanded on this in his influential theory of the 'experiential learning cycle': concrete experience–reflection–abstract conceptualization–active experimentation. A similar cyclical process of plan–act–observe–reflect–plan, etc. forms the basis of most theories of action research or action learning, so called because it places emphasis on research and learning as a way of modifying action, not just as a means of generating knowledge. As a means of improving action through changing behaviour, the construction of action research cycles within learning programmes for rural innovation is particularly critical.

Lessons from Kenya

In 2004, Kenya launched its ten-year strategy for revitalizing agriculture (MoA, 2004), as well as the Kenya Agricultural Productivity Project. Both of these instruments emphasized recognition and support for coordinated pluralism in R&D, reformed agricultural extension and farmer empowerment.

To support these policies, a multi-institutional IAR4D initiative was formed with the objective of facilitating the integration of IAR4D concepts, approaches and methods within the national agricultural research, extension and education systems. The initiative involves nine institutions: the Kenya Agricultural Research Institute (KARI), the Ministry of Agriculture (MoA), the Ministry of Livestock and Fisheries Development, the University of Nairobi, Egerton University, Kenyatta University, Jomo Kenyatta University of Agricultural Technology, the Kenya National Federation of Agricultural Producers and ICRA. Varying capacities and experiences in IAR4D exist within these different institutes and organizations; but what was deemed to be lacking was a mechanism to link them up, expand and use this knowledge base.

The initiative built upon a long history of engagements between ICRA and Kenya, during which many Kenyans attended ICRA's European ARD programmes. A number of ICRA field studies were conducted in Kenya as part of these programmes.

To manage the initiative, a task force comprising representatives of each of the member organizations was established and is defining a national IAR4D plan. The task force also took the approach of forming and capacitating a strong core team of IAR4D facilitators and advocates. This was achieved through teams of five and eight people from the member organizations who attended the ICRA programme in Wageningen, The Netherlands, in 2005 and 2006, respectively. Apart from the knowledge-acquisition component in The Netherlands, these programmes involved fieldwork conducted by the teams in East Africa.

In 2005, the team conducted fieldwork in Uganda, analysing the nascent apple cluster in Kabale (Turyamureeba et al, 2006). In 2006, the team carried out fieldwork in Kenya, focusing on improved use and management of water, the key natural resource in Katulani, Kitui District. This work attempted to establish linkages between key stakeholders at local and national levels, and both the task force and the members of the 2005 and 2006 teams intend to establish this collaborative initiative as a pilot site for IAR4D, which all the member organizations will use as a continuing learning site.

The core team is currently engaged in a number of activities to increase awareness and understanding of IAR4D among senior managers and policy-makers. It is also studying the administrative, financial and planning modalities of each member organization to gain a fuller and collective understanding of how each one operates. Particular attention is being paid to the university members and how best to introduce IAR4D into their programmes, both individually and collectively.

LESSONS LEARNED: PARTNERSHIP STRATEGY FOR CAPACITY STRENGTHENING

While there are differences in the experiences in the three countries, the partnership strategy for CS in rural innovation is broadly similar. This strategy is based on working towards a number of intermediate outcomes that are seen as necessary in order to achieve the goal of collective action and positive rural change (see Figure 20.1).

Figure 20.1 *Outcomes of capacity strengthening for rural innovation*

At the same time, we have learned a number of lessons relevant to the different outcomes, as presented below.

Creating a national vision for rural innovation

Strengthening capacity in rural innovation requires collaborative action and, therefore, shared objectives by diverse institutional actors at communal, district/provincial and national levels. The creation of some sort of national task force (South Africa, Kenya) or steering committee (Uganda) was central to developing a national effort for collective CS and overseeing the various activities.

While agreeing on a formal memorandum of understanding (MoU) between the three organizations in Uganda was a rapid process (given support from key decision-makers), formalizing arrangements in Kenya and South Africa, where more and varied partners were involved, is taking much longer. In these cases, something more than simply an MoU is required.

The difficulties of forming and sustaining multi-institutional partnerships should not be underestimated. The basic essentials of partnership are often lacking: open communication, attention to process and content, involvement of all concerned, mutual trust, respect for differences, arriving at an understanding of common goals, leadership, joint decision-making, constructive conflict resolution, building of individual self-esteem, etc. In all of the partner countries, the adaptation of structures and cultures to facilitate partnerships needs further dialogue and effort.

Different institutions have different administrative, financial and management systems. Learning to integrate these takes time. Because different amounts of resources are available from different sides, it is rarely possible to bring institutions together on a basis of equal power and influence; inevitably, one organization has more 'convening power' than the others.

In two of the three countries discussed here, special funding was obtained to advance the capacity-strengthening initiatives (one-year funding in Kenya and multi-year funding in South Africa). Separate dedicated funding obviously has advantages; but we think it is also necessary to embed CS initiatives in mainstream national R&D programmes to ensure sustainability. This implies considerable discussion and lobbying with the managers and donors of these programmes, and integration of CS activities within their work plans and budget cycles.

Focusing on learning cycles and then, subsequently, on institutionalization (as originally intended in Uganda) proved to be too simple and linear a process. In reality, raising awareness, in-service learning programmes and mainstreaming need to proceed simultaneously. Successful learning programmes and examples of collective action in the field provide focused discussion opportunities and motivation.

Forming a critical mass of learning facilitators

The design, implementation and evaluation of in-service and tertiary learning programmes require a critical mass of facilitators at national level. Again, the inter-institutional nature of the overall programme requires that these facilitators be drawn from the diverse institutions. Collaborating and creating a shared vision among these different individuals are key to creating an effective national group of facilitators.

The interaction of facilitators from different institutions (research, universities, development departments, etc.) helps to combine the strengths of different partners and is preferable to assigning the responsibility to just one organization. Normally, each institution has its individual strengths and particular capacities. In Uganda, for example, the joint mentoring by both the National Agricultural Research Organization (NARO) and Makerere University (MAK), with facilitators working in pairs, greatly enriched mutual learning.

Forming CS core teams through external and intensive programmes such as ICRA's Wageningen programme (three months of workshops in Europe and three months of fieldwork in the collaborating country) undoubtedly has benefits in creating a team from individuals of different organizations, but is relatively expensive. Despite prior inter-institutional agreements, it was difficult for all team members in Kenya, for example, to continue to work together and to organize in-country CS activities after the immediate programme had finished.

An alternative strategy is to strengthen national groups of facilitators *in situ* – through learning by doing, or organizing and implementing in-country learning programmes. In Uganda, NARO, MAK and ICRA facilitators agreed to cooperate to organize the first IAR4D learning cycle. While there were many heated discussions over content and process, the interaction over the year-long learning cycle was very effective in creating a shared vision of what needed to be done in Uganda to promote rural innovation.

Staff turnover (promotions with different responsibilities, new job opportunities with organizations not represented within the institutional partnerships, or further education outside the country) has resulted in the effective loss of about 10 to 15 per cent of the facilitators from the national CS core teams within two years after their formation. It is also often difficult for one or two individuals within a given institution to create a momentum for change within that institution. A larger group is more likely to have the critical mass needed. For all of these reasons, we suggest that a national capacity-strengthening core team needs to include about 15 to 20 individuals from institutions that are available to participate in organized learning programmes.

In-service professional development

Kolb's theory, as mentioned above, emphasizes the use of concrete experience as a facilitation and teaching method; the implications of this are that facilitation of learning needs to go beyond the organization of training courses or events, towards a long-term or even life-long learning. Creating such a continual process is currently not very evident in most professional capacity strengthening.

Learning programmes (or cycles) – consisting of a mixture of workshops for knowledge acquisition, planning and reflection, combined with fieldwork – need to be integrated within the context of specific development challenges in which participants in the learning programme have a collective stake. In the first learning cycle in Uganda and the first two in-service learning programmes organized by the Agricultural Research Council in South Africa, for example, participants were in some cases assigned to teams that were working outside their own geographic or programmatic area. Apart from the logistical, motivational and budget problems, these teams were not sustainable after the immediate learning programme, and the collective action with other stakeholders begun during the programme was usually not continued. Based on this experience, future learning programmes in both countries are planned around actual or potential *in-situ* teams, rather than on bringing in participants from other areas. The implication of this is that planning such an action-learning programme is more complicated and time consuming than a normal course, requiring considerable prior discussions and agreement between local stakeholders, and commitment to allowing the time and space for learning within the ongoing R&D or commercial activities.

Including participants from a broad range of institutions (research institutes, universities, provincial departments of agriculture, local government and NGOs) within learning teams improves mutual motivation and learning experience. In Uganda, for example, it was noticeable that teams who included local government professionals had a different 'outlook' and were more successful in establishing local stakeholder groups (or 'platforms') than teams comprising only researchers from NARO or MAK.

Choosing the entry point or specific development challenge around which to organize the learning programme affects the type of stakeholder platform constructed and the difficulty in bringing stakeholders together. Entry points organized around natural resource issues (e.g. water use in Kitui, Kenya) involved a wider variety of local government stakeholders. Those organized around market opportunities for specific products (e.g. apples in Kabale, Uganda) included more stakeholders, such as traders and market outlets at national or international level.

Including rural innovation concepts and practices in academic programmes

Universities have long been criticized for emphasizing teaching and not learning, and for neither promoting interdisciplinary work nor producing team players who can facilitate social processes and deal with the complex problems of rural innovation (Ison, 1990; Idachaba, 2003, Muir-Leresche, 2004; Wals, 2005). Over the last few years, the relevance of university curricula has increasingly been questioned by the rural development community, with universities being seen as 'ivory-tower' institutions.

In addition to the 'normal' or specialist professional capacities that each graduate needs, a capacity for promoting rural innovation requires competencies in a range of analytical, process and technical skills, as well as a professional ethos. A tentative framework or set of indicators for the integration of rural innovation competencies in curricula, developed during a recent workshop in South Africa, included the following elements:

- competencies of staff in facilitating learning (facilitating group and experiential learning, mentoring, building confidence of learners);
- inclusion of new competencies in the syllabus:
  - meta-disciplines such as process facilitation, systems thinking, planning, epistemology (learning how to learn), promoting equity, and rural innovation processes and approaches, etc.; and
  - personal and social skills, such as empathy, sensitivity, self-awareness, self-regulation, social and gender awareness, interpersonal communication abilities, and being a team player, etc;
- enabling interdisciplinary studies (facilitating students of different disciplines to work together on a common problem or research theme, both in classroom-based case studies as well as in practical work with non-academic stakeholders);
- embedding practical work within real-world action research projects (involvement of a diversity of non-academic rural stakeholders, including communities, farmer organizations, traders, local government, NGOs, etc.);
- assessment (for teamwork, personal and social skills, and problem-solving rather than knowledge *per se*); and
- management of the curriculum development process (involvement of non-academic stakeholders in determining the curricula).

There are, of course, many constraints to such an agenda. Teaching staff are often reluctant to analyse their own teaching abilities and methods. Adult education (e.g. stressing andragogy instead of pedagogy, as discussed by Green, 1998) is a discipline usually found in faculties separate from those of agricultural and rural development. Incorporating fieldwork outside the university increases costs and time for organization – and can be unpredictable in its outcomes, even though many universities have 'centres for development' that include outreach programmes. Assessment based on group outputs, using, for example, peer assessment, raises issues of subjectivity. Accrediting new courses often takes a minimum of 18 months to pass through curriculum-review committees and national qualifications authorities.

CONCLUSIONS

The change in institutional strategy from one of a training institute, accepting individual professionals from a wide range of countries, to a capacity-strengthening institute focusing on institutional partnerships in targeted countries has led to a new range of activities and experiences. ICRA and its partners in Uganda, Kenya and South Africa continue to learn and reflect on how best to strengthen capacity for rural innovation.

The progressive move towards more local and more in-service learning programmes has allowed the theoretical aspects and skill development to be more closely related to the real world situation faced by participants. This has led to the progressive change of starting point of these learning programmes from one based on a need to train a certain number of individuals in certain selected organizations to one based on the R&D needs of a certain group of stakeholders in a particular area (who may not yet be organized). The paradox to reconcile is that certain organizations (e.g. the national agricultural research organizations) may recognize the need to strengthen their own staff capacity and be willing to finance this; but learning programmes are most effective when these staff form only a minority in broader inter-institutional teams. Arranging agreement and finance in such multi-institutional partnerships has not been easy.

Perhaps the main lesson that we have learned so far is that CS for rural innovation involves much more than adopting suitable procedures, or improving the knowledge and skills of individuals in meta-disciplines such as systems thinking, planning, teamwork and networking – the importance of these aspects notwithstanding. It requires a different ethos than that often produced by professional formation: one that values the different aspirations of the various stakeholders and accepts that rural innovation involves social change. It requires research, development and educational institutions to provide the enabling environment by negotiating and compromising on collective challenges with other stakeholders rather than focusing on narrower institutional objectives and loyalties.

Promoting change on such a broad front requires activities at individual, operational and institutional levels. Pre-planned strategies for capacity strengthening in these complex institutional environments do not always lead to expected and successful outcomes. Rather, strategies need to be flexible in order to build on and add value to existing programmes, rather than trying to create separate programmes, and to build on emerging successes and opportunities.

ACKNOWLEDGEMENTS

Many people were involved in the experiences described in this chapter and have contributed in one way or another to many of the ideas expressed here; unfortunately, it is not possible to list them all. The financial support of the Netherlands Organization for International Cooperation in Higher Education (Nuffic) for part of the work in South Africa and Kenya is also gratefully acknowledged.

NOTE

1   ICRA changed its name from 'Course' to 'Centre' in 1990 when activities expanded.

REFERENCES

AISSA (Network for Agricultural Intensification in Sub-Saharan Agriculture) (2005) *Competitive Agricultural Systems and Enterprises (CASE)*, AISSA, www. aissa.org/case.asp, accessed 7 November 2006

Collinson, M. P. (ed) (2000) *A History of Farming Systems Research*, FAO, Rome/CABI, Wallingford, UK

Engel, P. and Salomon, M. (1997) *Facilitating Innovation for Development: A RAAKS Resource Box*, Royal Tropical Institute, Amsterdam, The Netherlands

Green, J. (1998) 'Andragogy: Teaching adults', in B. Hoffman (ed) *Encyclopedia of Educational Technology*, http://coe.sdsu.edu/eet/Articles/andragogy/start.htm, accessed 7 November 2006

ICARDA (International Centre for Agricultural Research in Dry Areas) (2005) *Integrated Natural Resource Management*, ICARDA, Aleppo, Syria, www.inrm. cgiar.org

ICRA (International Centre for development oriented Research in Agriculture) (1979) International Course for development oriented Research in Agriculture, ICRA, Wageningen, The Netherlands

Idachaba, F. S. (2003) 'Creating the new African university', Paper prepared for the international seminar series on Sustainability, Education and the Management of Change in the Tropics (SEMCIT), accessed from www.changetropics.org on 6 June 2006

Ison, R. (1990) 'Teaching threatens sustainable agriculture', Gatekeeper Series 21, International Institute Environment and Development, London

Kolb, D. A. (1984) *Experiential Learning*, Prentice Hall, Englewood Cliffs, NJ

MoA (Ministry of Agriculture) (2004) *Revitalizing Agriculture, 2004–2014*, MoA, Government of Kenya, Nairobi, Kenya

Muir-Leresche, K. (2004) 'Transforming university agricultural education', in A. Temu, S. Chakeredza, K. Mogotsi, D. Munthali and R. Mulinge (eds) Rebuilding Africa's Capacity for Agricultural Development: The Role of Tertiary Education, ANAFE Symposium on Tertiary Agricultural Education, April 2003, Nairobi, Kenya, pp420–432

Okali, C., Sumberg, J. and Farrington, J. (1994) *Farmer Participatory Research: Rhetoric and Reality*, Intermediate Technology Publications, London

Spielman, D. J. (2005)' Innovation systems perspectives on developing-country agriculture: A critical review', ISNAR Discussion Paper 2, International Food Policy Research Institute, Washington, DC

Turyamureeba, G., Cheminingw'a, G., Tum, J., Mwonga, S., Ndubi, J., Mulagooli, I., Kashaija, I. and Hawkins, R. (2006) 'Boom or bust: Strategies to exploit market opportunities for apple farmers in Kabale, Uganda', Innovation Africa Symposium, 20–23 November, Kampala, Uganda

Wals, A. E. J. (ed) (2005) *Curriculum Innovations in Higher Agricultural Education*, Elsevier, The Hague, The Netherlands

World Bank (2006) *Enhancing Agricultural Innovation: How to Go Beyond the Strengthening of Research Systems*, World Bank, Washington, DC

Building Competencies for Innovation in Agricultural Research: A Synthesis of Experiences and Lessons from Uganda

Diana Akullo, Arjen Wals, Imelda Kashaija and George Ayo

INTRODUCTION

Since the 1990s, public agricultural research organizations in Uganda have been increasingly challenged by the national government and international donors to be more oriented towards markets and clients. The National Agricultural Research Policy (MAAIF, 2003) required change in practices and procedures from research *and* development (R&D) to research *for* development (R4D). Before the new policy, agricultural R&D focused mainly on farm-level productivity by introducing new technologies developed by scientists. The applied methodology followed the farming systems research (FSR) format, where primarily scientists set the research priorities, although much of the actual research took place on farms (Collinson, 2000). The major weakness of the FSR approach was that it was based on quantitative models biased towards preset problem definitions and solutions. Most interventions were limited to biophysical characteristics and were implemented in a linear way. Moreover, a problem for Uganda could be traced to university curricula and capacity-building within the research system, which was mainly through projects and one-off training programmes, rarely conceived with the aim of influencing researchers' attitudes (Stroud, 2003). Consequently, little had been gained because of limited sharing of the useful but isolated information.

An innovation systems approach is now preferred to the science-based linear (top-down) implementation model (Hall and Nahdy, 1999; Daane and Booth, 2004). The innovation systems approach differs from the linear model in that the latter is based primarily on on-station research regarded as the unique source of innovation, while the former has a less clear-cut origin: innovations are seen as the result of shared efforts. This calls for an emphasis on communication, organizational change and other social factors, elements that are not entirely ignored in the linear model but considered important only during the dissemination and application stages.

When the National Agricultural Research Organization (NARO) was being reformed from 2001 onwards, the government and international donors gave it a new mandate, which included contributing to the national goal of improving the livelihoods of the poor in Uganda by adopting an integrated approach. In response to this and to the Sub-Saharan Challenge Programme of the Forum for Agricultural Research in Africa, NARO started building capacity in IAR4D to raise the level of agricultural performance in the rural economy through optimal use of natural resources (Wals et al, 2004).

METHODOLOGY

The 'self-help principle' of IAR4D was collaboratively used by NARO, Makerere University (MAK), the International Centre for development oriented Research in Agriculture (ICRA) and the African Highlands Initiative (AHI). These four entities identified one or two individuals from each to form an eight-member project implementation team (PIT). The PIT developed a common understanding of what the task of building competencies for innovation in agricultural research, code-named 'learning together for change through an IAR4D approach', would entail. It defined the purpose of the training programme and how the members would contribute to the process. It then outlined outputs, their indicators and how to measure them. The researchers trained and then formed smaller research teams, which were hosted at Abi, Bulindi, Kachwekano, Mbarara, Mukono, Ngetta and Serere Zonal Agricultural Research and Development Institutes (ZARDIs) of NARO. Two teams (Ngetta and Serere) were selected for focus group discussions. A total of five sets of workshops were organized in which ICRA training modules and personal development concepts were used for enhancing researchers' competencies in taking an IAR4D approach.

ENGAGING WITH INTEGRATED AGRICULTURAL RESEARCH FOR DEVELOPMENT

The IAR4D training is intended to build competencies among agricultural researchers to engage in participatory research. These competencies include:

- establishing, maintaining and working in interdisciplinary teams;
- involving clients and stakeholders in the agricultural research and development (ARD) process;
- seeking and applying both formal and indigenous knowledge;
- adopting systems-thinking approaches;
- working with multiple stakeholders to prioritize R&D opportunities; and

- deciphering research results into products applicable by clients according to their needs (Daane and Booth, 2004).

The IAR4D approach also includes organizational (or managerial) competencies (i.e. the knowledge and skills to engage relevant stakeholders in research processes).

The assumption is that skills training leads to institutional change. The activities initiated by NARO, MAK, AHI and ICRA should therefore also have an effect on these organizations. Applying the IAR4D approach requires a paradigm shift in institutional and researchers' thinking and approach to create an impact on research outputs (Kimmins and Sutherland, 2004; Leeuwis and Van den Ban, 2004). Reflection is central in this process (Stroud, 2003).

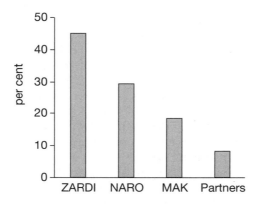

Figure 21.1 *Distribution of participants in integrated agricultural research for development training from different institutions in Uganda*

Note: NARO is comprised of the National Agricultural Research Institutes (NARIs) and the Zonal Agricultural Research and Development Institutes (ZARDIs); 'partners' refers to local governments and NGOs.

Where the PIT acted as 'flight control', the 'traffic' consisted of several training sessions (workshops) for the 54 researchers involved. They comprised seven action research teams. A total of five training sessions, each lasting five days, were alternated with four field visits that lasted three to six days, depending upon each research team's objectives. The field phases enabled the researchers to practise what they had learned.

IMMEDIATE IMPACT OF INTEGRATED AGRICULTURAL RESEARCH FOR DEVELOPMENT

The IAR4D initiative of NARO, MAK, AHI and ICRA introduced researchers to new work procedures and skills for undertaking agricultural research. The researchers were exposed to different methods of collecting data, including participatory rural appraisal (PRA), stakeholder analysis, scenario and strategy analysis, market opportunity identification, zonation and typology analysis, observation and facilitation. Researchers who engaged in the IAR4D training workshops developed new ideas for research. Some of these ideas were turned into concrete proposals that are now being implemented.

The effects of IAR4D mentioned so far are all on the level of researchers and their organizations. But these effects are supposed to contribute to the overall goal of a better-equipped and better-functioning innovation system. Evaluating the effects of IAR4D on the efficacy of innovation requires a different type of evaluation: how researchers involved in the IAR4D process experienced it and whether or not they are better equipped to serve the areas where they work. This was, indeed, what most researchers expressed. Specifically, they mentioned clarifying priorities and earmarking research activities to help improve the livelihoods of the users of research products in the agro-ecological zones in which they worked. Before IAR4D was introduced, all ZARDIs, except Serere, had carried out farming systems and livelihood analysis studies. Concerns identified in these studies provided a baseline on production and marketing of all potential commodities for intervention to solve communities' problems. Building competence in IAR4D led to more concrete utilization of these results. Six out of seven research teams adopted key issues from the studies in developing their R&D challenges. Research managers built on previous and ongoing work with the IAR4D approach, while researchers developed a better understanding of the systems in which they work.

The integrated agricultural research for development process

The 54 participants had varying degrees of research and training experience. Seven had attended the IAR4D training at ICRA, while others had attended short courses on topics such as participatory research approaches, monitoring and evaluation, proposal and scientific writing, mainstreaming gender, HIV/AIDS, and natural resource management (NRM). In other words, participatory research methods as presented by IAR4D were not entirely new to most of the researchers. A major difficulty, however, was translating the general principles into specific activities in the field, particularly to realize the 'learning together' component. In most cases, the identification phase was done as a team, but follow-up was not. Priority learning areas were identified at the beginning of the joint learning; but prioritizing the content for such a diverse group was a challenge.

IAR4D emphasized the need to involve all relevant stakeholders in the joint learning process; but it meant high costs. The organizations involved were particularly encouraged to support stakeholder participation. All partners were in principle and on record expected to contribute financial, technical and other resources to the process. However, this was not possible for MAK and the local government, who rely on the central government's budgeting and prioritization framework. IAR4D became an expensive project for NARO as it almost solely sponsored the process. Some activities were either delayed or skipped altogether because of financial stress.

CONCLUSIONS

Participants in IAR4D generally considered the process a valuable contribution to their work and an improvement compared to the previous situation. NARO and its partners now have additional skills that enable them to work towards a more demand-driven, market-oriented participatory form of agricultural research. The basic principles and objectives of IAR4D, however, are wider than that.

IAR4D is embedded within and emerges from a variety of policies and objectives formulated by a mixture of national and international administrative bodies, organizations and forums. It therefore becomes difficult to determine exactly what the objectives and ambitions are. When evaluating IAR4D in terms of described objectives, we can see two 'extreme versions'. One way to formulate the aim of the programme is to provide an avenue for researchers to contribute to NARO's new research strategy, which emphasizes inclusion of more end users in research and development processes. This we can call the minimum objective. The maximum or full objective is that NARO is fully geared to an integrative and participatory working method, with all programmes and projects based on close collaboration with a variety of partners and stakeholders. Thus far, the minimum objective has been achieved.

IAR4D is based on key areas in which interactions in an innovation system can be observed: technology, organization and methods. For each of these, change in working methods has been promoted through development of competencies or learning. Several forms of learning and learning events are distinguished; but the details about learning process and effects are left open. Individuals might state that they learned a lot; but, no matter how sincere and justified the statement might be, what matters from a practice-based perspective is whether the learning leads to observable changes in interaction and 'ways of doing'.

In IAR4D, technological improvement (innovation) provides the evidence of learning. From the data, we can say two things about this. Firstly, realizing innovation is a time-consuming process. Initial results show that some innovation was achieved with the new method; but, overall, our evaluation was not adequate to determine the effectiveness of the IAR4D approach in these terms. Second, what was observed in the process was a tendency of researchers to see the new participatory approach primarily as way of collecting more (new) information. In some cases (rice and cassava), this information was used for writing research proposals. There is a risk that, through the introduction of the IAR4D approach, researchers use the participatory methods primarily to gain a competitive advantage for research funding.

Like the other elements, the organization can be perceived as both mechanism (an organizational format conducive to IAR4D principles) and outcome (IAR4D pushing the organization in a new direction). What emerges from our data is that the organization (here primarily referring to NARO) is also an important context into which the IAR4D approach has to be integrated. NARO put much effort into making the IAR4D approach a success. At the same time, it was observed that, because of organizational constraints in the form of workloads or planning procedures, researchers often had to pull out from participating in the IAR4D training sessions. This shows that more effort is required before NARO and its partner organizations have fully internalized an innovation systems approach as envisioned by IAR4D. Again, the snapshot of the initiative as presented here probably comes too early to make any definitive claims. On a more general level, it brings us to our concluding point about the importance of methods, technologies and organization.

The introduction of IAR4D set in motion a series of changes that put on course a more participatory approach in doing research and being part of innovation processes. The analytical framework introduced shows that the incorporation of context is crucial in both setting goals and evaluating the IAR4D process in NARO. By being more specific about what outcome is expected in what context, the programme could create more effective feedback loops in order to continue with participatory research and to work according to an innovation systems approach.

REFERENCES

Collinson, M. P. (ed) (2000) *A History of Farming Systems Research*, CABI, New York, NY/FAO, Rome, Italy

Daane, J. and Booth, R. (2004) 'Integrated agricultural research for development: Lessons learnt and best practices', *Agricultural Sciences*, vol 9, pp126–131

Hall, A. and Nahdy, S. (1999) 'New methods and old institutions: The systems context of Farmer Participatory Research in national agricultural research systems: The case of Uganda', Agricultural Research and Extension Network Paper 76, Overseas Development Institute, London

Kimmins, F. and Sutherland, J. (2004) *COARD Project Review: Integrated Agricultural Research for Development*, Serere Agriculture and Animal Production Research Institute, Soroti, Uganda

Leeuwis, C. and Van den Ban, A. W. (2004) *Communication for Rural Innovation: Rethinking Agricultural Extension*, third edition, Blackwell, Oxford, UK

MAAIF (Ministry of Agriculture, Animal Industries and Fisheries) (2003) *The National Agricultural Research Policy*, MAAIF, Kampala, Uganda

Stroud, A. (2003) 'Transforming institutions to achieve innovation in research and development', in B. Pound, S. Snapp, C. McDougall and A. Braun (eds)

Managing Natural Resources for Sustainable Livelihoods: Uniting Science and Participation, Earthscan, London, pp88–112

Wals, A. J., Caporali, F. and Pace, P. (2004) 'Education for integrated rural development: Transformative learning in a complex and uncertain world', *Agricultural Education and Extension*, vol 10, no 2, pp89–100

Shaping Agricultural Research for Development to Africa's Needs: Building South African Capacity to Innovate

Aart-Jan Verschoor, Thembi Ngcobo, Juan Ceballos, Richard Hawkins, Colletah Chitsike and Petronella Chaminuka

INTRODUCTION

South Africa's main socio-economic challenge relates to inequality. An equitable society based on economic growth, in which agriculture plays a catalyst role, is a policy aim (Verschoor, 2003). Entrance of resource-poor farmers into mainstream agriculture is a government priority; but while a favourable policy environment has been established during the last decade, practical empowerment remains rare (Aliber et al, 2006).

This situation is mirrored in many African countries. A strategy to address these constraints would have widespread value. For Africa to achieve the Millennium Development Goals and the objectives of the Comprehensive Africa Agriculture Development Programme, research and development efforts on the continent need to improve. The conventional linear model of agricultural technology development and transfer is a poor representation of reality and is being replaced by an innovation systems model in which stakeholders interact towards a common objective (Spielman, 2005; see also Chapter 2 in this volume). Promoting the competitiveness of African agriculture in a global economy – a major objective of the New Partnership for Africa's Development (NEPAD) – requires such a shift towards a more systemic approach to research and development.

Here, we present how agricultural research for development (ARD), as an approach to action learning and research through collective innovation, is being introduced in South Africa. We elaborate on the potential role of ARD in empowerment, focus on experiences in establishing an enabling environment, describe initiatives in ARD capacity-building and institutional integration, and summarize the lessons learned.

BACKGROUND

Roughly 70 per cent of the poor in South Africa live in rural areas, with the rural economy providing insufficient remunerative opportunities. A unique rural–urban continuum is manifested because of the country's history. The apartheid system curbed viable small-scale farming; specifically, the land segregation laws of 1911, 1913 and 1932 effectively eliminated small-scale farmers' competition from the market. Government support for commercial farmers facilitated increased national output and self-sufficiency, at the cost of higher inequity and decreased food security (Aliber et al, 2006). Small-scale enterprises today are constrained by limitations in quality, quantity or accessibility of key inputs. Available technology often fails to match their constraints, environment and managerial ability (Verschoor, 2005). Many of these problems are faced by farmers all over Africa. Colonialism is hardwired into society, resulting in a major part of the population not participating fully in the economy (Aliber et al, 2006).

Most rural livelihoods in South Africa today depend upon non-farm incomes, including social government grants and remittances from urban industry and mining. More viable opportunities in urban South Africa result in massive migration. However, rural ties remain strong, resulting in high capital flows into poor rural areas. A social welfare system unequalled in Africa leads to lower reliance on agriculture than in most African countries. Most rural households in South Africa (75 to 85 per cent) use agriculture minimally to supplement larger, more stable income sources from elsewhere (Verschoor, 2005).

The agricultural policy framework focuses on strategic alliances, integrating financial services, lower production costs and supply-chain agreements (Verschoor, 2005; Aliber et al, 2006). The objective is equitable access and participation in a globally competitive and sustainable agricultural sector, with priorities of transformation of research, technology transfer and, specifically, human capacity development (NDA, 2001). The Agricultural Research Council (ARC) provides a scientific base and technology-transfer capacity to South Africa's agricultural industry. Provincial Departments of Agriculture (PDAs), guided by the National Department of Agriculture (NDA) and the ARC, form the mainstay of support to the diverse farming sector.

However, the gap between white and black producers is slow in closing and no significant improvement in rural livelihoods is evident. Generally, agriculture in South Africa has not fulfilled its potential as a catalyst for economic development. One reason is that support services lack the skills to deal with resource-poor farmers, who differ from the commercial clientele in terms of priorities, scale, capacity and experience.

An approach to agricultural research, development and education that meets broader economic and social objectives, aligned with policy priorities, is required. Skills to engage farmers effectively, to achieve collaboration and to try out new options collectively are needed. We argue that capacity development can be addressed by integrating ARD within agricultural research, development and tertiary education. ARD is used as an umbrella name for a range of approaches to collective rural innovation that:

- respond to the needs of clients;
- facilitate teamwork and communication across disciplines and institutions;
- combine scientific and local knowledge in a systemic perspective;
- integrate a set of participatory action research tools within pragmatic experiential learning; and
- enhance meta-disciplinary analytical skills.

AGRICULTURAL RESEARCH FOR DEVELOPMENT INTEGRATION

South Africans were exposed to the ARD approach in 1995, when two citizens took part in the International Centre for development oriented Research in Agriculture (ICRA) training in Wageningen, The Netherlands, with fieldwork in Uganda. ICRA recognizes that agricultural research responsive to the needs of resource-poor farmers demands competencies other than those acquired through discipline-oriented education. It offers learning programmes that integrate inter-institutional transdisciplinary contributions within collective innovation processes.

Subsequently, the ICRA alumni established ARD initiatives in South Africa. In the North-West Province, a multidisciplinary discussion forum, surveys into resource-poor farmers' enterprises and an inter-institutional on-farm project were initiated. Farmer study groups were established and interaction facilitated. This led to a series of ICRA field studies from 2001 onwards. In collaboration with the ARC, PDAs hosted teams in Wageningen that included local scientists, using their new ARD skills to address a development-oriented research topic. Between 2001 and 2004, five studies were completed with stakeholders (PDAs in Eastern Cape, North-West Province and Limpopo, the ARC and ICRA). Eight PDA and six ARC officials took part. These studies dealt with communal land-use systems, sustainable management of land and water use, and animal production. Analytical reports with recommendations for research and development were submitted and often the recommendations were reflected in the following planning cycle of the department. However, there was a lack of implementation capacity.

The ARC has been criticized since its inception in 1992 for its lack of impact on resource-poor farmers. Its scientists are well-trained technically, but deficient in the integrative and holistic skills and mindsets needed for innovation. The ARC created the Sustainable Rural Livelihoods Division in

2003 to facilitate appropriate research for resource-poor farmers. Since 2004, this division – with assistance from ICRA and in collaboration with other stakeholders – trained ARC and PDA staff in ARD. The ARC Technology Transfer Academy (ATTA) was established to provide a conduit for ARC services in engaging resource-poor farmers and other stakeholders and in facilitating human capacity development. Since collaboration within the research, development and higher education environment was clearly required, ATTA initiated a National ARD Task Team (NARDTT) to oversee national ARD processes. This team consists of the ARC; the Universities of KwaZulu Natal, Fort Hare, Limpopo, Venda and the Free State; the Tompi Seleka and Madzivhandila Agricultural Colleges; and PDAs in Limpopo, Eastern Cape, KwaZulu Natal, Northern Cape, Mpumalanga and the Free State. Common premises guiding the team are that:

- A lack of insight limits rural development.
- Current systems are too focused on disciplinary expertise.
- Soft skills to address resource-poor farmers' needs must be enhanced.

The team guides the tailor-made ARD learning programme for research and development practitioners, facilitates participation in annual ARD training in Wageningen and oversees integration of ARD within the curricula of the universities.

The first local initiative in building ARD capacities was a 15-week in-service training programme developed in collaboration with NARDTT. Implemented by ATTA in 2004, 2005 and 2007, it enabled 73 participants to apply ARD principles in collective problem analysis, planning and prioritizing research, and development opportunities. Field studies dealt with aspects of the national Land Reform Programme, conservation agriculture and indigenous knowledge.

CHALLENGES

The in-service ARD training programme is a key initiative and has received many accolades. It established working relationships, and its principles have become ingrained in some activities of the ARC and its partners. Resulting field study reports are deemed useful to the PDAs that integrate its recommendations in subsequent planning processes.

However, bottlenecks became apparent. The programme takes three months, which strained participants and facilitators. Logistical arrangements were detailed and time consuming. Ensuring participants with the aptitude and attitude required is problematic, while support from stakeholders has been varied. Significant pressure to increase trainee numbers is countered by a demand to reduce training time. A key constraint is the limited implementation of recommendations from the field studies.

Most PDAs do not have the skills to deal with the demands of a collective innovation process. While the ideal would be that the trained inter-institutional teams would lead such a process, these teams had limited incentive to remain involved. Performance evaluation of researchers does not value participatory processes. Decision-makers' appreciation of the value of such collective innovation is also still limited. A critical mass of ARD practitioners has not yet been reached.

A more participatory model was implemented in 2007. A provincial ARD forum was set up in Limpopo Province to act as a platform for stake-holders from government, the ARC and universities to initiate collaboration on common research and development issues. These parties agreed on priority topics, and interdisciplinary inter-institutional teams determined terms of reference on these priority projects. Participants were selected from the inter-institutional group, who formed teams that followed a staggered programme of knowledge acquisition and practice, providing a report with recommendations to be implemented by the larger group. Collaborating institutions provided guidance and invested in terms of manpower, logistical support, access to information and communication. The programme of three workshops and three fieldwork periods was interspaced with 'normal work' periods. Its iterative design allowed for assessment and for concepts to be tied to fieldwork. Field studies are integrated with activities of the participants and the other stakeholders.

In a survey held among alumni of the in-service ARD course, 97 per cent perceived the ARD training as advantageous. Its value was described in terms of enhanced ability to function in interdisciplinary, intercultural teams; improved understanding of livelihood strategies; the effective combination of social and biophysical factors; and the broadening of interpersonal and social skills. Alumni, in general, stated that they continued to use the ARD tools in their work to communicate, analyse situations, facilitate stakeholder engagements and prioritize research and development options. More than 90 per cent of former participants felt that ARD should be integrated within mainstream research and develop-ment. Many respondents mentioned that these skills are not taught at universities, but are critical for the ability to work with people and would improve the quality of research output.

INSTITUTIONALIZING AGRICULTURAL RESEARCH FOR DEVELOPMENT

A breakthrough in establishing ARD in South Africa was obtaining fund-ing for an ARD institutionalization project from the Netherlands Organ-ization for International Cooperation in Higher Education (Nuffic). This resulted in more focused interaction between participating institutions, broader understanding and exposure. As a key employer of graduates, the ARC has a keen interest in curriculum development and therefore initiated this project, aimed at enhancing the skills of its staff in dealing with resource-poor farmers' priorities. Three strategies were developed:

1   enhancing inter-institutional linkages;
2   training tertiary education staff in ARD; and
3   augmenting the curricula of learning institutions to integrate ARD within the education of research and development professionals.

Various committees were constituted to link partners and to handle project activities. The aim is to create a national critical mass of ARD practitioners and teachers across institutions. The project facilitates various discussion opportunities, capacity development of national teams representative of the NARDTT through ICRA's programme, and engagement with universities regarding curriculum development.

As part of the project, various capacity development workshops are organized for 20 NARDTT institution staff members. Others attend short courses in Wageningen (e.g. on facilitating multi-stakeholder processes). Throughout the project, these short courses in The Netherlands will address missing competencies in the capacity to teach ARD. An ARD-based MSc programme for a group of NARDTT staff members is also being refined. The final objective of the project is integrating ARD concepts within the curricula of partner learning institutions. A curriculum development group has defined key ARD elements required at each participating university through a curriculum audit. This informs attendance at the short courses in Wageningen in line with the identified ARD capacity-strengthening needs. The audit focused on content and is now augmented with an analysis of process, the extent of integration of practical work and the transdisciplinary multi-stakeholder processes involved. The focus is on integration within existing courses, ensuring multi-stakeholder interaction in fieldwork.

CONCLUSIONS

A paradigm shift towards increasingly collaborative modes of engagement among actors is evident in agricultural research and development, driven by recognition of the contributions that the various actors in the value chain bring to the process. It is also informed by the limited impact of decades of linear and largely disjointed research efforts, especially in heterogeneous environments, as in developing countries. Participatory processes are needed to promote the relevance of science for development and impact. The South African ARD programme represents an attempt to address this. It established a broader awareness of the need for collaboration to improve research impact on development. An NARDTT mandated by the leadership of the partner institutions has been established, and working procedures and strategies have been defined. A provincial ARD hub has been initiated. Roughly 120 research and development practitioners, trainers and lecturers have thus far been trained through the ARC's in-service course, short courses and the ICRA course. Outputs include a webpage, a brochure detailing NARDTT activities and several papers.

Integrating ARD in research, development and tertiary education requires a revolution, as collective innovation is foreign to the local bureaucratic systems. A diversity of interests – also in terms of organizational politics – is recognized, as is resistance from people not inclined towards collective innovation. However, the initiative described will, in the long term, enhance farmer support in South Africa through the recognition of the social dimensions of development and the need for more collective synergetic action. It addresses the challenge of science having to contribute more to development. Decision-makers need to deal with incentives promoting innovation rather than mere research.

ACKNOWLEDGEMENTS

The contributions of all partners including ICRA, Nuffic and partners in NARDTT are gratefully acknowledged.

REFERENCES

Aliber, M., Kirsten, M., Maharajh, R., Nhlapo-Hlope, J. and Nkoane, P. (2006) 'Overcoming underdevelopment in South Africa's second economy', *Development Southern Africa*, vol 23, no 1, pp45–61

NDA (National Department of Agriculture) (2001) *The Strategic Plan for South African Agriculture*, NDA, Pretoria, South Africa

Spielman, D. J. (2005) 'Innovation systems perspectives on developing-country agriculture: A critical review', ISNAR Discussion Paper 2, International Food Policy Research Institute, Washington, DC

Verschoor, A. J. (2003) 'Agricultural development in the North West Province of South Africa through the application of comprehensive project planning and appraisal methodologies', PhD thesis, Department of Agricultural Economics, Extension and Rural Development, Faculty of Natural and Agricultural Science, University of Pretoria, South Africa

Verschoor, A. J. (2005) 'New agricultural development criteria: a proposal for project design and implementation', *Development Southern Africa*, vol 22, no 4, pp501–514

Going to Scale with Facilitation for Change: Developing Competence to Facilitate Community Emancipation and Innovation in South Africa

Hlamalani Ngwenya, Jürgen Hagmann and Johannes Ramaru

Enabling communities to become drivers of their own development has become a major focus for many participatory development efforts in the past decade. This move from 'participation' to 'emancipation' of communities requires a drastic transformation in how development is perceived by all actors and how support systems such as rural services are structured and operate. This also requires new competencies to facilitate the transformation processes in communities.

The deeper dimension of 'facilitation' is, however, underestimated and open to many interpretations. The word is used to refer to 'bribing', 'paying *per diems*', 'chairing' meetings or 'transformative learning processes'. Here, facilitation is understood in the last-mentioned way and is referred to as facilitation for change (F4C). This kind of facilitation aims at stimulating fundamental change in both individuals and organizations (see Hagmann et al, 1999; Groot, 2002; Rough, 2002). It is inspired and organized on the basis of organizational change and/or development through action learning and learning organization theories (Schein, 1992) and systemic approaches (Senge, 1990).

Starting in 1998, the Limpopo Province Department of Agriculture (LDA) in South Africa, supported by the German Agency for Technical Cooperation (Deutsche Gesellschaft für Technische Zusammenarbeit, or GTZ), engaged in an action-learning process to develop and institutionalize the participatory extension approach (PEA). PEA is a facilitative intervention approach that focuses on community organization and innovation. It depends largely on the facilitation capacities of those involved in catalysing and managing the process. Hence, deliberately developing facilitation competence of extension officers to mobilize communities to better articulate their demands and to strengthen local organizational capacities has been central to PEA. It was initially developed in Zimbabwe during the 1990s (Hagmann et al, 1998) and was then adopted, adapted and further developed as an alternative approach to delivering rural extension services in South Africa.

The aim is to challenge extension officers to shift their paradigm radically in terms of personality and professional attitude. This means de-learning the top-down mode of engaging with farmers, where they are supposed always to have answers to farmers' problems, and assuming a role of catalyst for social change in the sense of what Hagmann (1999) calls 'learning together for change'.

THE CONTEXT OF FACILITATING THE PARTICIPATORY EXTENSION APPROACH

We differentiate between two models in extension and rural service delivery: technical advisory services and social extension. In the first model, the role of extension services is to provide technical advice on enhancing production of specific commodities and all of the related service functions, including input and output markets. The knowledge on those commodity packages is clearly with the experts who can provide the advice required by the clients. This model does not reflect the social dynamics of a society or community. It often results in a minority of farmers benefiting from these extension services.

The participatory extension approach (PEA) reflects the second model in trying to deal with social dynamics, looking at service functions required in an innovation system based on solving problems in smallholder farming. It focuses on establishing a common platform for trying out new things, including the majority of community members in the process. It aims at enhancing the adaptive capacity of rural people, enabling them to manage better a changing economic, social and ecological environment, to adapt their practices and the way in which they are organized, etc. To a large extent, this depends upon collective capacity rather than individual capacity.

Both models are required to support communities in their own development. It is therefore not about 'either/or'. Rather, the key is successful integration of technical advice into a sound social process. This integration is at the heart of PEA.

In this example from South Africa, the objectives of PEA were to:

- develop the individual and organizational capacities of rural people and their communities to be able to deal with the dynamic challenges and changes in development;
- facilitate a process of self-organization and community emancipation to enable people to articulate and represent their needs better *vis-à-vis* a wide range of service providers;
- develop and spread technical and social innovations in a process of joint learning that builds on the rural people's 'life world' and local

knowledge of agriculture and can then spread to other fields of rural development, closely connected to decentralization, municipal development and service delivery; and

- link rural people and organizations to external service providers, input and output markets and sources of information in order to create a functional innovation system in which both the demand and the service-supply side are well developed.

Implementation of PEA in the communities was structured along the operational steps described in the learning cycle shown in Figure 23.1. This integrates various extension methodologies and tools in a consistent and rigorous learning process in order to deal with different topics in rural development (Hagmann et al, 1998). Its initial focus is on agriculture; but, because it builds a foundation capacity for rural communities to deal with their challenges, it becomes applied more broadly (Ramaru et al, 2004).

The learning cycle comprises six aspects: initiating change; searching for new ways; planning and strengthening local organizational capacity; experimentation while implementing action; sharing experiences; and reflecting on lessons learned and re-planning. Local organizational change is the backbone for all phases as a continuous process.

COMPETENCIES REQUIRED FOR FACILITATING THE PARTICIPATORY EXTENSION APPROACH

Facilitation for change plays a significant role in catalysing the PEA process. We distinguish between two levels of facilitation competencies: trainers' facilitation competence and the community's facilitation competence. Here we focus on the latter – the competence needed for scaling up – which comprises four broad competence areas:

1 competencies regarding vision and values;
2 competencies regarding personal development;
3 competencies regarding facilitation; and
4 competencies regarding conceptual and methodological aspects.

Competencies regarding vision and values for oneself and for development

F4C requires a strong emancipative vision for oneself and for agricultural development to be able to provide orientation for others. Extension facilitators need to understand fully and orient themselves towards a vision of participatory development in which social development and personal self-development are the ultimate goals of extension, rather than only technical development (Moyo and Hagmann, 2000).

As a process-oriented learning approach, PEA is not a blueprint with fixed goals and time-frames. However, having a clear 'guiding star' helps

Figure 23.1
The participatory extension approach learning cycle as developed in South Africa

Source: Hagmann et al (2003)

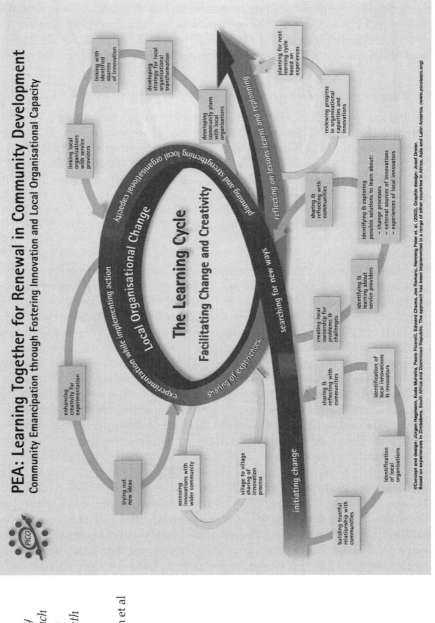

PEA: Learning Together for Renewal in Community Development
Community Emancipation through Fostering Innovation and Local Organisational Capacity

planning and strengthening local organisational capacity

Local Organisational Change

experimentation while implementing action

The Learning Cycle
Facilitating Change and Creativity

sharing of experiences

searching for new ways

reflecting on lessons learnt and replanning

initiating change

linking with identified sources of innovation

developing strategy for local organisational transformation

linking local organisations with service providers

developing community plans with local organisations

planning for next learning cycle based on experiences

reviewing progress in organisational capacities and innovations

enhancing creativity for experimentation

sharing & reflecting with communities

identifying & exploring possible solutions to learn about:
– change processes
– external sources of innovations
– experiences of local innovators

trying out new ideas

creating local ownership for problems & challenges

identifying & learning about service providers

assessing innovations with wider community

village to village sharing of innovation process

sharing & reflecting with communities

identification of local innovations & innovators

building trustful relationship with communities

identification of local organisations

©Concept and design: Jürgen Hagmann, Paolo Ficarelli, Edward Chuma, Joe Ramaru, Henning Peter et. al. (2002), Graphic design: Josef Dorler. Based on experiences in Zimbabwe, South Africa and Dominican Republic. The approach has been implemented in a range of other countries in Africa, Asia and Latin America. (www.picoteam.org)

the facilitators to navigate amidst complexity and allows for a systematic flow of processes. The 'guiding star' includes a vision and some core values that guide the implementation process. These have to be internalized and made transparent in order to minimize suspicion about hidden agendas.

Competencies regarding personal development

An insecure person is seldom a good facilitator. Hence, in F4C, special attention is given to the extension officers' self-development. This aims at stimulating and enhancing the cognitive, behavioural, attitudinal and emotional levels simultaneously in order to build their capacities to act differently (Hagmann et al, 2003). Such a holistic approach to personal development recognizes that human beings have a potential for multiple intelligences, at different levels (see sub-sections below), and have to learn to exercise them simultaneously.

Cognitive level
The complexity and interconnectedness of development issues require a radical change in thinking to becoming more adaptive and systemic, in contrast to conventional linear thinking with fixed procedures and routines. Most extension officers thought in this conventional way and lacked the capability to deal effectively with the complexity of development challenges. The process of developing competence in PEA focused on intensive cognitive change and stimulated them to think laterally in terms of system perspectives and processes (Moyo and Hagmann, 2000).

The extension officers first had to open their minds to a type of learning that was based on self-awareness and on critical thinking and reflection. This meant exposing them to various alternative concepts and paradigms, as well as stimulating their creativity by giving them space to try out and experiment with new ideas.

Behavioural and attitudinal level
The ability to think in a different way often requires a radical swing in attitude and behaviour that allows de-learning of some historical patterns. Facilitating PEA means challenging the deep-rooted prevailing values and social norms that affect people's perceptions. For example, formal education has long been seen as the key to development. This attitude of valuing formal education more than experiential knowledge and non-formal education was reflected in how extension officers and farmers related to each other. Farmers perceived extension officers as superiors who cannot be challenged – because they have formal education – while they regarded farmers as backward and incapable of making any valuable contribution, despite their vast experience and local knowledge.

A PEA process, however, requires a less hierarchical mode of learning where both extension officer and farmer engage in joint learning. This means the extension officer must relinquish power, while the farmer becomes emancipated to challenge the officer and both begin to learn together for change.

Emotional level
Managing complex social processes in communities characterized by continuous uncertainty requires some level of confidence on the part of the facilitator. To be able to 'read' a process and thus reduce the uncertainty and create a reference base for making decisions, facilitators need a sound degree of common sense, empathy, self-awareness and self-regulation – in other words, 'emotional intelligence' (Goleman, 1995).

There is no clear divide between these three levels. In practice, it is difficult to address them separately; they need to be addressed holistically. A reduced approach that does not take into account the importance of personal development is less likely to be effective and sustainable.

Competencies regarding facilitation

Facilitation skills comprise the ability to observe processes as well as various techniques in using different types of tools.

Process-related skills
The unpredictable nature of PEA processes requires strong observational skills to understand the environment better. Intuition is also essential to sense how one's own thinking, attitude and behaviour influence the group. This means finding the right balance between addressing individuals and the group. These skills allow the facilitator to adapt continuously to changes in the environment. Adaptive capacity also means that one is aware of several options and dares to choose from among them.

The art of questioning and probing
The ability to ask relevant questions that stimulate people's ability to think below the surface is a crucial skill in facilitating PEA. Questioning and probing is in contrast to the conventional method of providing supposed solutions to problems. Through questions, the PEA facilitators continuously stimulate the local people to discover what the facilitators may already see, but without imposing this view. To be able to use this technique effectively, a facilitator should be broadminded and able to see how different issues are interconnected.

Managing facilitation tools such as codes and simulations
Facilitation of PEA derives its strength from the use of codes, simulations, songs and proverbs. However, the effectiveness of these tools depends greatly upon how they are applied. Having a 'toolbox' of codes and simulations is not enough; the facilitator needs to know when and how to use a certain tool and what questions to ask in order to enable effective 'decoding'.

Visualization skills
The use of visual language helps people to focus and makes communication more effective. Facilitators are challenged to be creative in making visuals clear, interesting and humourous. For some facilitators, this 'artistic' expression is inborn; but for most it is a real challenge, making them hesitant to use visualization.

Giving and receiving feedback
In PEA, the 'feedback culture' is a core value and is consciously instilled from the outset at both extension and community level and is maintained throughout the process. The facilitators continuously have to encourage openness and constructive feedback among farmers. However, facilitators cannot hope to foster these skills in the group if they cannot deal with feedback.

Managing group dynamics and team-building techniques
One value that PEA promotes is the move from an individualistic to a collective approach of dealing with development challenges. Working collectively implies that people with different backgrounds, needs, attitudes, etc. have to find some common ground. The challenge for the facilitator is to find ways of ensuring that the diversity is not detrimental to the group, but rather enriches group performance.

The 'toolbox' is a collection of different tools needed to facilitate PEA. During competence development, extension officers are introduced to as many tools as possible so that they can make their own collection – 'a basket of options' – to complement their facilitation skills. In the course of the training, most of the tools are practised so that the officers can experience how the tools work and can understand the effects of each one. In this way, they can internalize the tools and are better able to use them later in the field.

Competencies regarding conceptual and methodological aspects

This involves broader technical, conceptual and management knowledge in relation to the context of extension organization and community development, as well as operational and process-management aspects.

Extension organizational context
Critical analysis of the current extension situation in terms of its successes and constraints, reflection on and analysis of the history of extension approaches, and articulation of a vision for effective extension form the basis for discussion of ways to improve it.

Community development context
Facilitators need to gain a good understanding and internalize concepts related to community development, such as local organizational

development, rural livelihood systems, sustainable agriculture and related fields.

Operational and process management
In order to operationalize and manage the PEA process, extension officers need to be exposed to concepts of change and managing change; facilitation for change; design and management of learning process intervention; and mentoring and coaching.

HOW TO DEVELOP THESE FACILITATION COMPETENCIES?

The nature of the key competencies required to facilitate PEA demands an integrated approach. The process was not a 'one-off' training exercise. It was organized in a series of five learning workshops spread over 18 months. Each workshop was followed by two to four months of practice in the field: in selected villages, the trainees applied what they had learned, and periodic mentoring and coaching in the field as well as peer learning in groups and self-learning by the trainees took place (see Table 23.1).

The learning workshops

The learning workshops were focused on exposure to concepts and reflection on practice. The first workshop introduced the basic concept of PEA and other concepts related to development as a basis for initiating change. This was the longest workshop and laid a good foundation for

Table 23.1 *The participatory extension approach strategy for developing competence*

Phase	Activity	Duration
1	Orientation learning workshop	Fifteen days
	Field practice: initiating change	Two months
2	Second learning workshop	Ten days
	Field practice	Four months
3	Third learning workshop	Ten days
	Field practice	Four months
4	Fourth learning workshop	Ten days
	Field practice	Four months
5	Fifth and final learning workshop	Five days

Source: adapted from Moyo and Hagmann (2000)

sharing and feedback, which is crucial for the entire process of competence development.

The following workshops built on each other and provided space for reflection at both individual and team level, and for sharing experiences from practice in the field. This made possible a continuous monitoring and evaluation process. Each workshop also deepened certain concepts (i.e. vision, personal development, facilitation skills, technical and method-ological aspects) that had been introduced in the previous workshop, while

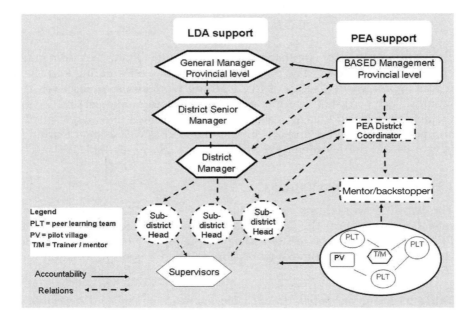

Figure 23.2 *Organization of the mentoring process and support structure*

introducing new ones according to the different phases in the learning cycle.

In addition to this series of workshops, trainees also went through specific technical workshops in which they learned and deepened technical issues. The PEA process in South Africa focused on four major technical areas, based on the farmers' needs – namely, soil fertility management (SFM), soil and water conservation (SWC), small-scale seed production (SSSP) and livestock production (LP). Various other institutions (i.e. local universities, colleges and research institutions) were involved in these technical workshops in order to provide support in terms of technical expertise.

Organizing the field practice and mentoring

In order to manage the integration of PEA within the existing extension system, a temporary structure was established to support the processes of developing competencies and implementing PEA. The Broadening Agricultural Services and Extension Delivery (BASED) management team, with a provincial coordinator, district coordinators and sub-district mentors, each with peer-learning teams working at ward level, was set up alongside the existing line structure of LDA (see Figure 23.2).

The mentoring and coaching process was designed to support the trainees in the field. The trainees formed peer-learning teams (PLTs) that provided support to each other during field practice in planning together, giving each other feedback and providing moral support when facilitating community meetings. Each PLT consisted of three to four trainees, who each had full responsibility for one village. However, because of peer learning, trainees are also familiar with what is happening in the villages of the other PLT members. Each mentor was responsible for giving guidance and support to three to four PLTs, amounting to 9 to 16 trainees altogether.

GOING TO SCALE WITH FACILITATION FOR CHANGE

The broader concept of scale

In referring to the issue of 'scale', several authors differentiate between 'scaling up' and 'scaling out'. Some authors (e.g. Edwards and Hulme, 1992; Howes and Sattar, 1992; Blackburn and Holland, 1998) use only the term 'scaling up' to refer to what the others call 'scaling up and/or out'. Among the second group are Uvin and Miller (1994), who further categorize scaling up as quantitative, functional, political or organizational. In this chapter, we use 'scaling up' in the latter sense, including horizontal expansion, intensification and vertical integration.

The triggers for going to scale

A combination of factors triggered the decision to scale up PEA. After three years of piloting, LDA senior management made an intensive impact assessment at different levels. This showed the high impact of PEA in the pilot villages in terms of, among other factors:

- self-organizational capacity and linkages with service providers, including input and output markets;
- inclusiveness, collective action, bargaining power and economies of scale;
- enhanced leadership skills and recognition of women leaders;

- mobilization of own resources for services (self-reliance);
- learning through experimentation and farmer-to-farmer sharing; and
- increased production, market orientation and income generation.

The LDA saw PEA as addressing its focal areas and therefore decided to scale up and institutionalize the approach.

Going to scale: What it meant in this context

Horizontal expansion of the participatory extension approach
Going to scale meant expanding or spreading the application of PEA. LDA adopted two approaches to horizontal expansion. The first one involved developing more facilitation competence of extension officers to take the process further in more villages. This was envisaged even at the beginning of the piloting phase.

The second approach involved providing farmer trainers with the skills to facilitate the spread of technologies beyond the PEA villages. This was not envisaged at the beginning, but came as a response to the spread of technologies from the pilot villages to others. While the extension officers were going through the lengthy learning process, the technologies were spreading quickly from farmer to farmer. Selected farmers in the pilot villages were trained in different technical areas in order to enhance their technical and facilitation capacities and thus improve farmer-to-farmer extension.

Extension officers trained in the participatory extension approach
From 1998 to 2000, 35 extension officers working in six villages in two sub-districts (Vhembe and Capricorn) were trained. This was the 'first generation'. In the second-generation phase (2001 to 2003), the number of trainees increased to 103 in the two initial districts, working in more than 80 villages. After 2003, the process of competence development spread to the other four sub-districts in Limpopo Province.

According to the project process review (PPR) (BASED, 2005), by June 2005, 389 extension officers had been trained in the five phases of the PEA learning cycle and were applying the approach in 211 villages in five of the six districts of the province. The PPR also revealed that 142 (37 per cent) officers had been trained in SWC, 109 (28 per cent) in SFM, 71 (18 per cent) in LP and 67 (17 per cent) in SSSP. Each officer was trained in only one technical area. The idea was that the PLT members would be trained in different areas to allow for complementarity.

Farmers trained as farmer trainers
The PPR (2005) revealed that about 200 farmer trainers had been trained in the same four technical areas to support the spread of technologies from farmer to farmer. It also revealed that 105 villages in five districts of

Table 23.2 *Number of villages in Limpopo Province practising different technical areas, 2005*

Technical area	District					Total
	Vhembe	Capricorn	Mopani	Sekhukhune	Bothlabelo	
Soil fertility management (SFM)	74	24	2	2	3	105
Soil and water conservation (SWC)	68	21	3	3	4	99
Small-scale seed production (SSSP)	68	23	0	4	3	98
Livestock production (LP)	75	12	1	4	3	95

Source: BASED (2005)

Limpopo Province were by then innovating in SFM, 99 villages in SWC, 98 villages in SSSP and 95 in LP. Table 23.2 shows the number of villages that were practising different technical areas in Limpopo Province in the five districts. The figures indicate that some villages were applying more than one type of new technology.

The participatory extension approach beyond Limpopo Province
Beside the horizontal expansion within Limpopo Province, in 2001 PEA was also initiated in Eastern Cape Province and in 2002 in Mpumalanga Province. This was facilitated by the BASED team in partnership with the respective provincial GTZ-funded programmes: Promotion of Rural Livelihoods (RULIV) in the Eastern Cape and Mpumalanga Rural Development Programme (MRDP) within the Department of Agriculture, Conservation and Environment (DACE). Thus, when the BASED team started larger-scale implementation of PEA in Limpopo Province in 2001, it was also establishing the process in two other provinces.

Intensification of activities
As the PEA learning cycle suggests (see Figure 23.1), action and learning form the main mode of operation. By using short, reflective cycles, the BASED team could continuously assess its activities and adapt the focus of its work accordingly. Along these lines, the scaling-up programme involved not only expansion to cover a larger constituency, but also an

increase in, or diversification of, its activities. Thus, the 'intensification' (Howes and Satter, 1992) occurred in two ways:

1   changing the focus and strategies in the PEA operational areas, as the process moved from a pilot to the entire province; and
2   intensifying within different technical areas.

As an example of the latter, in small-scale seed production, different activities were added to intensify the value chain. The farmers had started simply by testing different maize varieties. They selected the best-performing ones and produced seed. High demand for the seed prompted a need to produce more. To be able to market the seed nationally and internationally, it had to be certified by the national regulatory authority. A seed unit had to be established for treating and packaging seed, with regular inspection to ensure that the seed would meet the quality standards.

Vertical integration
After the LDA management had assessed the impact of the PEA pilots and decided to adopt the approach, it had to find mechanisms for integrating PEA within its existing structure and system. It did so by mandating the senior manager of extension to be the 'champion' to oversee the overall PEA integration process and by establishing a provincial change management team to facilitate the PEA integration activities.

MAJOR LESSONS AND INSIGHTS

Lessons in terms of developing facilitation competence

Learning workshops

- *Learning versus training workshops.* The emphasis on learning rather than training suggests that the learning involves co-generating knowledge grounded in people's experience, rather than receiving it from one who knows better. While the principle worked very well, it makes high demands on the quality of trainers. It is easy to take people through modules; but this may not lead to the desired outcome. Exposing the real issues, and confronting and provoking people, require deep experience and good orientation of the trainers. This has been a major challenge in scaling up.
- *Systemic nature of competence development.* Facilitation of competence development and of the PEA process as a whole is a systemic intervention based on principles of process orientation and strategic thinking. When one part in a system is moved, many other parts will move

as well. Often this is unpredictable and must therefore be observed and analysed closely. The intervention needs to be adapted step by step. Facilitation of this flexible process with its interconnected parts is a great challenge, and the trainers struggled with it. One should not expect quick success through a training-of-trainers approach, but rather regard development of trainers as a longer-term coaching process.

- *Appreciation of current success as a starting point.* At the outset, it is crucial to create awareness about the current situation of extension services in terms of roles, responsibilities and vision, and to appreciate the successes and failures of current approaches. This lessens resistance. Instead of giving people the impression that the introduction of PEA replaces the previous approach, it tries to recognize the good things about the old ways and then seeks to add value by providing strategies for dealing with the remaining challenges that people face. Exposing the trainees to concrete cases during the first workshop helped to stimulate them to imagine alternatives.

- *Learning through self-reflection.* The short iterative cycles in the PEA process were crucial in enabling action learning and reflection, making the process more manageable and fuelling the energy. The longer the time without contact with the learners, the more the energy waned. The approach allowed for flexibility and adaptation to accommodate emerging issues along the way, while stimulating capacities to grow and enabling a better understanding of the process.

- *Feedback and sharing trainees' field experiences.* Laying a good foundation for sharing by consciously promoting a feedback culture from the outset was vital in stimulating debate and experiential learning. During the sharing, trainees questioned each other and demanded transparency and evidence of the progress that their fellow trainees had made. This created peer pressure for the trainees to be active during their field activities so that they did not lose face when they had to report on their progress. In the case of public servants who receive no additional incentives, this was important. The sharing also served as a platform for developing a pool of possible solutions to the challenges faced by the trainees.

- *Using codes, role plays, simulations, proverbs and songs.* PEA draws its strength from the use of such communication tools at all levels. These tools were very effective in instilling a culture of sharing, self-reliance, cooperation and self-organization. People were challenged to reflect critically on their situation and patterns of behaviour, 'de-politicizing' issues and engaging in a learning process geared towards their own development. More important than the tool *per se* was the extension officer's ability to use it appropriately and to facilitate 'decoding' in a way that people could learn from it.

Field practice

- *Peer learning.* The formation of PLTs during field practice provided a strong support base, especially for new extension officers. 'Knowing that I am not alone helped boost my confidence when addressing the entire community for the first time,' one extension officer said. The PLTs also increased collaboration and co-learning among extension officers. The fact that PLT members become acquainted with the villages in which their members worked ensured continuity of activities even if one team member was absent. The PLT members specialized in different technical areas, which compelled them to collaborate in order to complement each other. However, this had limitations in the sense that some extension officers tended to promote their specific technical areas in the communities where they worked.
- *Mentoring and coaching.* This was important during field practice to help guide the PLTs in operationalizing PEA. The PLTs that reported having received support from their mentors in terms of regular joint planning and feedback meetings outperformed those that complained about not having enough support from their mentors.

Lesson in terms of going to scale

Going to scale through horizontal expansion, intensification of activities or vertical integration had many implications. As the PEA process expanded to involve more people at different levels and as the activities intensified, it became more complex and required complex measures to manage it. Some trade-offs that came with going to scale are outlined in the following sub-sections.

Quality assurance
This was a big challenge as PEA unfolded from one generation to the next in the process of scaling up. Dilution of the approach is inevitable because of the different contexts in which each generation of trainees implemented it.

During the third-generation training, some dilution effects were observed. Because the government exerted pressure to train more extension officers, but financial resources were limited, all learning workshops were cut back to a maximum of five days per phase. This meant that what had been done in 10 to 15 days in the first and second generations had to be squeezed into half or one-third of the time. This affected the content and quality of the training. The process of generating knowledge based on experience, which the trainees in previous generations had gone through, was lost along the way. Issues that were generated and documented in the earlier workshops tended to be copied and presented as standard knowledge in a 'cut-and-paste' manner, instead of taking the trainees through a learning process. This raised questions of whether future trainers would emerge from this third generation.

Process documentation was another weakness as it became very shallow during the third generation of competence development.

Technology versus process
Because processes are intangible, knowledge about them travels more slowly than knowledge about technologies. People remember what they see and what they have achieved, but tend to forget how they got there. In some cases, extension officers continuously reminded the communities about the process they had gone through and the benefits attained. This helped the communities to internalize the PEA process and its values.

This issue also arose when neighbouring communities adopted technologies that had been developed in the pilot villages, without having gone through the learning process themselves and without the accompanying organizational development that is key for lasting success. Disseminating technology is a great achievement; but developing farmers' capacities to conduct farmer-to-farmer training in technical areas not only helped to spread PEA in an organized manner, but also encouraged farmers to learn since they learn better from their fellow farmers.

Larger scale versus inclusiveness
As the process moved from one generation to the next – expanding in size and increasing in complexity – inclusiveness also suffered. Those who became trainers naturally gained recognition, while the majority of extension officers who worked only in their communities became less visible. The least included were those PEA learners who did not manage to complete all of the steps in the operational framework during the formal learning cycle. This increased the likelihood of an unseen collapse of the process. There is a need to find a balance between the so-called 'super facilitators' and the mass of PEA practitioners in order not to create jealousy among the latter. The challenge is how to keep the majority on board while creating champions to take the process further.

Lessons in terms of institutional response to pressures to scale up

Getting the buy-in from the Limpopo Province Department of Agriculture
The involvement of the LDA senior management in designing and conducting the impact assessment and their exposure to the pilot cases played a significant role in encouraging them not only to appreciate the contributions made by PEA, but also to adopt it as a promising approach to improving the delivery of extension services.

Integration of the participatory extension approach in Limpopo Province Department of Agriculture
Adoption of the approach forced the LDA to radically adjust its service delivery system and budget allocation process. In addition to establishing

a provincial change management team to facilitate the integration, PEA competencies were included in the contractual agreement within LDA's performance management system. This recognized that while the facilitation competence of extension officers is the centre of PEA, its success depends upon other competencies needed to support the process. In this light, LDA identified the minimum competence requirements of various actors ranging from the top management to the lowest level in the hierarchy in order to strengthen support for PEA integration.

LDA also allocated funds for competence development in the districts and other activities related to the integration of PEA within its system.

FUTURE CHALLENGES

The challenges for the future include:

- *Keeping momentum in initial areas while operating in new areas.* The pilots must continue to be supported and further developed as a source of inspiration for new trainees being exposed to the approach in practice and for further concept development and learning for the future. The pilot cases remain the forerunners of the approach.
- *Maintaining quality of PEA while scaling up.* The challenge was to maintain high quality of learning matched with the available human and financial resources of LDA (i.e. to avoid dilution and blueprinting of the approach while scaling up). It was difficult to build up a quality assurance system to continue and further develop the approach. Without this, the training will probably be the same ten years later and will have lost its energy. New ideas, concepts, methods and tools are needed to keep an approach alive.
- *Maintaining critical mass.* The more that people become 'capacitated', the more attractive they become for other organizations, leading to higher staff turnover. A strategy for building and maintaining high critical mass therefore needs to be developed in order to keep the process going. Without a continuous nurturing and grooming of new people, the competence will soon be exhausted because of staff turnover or – as we observed – good staff being promoted to management positions and removed from the field.
- *Harmonizing PEA with other departmental programmes and projects.* This remains a major challenge for LDA. There are many other donor programmes within LDA, with their own mandates and differing approaches. Some still operate in a mode that promotes dependency, destroying the value of self-reliance promoted by PEA.

These are enormous challenges in public service institutions, where human resource development is bound by many regulations. Thus, the process of competence development needs to be continued and embedded within a quality assurance system for facilitating learning and implementation.

REFERENCES

BASED (Broadening Agricultural Services and Extension Delivery) (2005) Project Progress Review by LDA Senior Management, Unpublished Report for BASED, Polokwane, South Africa

Blackburn, J. and Holland, J. (1998) *Who Changes? Institutionalising Participation in Development*, Intermediate Technology Publications, London

Edwards, M. and Hulme, D. (1992) 'Scaling up NGO impact on development', in M. Edwards and D. Hulme (eds) *Making a Difference: NGOs and Development in a Changing World*, Earthscan, London

Goleman, D. (1995) *Emotional Intelligence: Why It Can Matter More than IQ*, Bantam Books, New York, NY

Groot, A. E. (2002) 'Demystifying facilitation of multiple-actor learning processes', PhD thesis, Wageningen University, Wageningen, The Netherlands

Hagmann, J. (1999) *Learning Together for Change: Facilitating Innovation in Natural Resource Management through Learning Process Approaches in Rural Livelihoods in Zimbabwe*, Margraf, Weikersheim, Germany

Hagmann, J., Chuma, E., Murwira, K. and Connolly, M. (1998) *Learning Together through Participatory Extension: A Guide to an Approach Developed in Zimbabwe*, Department of Agriculture, Technical and Extension Services, GTZ/ITDG, Harare, Zimbabwe

Hagmann, J., Chuma, E., Murwira, K. and Connolly, M. (1999) 'Putting process into practice: Operationalising participatory extension', Agricultural Research and Extension (AGREN) Network Paper 94, Overseas Development Institute, London

Hagmann, J., Moyo, E., Chuma, E., Murwira, K., Ramaru, J. and Ficarelli, P. (2003) 'Learning about developing competence to facilitate rural extension processes', in C. Wettasinha, L. van Veldhuizen and A. Water-Bayer (eds) *Advancing Participatory Technology Development: Case Studies on Integration into Agricultural Research, Extension and Education*, IIRR, Silang, The Philippines/ETC EcoCulture, Leusden, The Netherlands/CTA, Wageningen, The Netherlands, pp21–38

Howes, M. and Sattar, M. G (1992) 'Bigger and better? Scaling-up strategies pursued by BRAC 1972–1991', in M. Edwards and D. Hulme (eds) *Making a Difference: NGOs and Development in a Changing World*, Earthscan, London, pp99–110

Moyo, E. and Hagmann, J. (2000) 'Facilitating competency development to put learning process approaches into practice in rural extension', in *Human Resources in Agricultural and Rural Development*, FAO, Rome, Italy, pp143–157

Ramaru, J., Hagmann, J., Chuma, E., Ficarelli, P., Netshivhodza, M. and Mamabolo, Z. (2004) 'Building linkages and bargaining power between smallholder farmers and service providers: Learning from a case on soil fertility inputs in South Africa', *Ugandan Journal of Agricultural Sciences*, vol 9, pp2004–2014

Rough, J. (2002) *Society's Breakthrough: Releasing Essential Wisdom and Virtue in All the People,* First Books Library, Bloomington, IN

Schein, E. (1992) *Organizational Culture and Leadership*, Jossey-Bass, San Francisco, CA

Senge, P. M. (1990) *The Fifth Discipline: The Art and Practice of the Learning Organization*, Bantam Doubleday Dell Publishing, New York, NY

Uvin, P. and Miller, D (1994) 'Scaling up: Thinking through the issues', The World Hunger Program, www.globalpolicy.org/ngos/intro/growing/2000/1204.htm

CHAPTER 24

Building Capacity for Participatory Monitoring and Evaluation: Integrating Stakeholders' Perspectives

Jemimah Njuki, Susan Kaaria, Pascal C. Sanginga, Festus Murithi, Micheal Njunie and Kadenge Lewa

The idea of participation in development and, with it, participatory methodologies have now become widely accepted internationally. The language of participation is used by multilaterals such as the World Bank and many United Nations agencies, as well as by bilateral donors. International non-governmental organizations (NGOs) have played an important role in catalysing the spread of participatory methodologies worldwide. In the recent past, research organizations have also embraced participatory approaches, innovation-system perspectives and systems thinking. Participatory approaches have the potential to substantially increase the downward accountability in the development process and to contribute to empowerment of civil society (Chambers and Guijt, 1995; Chambers, 1997, 2005). While many agricultural research and development (ARD) organizations have embraced these approaches, especially in planning and implementation (Maguire, 1987), they often revert to traditional models when the time for programme evaluation arrives (Abbot and Guijt, 1998). Most evaluations in ARD organizations are conducted, with some exceptions, by consultants or outside researchers, rather than through *participatory* evaluation involving project stakeholders (Grundy, 1997). These evaluations are driven more by donor criteria and demands for accountability than by local needs and long-term interest in programme improvement. The shortcomings of such traditional evaluations are well documented and include evaluation questions developed by consultants and donors with a narrow focus; limited use of evaluation results for programme improvement; lack of capacity-building for mainstreaming evaluation practice into organizations; lack of integration of local knowledge to achieve programme improvement; and collection of data not responsive to the needs of programme staff and stakeholders (Carden, 1997; Patton, 1997; Sanders, 2003).

Participatory evaluation has developed in response to the constraints of traditional evaluation models, as well as growth in participatory research methods (e.g. Chambers, 1997). As practitioners integrated local participation within project planning, the need for participatory strategies in monitoring and evaluation (M&E) became apparent (Feuerstein, 1986; Mulwa, 1993; Jackson and Kassam, 1998). Participatory evaluation builds on the importance of capacity-building and learning, the need to adapt methods to the local context and the involvement of multiple stakeholders, who negotiate issues to be monitored and the indicators and methods. It promotes the use of, and learning from, the evaluation findings and the process (Abbot and Guijt, 1998; Cousins and Whitmore, 1998; Davis-Case, 1990; Patton, 1997). Bringing stakeholders, including beneficiaries, together to undertake the analysis leads to practical and mutually acceptable solutions that can then be adopted more quickly by all concerned. There are many documented outcomes of participatory evaluation, which include but are not limited to increased validity and utilization of results, improved communication between funding agencies and local partners, enhanced local capacity for decision-making, and cost-effective strategies for ongoing assessment (Bryk, 1983; Feuerstein, 1986; Pfohl, 1986; Rugh, 1986, 1994; Narayan-Parker, 1993; Johnson, 1995; Cousins and Whitmore, 1998; Jackson and Kassam, 1998).

The literature now goes beyond participatory to empowerment evaluation. Cousins and Whitmore (1998) distinguish between two types of participatory evaluation: practical and transformative. They define the former as evaluation that supports programme, policy or organizational decision-making, with the primary function of involving stakeholders in order to foster evaluation use – mainly instrumental, conceptual and symbolic use. Increasingly, research has focused on these three dimensions of use (Knorr, 1977; Alkin et al, 1979) and more recently on process use (Patton, 1997; Henry and Mark, 2003). Transformative participatory evaluation, which grew out of participatory research and participatory action research, involves the application of participatory principles and action in order to achieve social change. Its primary function is to empower individuals and groups.

In current literature on empowerment evaluation (Fetterman, 1994; Fetterman and Wandersman, 2005), transformative participatory evaluation leans more towards empowerment evaluation, while practical participatory evaluation leans towards collaborative evaluation. Fetterman (1994) defines empowerment evaluation as the use of evaluation concepts, techniques and findings to foster improvement and self-determination. Wandersman (1999) and Wandersman et al (2005) expanded this definition to emphasize empowerment evaluation as an approach that helps to increase the probability of achieving results by providing practitioners with tools for assessing, planning, implementing and evaluating their programmes, thus mainstreaming evaluation as part of programme planning and implementation. Here, we expand further on this definition by focusing on the process of strengthening the capacity of programme staff and other stakeholders, including the communities with whom they work, to use participatory tools for planning, implementation and evaluation, and to use reflection and evidence-based strategies to increase the likelihood of achieving results that satisfy the different stakeholder groups.

The development of evaluation capacity had to start with the existing situation and a diagnosis of actual needs. It is a process that takes time and goes through a number of stages and in which the role of the facilitator changes as the capacity is built. The objective of this evaluation research and capacity-building programme was threefold:

1   Build the capacity of staff of the Kenya Agricultural Research Institute (KARI) in participatory monitoring and evaluation (PM&E).
2   Integrate other stakeholders within the PM&E process, including organizations collaborating in KARI's research programmes and community groups.
3   Achieve organizational learning and change through the mainstreaming of evaluation in KARI.

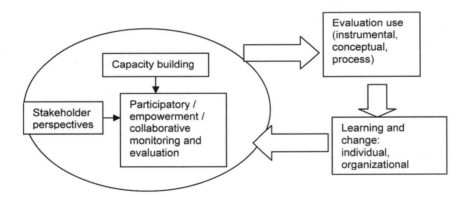

Figure 24.1 *The conceptual framework for developing evaluation capacity*

Kenya Agricultural Research Institute

KARI is the research organization in Kenya with a mandate for national research. It collaborates with other governmental, non-governmental and international organizations to fulfil this mandate. KARI manages and supports four main research programmes, within which many individual research projects are undertaken:

1   *Crops Programme:* focuses on profitability of farming, contribution to food security, improved livelihoods and providing a basis for industrialization.

2   *Livestock and Range Programme:* focuses on contribution of livestock to net farm incomes and national food self-sufficiency through developing economically efficient, socially acceptable and ecologically sound technological production packages.

3   *Land and Water Management Programme:* concentrates on improved soil and land productivity, optimum use of rainfall and other water resources, and integrated natural resource management to ensure the sustainable productive use of the natural resource base.

4   *Socio-Economic Programme:* focuses on methods of participatory research, technology transfer and M&E of technological packages with regard to adoption and impact, and contributions to policies affecting agricultural production.

In addition, KARI has a number of cross-cutting programmes, including the Seed Unit, the Agricultural Technology and Information Response Initiative (ATIRI), the Agricultural Research Fund and Agricultural Research Investments and Services. Administratively, KARI is organized into a headquarters and national and regional sub-centres that are located throughout the country.

The collaborative initiative on monitoring and evaluation

In an effort to promote the use of participatory approaches, especially in M&E, the Socio-Economics Programme and the International Centre for Tropical Agriculture (CIAT) developed a collaborative initiative to implement and mainstream M&E systems and processes into research. The pilot phase, which we describe in this chapter, involved five of the KARI centres and ten research projects. These were drawn from across the research programmes and cross-cutting programmes, involved multiple partnerships, were funded by different donors and were at different stages of implementation.

Assessment process and results

Processes become more sustainable if they build on existing structures and opportunities within organizations. Based on this premise, a facilitated self-assessment of M&E was the first step in building a robust M&E system within KARI. Over 100 researchers and extension staff involved in the research programmes did the assessment according to the strengths, weaknesses, opportunities and threats (SWOT) framework during five workshops in five KARI centres or sites.

Some of the strengths identified included the existence of an information management system and a commitment by both researchers and management to PM&E for learning and improvement. Identified weaknesses included the lack of a systematic process for monitoring, low skills in M&E, lack of in-built M&E during project development and non-involvement of stakeholders.

Based on this assessment, key intervention areas were jointly developed. These included:

- building the capacity of scientists to establish and support PM&E systems;
- facilitating scientists to build the skills of communities and other local stakeholders in PM&E;
- conducting training for attitude change; and
- implementing an action-learning process integrating PM&E in selected pilot projects.

Capacity-building would focus on:

- identifying different stakeholders and their roles in the PM&E process (including farmers and other community members, etc.);
- strategies to develop appropriate qualitative and quantitative indicators;
- integrating gender and equity issues within the PM&E process;
- skills to facilitate interaction between scientists, farmers and other stakeholders;
- data analysis in PM&E at different levels; and
- data management, analysis, interpretation and use.

This would include synthesizing PM&E data to facilitate its use for decision-making at different levels in order to provide feedback and to support learning.

THE CAPACITY-BUILDING PROCESS

Capacity-building was carried out in four phases:

1 awareness training for research managers and researchers aimed at creating buy-in to the process and changing long-held attitudes towards M&E;
2 training a core team (or the 'champions');
3 on-site training workshops; and
4 a two-year mentoring process with selected research projects as pilot learning sites.

Awareness and training for attitude change

The assessment revealed that most of the researchers viewed M&E as a policing tool, as something done to them by consultants sent by donors, project managers and/or research managers, mainly to check on whether they were doing what they were supposed to be doing. They therefore had a negative attitude to M&E and believed it was not meant to be done by programme staff, but rather by others. This implied that researchers regarded M&E from the perspective of accountability and not from the other perspectives of learning and programme improvement. Wandersman and Snell-Johns (2005) describe this as the greatest challenge to evaluation: fear and resistance because of what they describe as the 'gotcha' mentality associated with an audit-like approach to evaluation. However, evaluation does not always have to be about catching people doing something wrong, and it does not have to involve winners and losers.

It was apparent that a change in attitude was needed before starting to build technical capacity in evaluation and that this change had to occur at both levels:

- researchers involved in programme implementation; and
- research and project managers doing the M&E checks.

This start-up phase of the training was designed to demystify evaluation and to address the misconception that evaluation is an activity conducted only by outsiders and project managers. The training involved both researchers and management.

Using graphics and occurrences in the daily lives of people, such as planting a crop or building a house, the concepts of M&E and the purposes for which it can be used were discussed. The question as to who should do M&E, given its broad functions, became very evident (i.e. that researchers themselves had to be involved and, in fact, take charge of the M&E process in their programmes).

The challenges of doing this were identified. These included limited capacity for evaluation and finding ways of mainstreaming M&E into project implementation so that it was not a separate and marginalized activity, but rather an integrated part of the process. Including research managers in this assessment created the necessary support for making the programme changes that would become necessary. Project teams of researchers started seeing M&E as a learning process, as something they do for themselves to improve project implementation, achieve the desired outcomes and involve all interested parties.

Building capacity for technical skills in participatory monitoring and evaluation

This phase of the capacity-building had two aims:

1   Develop a strategy for collecting evidence on which changes and improvements should be made.

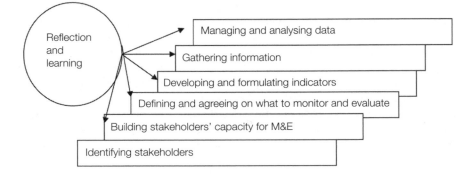

Managing and analysing data

Gathering information

Developing and formulating indicators

Defining and agreeing on what to monitor and evaluate

Building stakeholders' capacity for M&E

Identifying stakeholders

Reflection and learning

Figure 24.2 *The stepwise participatory monitoring and evaluation implementation process*

Source: adapted from Pali et al (2005)

2   Create a culture in which researchers reflect on the effectiveness of their work and make necessary adjustments in order to achieve better results.

The monitoring process is especially important in identifying barriers to achieving results and addressing these barriers on time (Wandersman and Snells-John, 2005). The technical capacity-building was organized around methodological steps in the PM&E process, as shown in Figure 24.2.

Building capacity to involve communities in participatory monitoring and evaluation

While this training goes hand-in-hand with the technical skills training, it is treated separately here for emphasis as it goes beyond technical skills. It involves training in the use of participatory approaches to M&E with communities and the facilitation skills required to engage stakeholders and, especially, communities in the process. The traditional role for communities in M&E has been to provide information required for the evaluation and, in some cases, to collect basic monitoring data. The main objective of this training was to look beyond this traditional role and integrate communities as strategic partners, contributing to the content and process of PM&E. Of course, capacity-building for communities requires a different set of skills and tools compared to capacity-building for researchers and extension agents.

Although the extension staff and even the researchers had been working with communities for years, one of the expressed needs during the self-assessment was for training in planning and facilitation skills. Working with communities requires not only having the technical capacity or technical messages, technologies and information to deliver to farmers, but also having skills to engage effectively with communities as partners in a way that allows mutual learning by researchers and communities. This engagement determines the extent to which communities feel part of the project. The hypothesis of this training was that 'process precedes content'. The training covered skills such as how to plan for field activities, good interpersonal communication (active listening, verbal and non-verbal communication, language), sensitivity to others' needs and culture, and demonstrating respect and building trust.

Bringing it all together: The action learning and mentoring process

The most critical part of the capacity-building was to bring these three processes together in an action-learning and mentoring programme, or what Freire (1987) calls 'learning to do it by doing it'. A cycle of workshops followed by implementation was designed that complemented technical skills in PM&E with community facilitation skills. Action learning, like action research, is a process through which participants learn by doing.

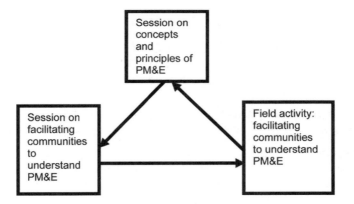

Figure 24.3 *The mentoring process: Understanding concepts of participatory monitoring and evaluation*

For example, a session for researchers on understanding the concepts and principles of PM&E was followed by a session on how to facilitate communities to understand PM&E and the tools used, and this was followed by fieldwork. After the fieldwork came a reflection process, the lessons from which fed into the next training session. Some concepts, such as developing the results chain, went beyond a three-stage process, as illustrated in Figure 24.4.

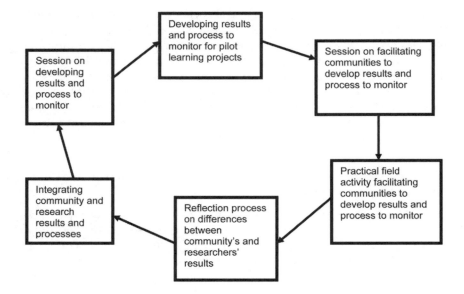

Figure 24.4 *The mentoring process: Defining what to monitor*

PM&E PROCESS OUTCOMES RELATED TO CAPACITY STRENGTHENING AND COMMUNITY INVOLVEMENT

Integrating community objectives within the participatory monitoring and evaluation framework

Involving different stakeholders led to broadening how the researchers viewed project results. While their focus had been mainly on project outputs (e.g. number of technologies that they developed, number of farmers who adopted them and number of papers published from their work), the communities viewed the results in terms of the extent to which these would improve their lives in different ways. This meant that researchers had to extend their results chain to go beyond the direct outputs to look at the implications of these outputs for farmers' lives and livelihoods. Table 24.1 provides an example of the expected results of projects as articulated by researchers and the expected results of the same projects as articulated by farmers involved in it.

During the reflection meetings on these different expectations with regard to results, we asked: 'Are these expected results realistic based on the current activities, or do we need to include other activities in order to meet them?' A yes to the first question led to broadening the scope for M&E to include these outcomes, while a no to this question led to:

- a rejection of some of the farmers' expected results; or
- a readjustment of activities to include activities that would lead to achieving these results.

Table 24.1 *Differences in results expected by researchers and farmers*

Type of project	Researchers' expected results	Farmers' expected results
Up-scaling soil and water management technologies using farmer field school (FFS) methodology	Increased adoption of soil and water management technologies Improved soil fertility (organic matter content, pH) Increased yields	Increased yields from major crops Improved food security Increased incomes Ability to do own experiments to solve farming problems Increased confidence to diagnose community problems and approach service providers
Improving banana through tissue-culture technologies	Farmers' increased adoption of improved banana varieties Increased availability of banana-planting material at community level	Improved banana production Increased income from banana sales Increased income from selling tissue-culture planting material Increased recognition within the village as sources of planting material and knowledge on the banana

Integrating community indicators within participatory monitoring and evaluation: Measuring the unusual

Another interesting result was the clear difference between researchers' and community indicators (see Table 24.2). While the former tended to be generic and mainly quantitative, community indicators were mainly local, qualitative and a reflection of the community situation. Integrating these indicators within the PM&E system provided the researchers with insights into changes important to farmers and allowed these changes to be measured from the farmers' perspectives.

Change in relationships between researchers and communities

In the past, most applied research projects have used farmers as sources of information. Our process involved all stakeholders, including farmers, in making decisions about desired changes, activities to be carried out

Table 24.2 *Differences between researchers' and community indicators*

Expected result	Researchers' indicators	Community indicators
Increased food self-sufficiency	Changes in expenditure on non-food and food items Maize yields	Good health: shiny faces, reduced skin diseases, reduced cases of malnutrition Number of months that the harvested maize lasts Reduction in hunger period Increased ceremonies Fruits ripening on the plants (pawpaw and banana)
Capacity-building	Number of people trained Number of workshops held Diversity of topics in which farmers are trained	Application of knowledge on their farms Ability to carry out research on their own Capacity to approach the extension worker Self-reliance in finding seed, market and services from other organizations
Improved soil fertility	Nutrient levels (carbon, phosphorus, macronutrients) Change in pH Organic matter content	Types and presence of certain weeds 'Feel' of the soil Capacity of soil to hold water Increase in yields

and indicators of change, as well as in collecting, analysing and using the PM&E information. This, in effect, transformed the relationships between the researchers and the communities with whom they work. As some community members noted: 'This is the first time we have been asked what results we want from these projects. Before, scientists came with questionnaires, collected information and then went away. Now we are sitting together to have discussions about the project.'

The process also helped researchers to better understand community priorities and to engage the communities in better planning of the project activities:

Now when we [researchers and communities] both understand what we want to achieve from the projects, we are better able to plan for activities. There is more ownership of the projects by communities and therefore more sharing of roles and responsibilities between us and the communities.

The training in facilitation skills increased the confidence of researchers and extension agents in working with communities and in facilitating communities to organize themselves better to achieve results. As one technical assistant remarked:

> Now when I go to work with the communities, I have more confidence; I
> have skills to deal with different situations and to facilitate the communities
> for different issues. When women do not speak in the meeting, I know some
> ways to facilitate so that their ideas are also heard.

The integration of planning within the PM&E process ensured that the
researchers and extension agents were better prepared for their field
activities:

> Now we know what things we should do to make sure our projects are
> successful. Before, we would just carry out the activities without thinking
> through whether these activities were really leading us to the desired results.
> We have even added some activities because the activities we were carrying
> out before were not enough to achieve the results we want to achieve.

A researcher noted:

> Before, we would leave the office without planning for what we were going
> to do. We would hastily plan in the car on our way to the field; this would
> disorganize us and disorganize the farmers as well. In the end, we would
> not achieve much. Now we plan adequately and we involve the farmers
> and we achieve a lot when we go to work with the farmers because we are
> prepared and the farmers are prepared.

Creating a culture of reflection, learning and improvement

One of the main goals of empowerment evaluation that distinguishes it
from other forms of participatory evaluation is the creation of a culture of
reflection, where programme staff continuously consider their effectiveness
and make adjustments to improve their programmes (Fetterman, 2005;
Lenz et al, 2005; Wandersman and Snell-Johns, 2005). Achieving improvement requires that the programme staff and the organization change their
practice, procedures and behaviour. These cycles of reflection, learning and
action were based on both the researchers' perceptions of what was going
well – or not – with the programmes and the evidence and data collected
from stakeholders. These reflections were both on the programme and on
the PM&E process itself.

During the early stages in the PM&E process, an issue that came out of
the reflection, learning and action was that other researchers and extension
agents were relying too heavily on the core team of researchers who had
been trained in PM&E, leaving them too constrained in time to perform
their other duties in their research programmes. Table 24.3 provides some
results of the reflection and learning process.

By building in cycles of reflection, learning and action, both single-loop
and double-loop learning were achieved. Single-loop learning is when
improvements are made to the way in which existing rules, procedures

Table 24.3 *Example of reflection, learning and action cycle*

What?	What was not going well	Why?	Recommendation for action	Who will do it?
At researcher level				
The participatory monitoring and evaluation (PM&E) process	The monitoring and evaluation (M&E) process is not yet well integrated within the programmes, especially at community level	Researchers not sure at what stages of farmer field school (FFS) process the M&E should be integrated	Training workshop on integrating M&E within the FFS methodology	FFS facilitator and M&E facilitator to design training with input from researchers and extension facilitators
Soil Management Project, Kenya Agricultural Research Institute (KARI) Kitale	Project aimed to increase the adoption of technologies, including soybeans; but few farmers adopting soybeans, although group discussions indicated farmer interest in the crop	Low availability of soybean seed, a relatively new crop in the area	Implement a seed-system component where farmers multiply soybean at community level for distribution to other farmers	Project manager to source for initial basic seed Project team to organize community groups for seed multiplication
At community level				
Tissue-Culture Banana Project	Community has joint plot from which they sell bananas, but no clarity as to who should sell and who keeps the money Only the treasurer knows how much money the group has	Group did not allocate responsibility for selling bananas No periodic review of accounts	Establish a committee responsible for selling bananas Treasurer to make monthly review and presentation of group accounts to the whole group All sales of bananas to be receipted and receipts audited	Group members to develop criteria for committee and elect it Group treasurer to prepare monthly accounts to present to the group A community member to be asked to do independent audit of accounts once a year

and activities are applied in the programme and within community groups. Single-loop learning is often called 'thinking inside the box' and poses 'how?' questions, but almost never the more fundamental 'why?' questions. The double-loop learning is often called 'thinking outside the box'. It questions the underlying assumptions and principles upon which the rules and procedures are based.

From monitoring and evaluation to planning and budgeting

A main concern in M&E is that the information provided by the process influences neither decision-making during project implementation nor the planning of ongoing project development and new initiatives. The cycles of reflection, learning and action described above led to the inclusion of changes within existing projects, and plans were adjusted based on the recommendations made during the reflection meetings.

The inclusion of these lessons in planning for new projects was also evident. A notable case was the inclusion of the PM&E process in other approaches that KARI was using for adaptive research and for working with communities. The ATIRI programme, aimed at accelerating the uptake of agricultural technologies and information by communities through community-managed funds, has now integrated PM&E within the methodology that it applies on a pilot scale in one of the centres.

CONCLUSIONS

The achievements made in this collaborative effort were a result of a long-term partnership-building process between CIAT and KARI and between the evaluation facilitator and the research and extension staff involved in the process, as well as the communities. Moving from M&E by outsiders and managers for purposes of accountability to an M&E process geared towards learning and improvement relies heavily on this kind of partnership, but also on intensive capacity-building for the programme staff to be able to carry out the process and to integrate and mainstream it within the research programmes.

While this collaborative effort has achieved a lot, it has not been without its challenges. Balancing between the M&E process and actual implementation of programme activities was delicate in the initial phase of the process and involved much capacity-building. During the later stages of the process, M&E became better integrated within the programme activities. Programme staff must be prepared to make these initial investments in time and resources as the process develops. Internalization of the M&E process into the research programmes has been slow. Initially, the researchers and extension staff saw it as something separate from the day-to-day programme planning and implementation.

Another major challenge has been due to the fact that the process was introduced to research programmes that were already under way. This posed problems, especially when there were recommendations for change, as long negotiations were needed to be able to make these changes. In some cases, the programmes were so rigid that changes could not be made. The scaling out process to other programmes and projects within KARI that were not initially involved in the process has been slow. This is because KARI has been decentralized and, in most cases, the individual centres act independently from each other, so that opportunities for learning across centres are limited. The next step in this process is to identity some of the facilitating and constraining factors to the mainstreaming (both institutionalization and out-scaling) of the PM&E process.

REFERENCES

Abbot, J. and Guijt, I. (1998) *Changing Views on Change: Participatory Approaches to Monitoring the Environment*, International Institute for Environment and Development, London

Alkin, M., Daillak, R. and White, P. (1979) *Using Evaluations: Does Evaluation Make a Difference?* Sage, Beverly Hills, CA

Bryk, A. S. (ed) (1983) *Stakeholder-Based Evaluation*, Jossey-Bass, San Francisco, CA

Carden, F. (1997) 'Giving evaluation away: Challenges in a learning-based approach to institutional assessment', International Workshop on Participatory Monitoring and Evaluation, 24–29 November, Cavite, The Philippines

Chambers, R. (1997) *Whose Reality Counts? Putting the First Last*, Intermediate Technology Publications, London

Chambers, R. (2005) 'New directions in impact assessment for development: Methods and practice', University of Manchester, 24–25 November, Manchester, UK

Chambers, R. and Guijt, I. (1995) 'PRA five years later: Where are we now?' *Forest, Trees and People Newsletter*, vol 26/27, pp4–14

Cousins, J. B. and Whitmore, E. (1998) 'Framing participatory evaluation', *New Directions for Evaluation*, vol 80, pp5–23

Davis-Case, D. (1990) *The Community's Toolbox: The Idea, Methods and Tools for Participatory Assessment, Monitoring and Evaluation in Community Forestry*, Community Forestry Field Manual 2, FAO, Rome, Italy

Fetterman, D. M. (1994) 'Empowerment evaluation', *Evaluation Practice*, vol 15, no 11, pp1–15

Fetterman, D. M. (2005) 'A window into the heart and soul of empowerment evaluation (Looking through the lens of empowerment evaluation principles)', in D. M. Fetterman and A. Wandersman (eds) *Empowerment Evaluation Principles in Practice*, Guilford, New York, NY, pp1–26

Fetterman, D. M. and Wandersman, A. (2005) *Empowerment Evaluation Principles in Practice*, Guilford, New York, NY

Feuerstein, M. T. (1986) *Partners in Evaluation: Evaluating Development and Community Programmes with Participants*, MacMillan Education, Hong Kong

Freire, P. (1987) 'Creating alternative research methods: Learning to do it by doing it', in B. Hall and J. G. Greene (eds) 'Stakeholder participation in evaluation design: Is it worth the effort?', *Evaluation and Program Planning*, vol 10, pp379–394

Grundy, I. M. (1997) 'Participatory monitoring and evaluation of natural resource use in communal farming areas of Zimbabwe', International Workshop on Participatory Monitoring and Evaluation, 24–29 November, Cavite, The Philippines

Henry, G. T. and Mark, M. M. (2003) 'Beyond use: Understanding evaluation's influence on attitudes and actions', *American Journal of Evaluation*, vol 24, no 3, pp293–314

Jackson, E. T. and Kassam, Y. (eds) (1998) *Better Knowledge, Better Results: Participatory Evaluation in Development Cooperation*, Kumarian Press, West Hartford, CT

Johnson, R. B. (1995) 'Estimating an evaluation utilization model using conjoint measurement and analysis', *Evaluation Review*, vol 19, no 3, pp313–338

Knorr, K. D. (1977) 'Policymakers' use of social science knowledge: Symbolic or instrumental?', in C. H. Weiss (ed) *Using Social Research in Public Policy Making*, Lexington Books, Lexington, MA, pp165–182

Lenz, B. E., Imm, P. S., Yost, J. B., Johnson, N. P., Barron, C., Lindberg, M. S. and Treistman, J. (2005) 'Empowerment evaluation and organizational learning', in D. M. Fetterman and A. Wandersman (eds) *Empowerment Evaluation Principles in Practice*, Guilford Press, New York, NY, pp155–182

Maguire, P. (1987) *Doing Participatory Research: A Feminist Approach*, Centre for International Education, Amherst, MA

Mulwa, F. W. (1993) *Participatory Evaluation in Social Development Programmes*, Premese Africa Development Institute, Nairobi, Kenya

Narayan-Parker, D. (1993) 'Participatory evaluation: Tools for managing change in water and sanitation, Technical Paper 207, World Bank, Washington, DC

Pali, P. N., Nalukwago, G., Kaaria, S., Sanginga, P. and Kankwatsa, P. (2005) 'Empowering communities through participatory monitoring and evaluation in Tororo district', African Crop Science Conference Proceedings, vol 7, pp983–989

Patton, M. Q. (1997) *Utilization-Focused Evaluation: The New Century Text*, third edition, Sage, Thousand Oaks, CA

Pfohl, J. (1986) *Participatory Evaluation: A User's Guide*, Pact Publications, New York, NY

Rugh, J. (1986) *Self-Evaluation: Ideas for Participatory Evaluation of Rural Community Development Projects*, World Neighbors, Oklahoma City, OK

Rugh, J. (1994) 'Can participatory evaluation meet the needs of all stakeholders? A case study: Evaluating the World Neighbors West Africa program', American Evaluation Association Conference, 2–5 November, Boston, MA

Sanders, J. R. (2003) 'Mainstreaming evaluation', in J. J. Barnette and J. R. Sanders (eds) *The Mainstreaming of Evaluation*, International Institute for Environment and Development, London, pp46–52

Wandersman, A. (1999) 'Framing the evaluation of health and human service programmes in community settings: Assessing progress', *New Directions for Evaluation*, vol 83, pp95–102

Wandersman, A. and Snell-Johns, J. (2005) 'Empowerment evaluation: Clarity, dialogue and growth', *American Journal of Evaluation*, vol 26, no 3, pp421–428

Wandersman, A., Snell-Johns, J., Lentz, L., Fetterman, D. M., Keener, D. C. and Livet, M. (2005) 'The principles of empowerment evaluation', in D. M. Fetterman and A. Wandersman (eds) *Empowerment Evaluation Principles in Practice*, Guilford, New York, NY, pp27–41

CHAPTER 25

Innovation Africa: Beyond Rhetoric to Praxis

Pascal C. Sanginga, Ann Waters-Bayer, Susan Kaaria, Jemimah Njuki and Chesha Wettasinha

PARADIGM SHIFTS AND NEW PARADOXES

Agricultural research and development in Africa is undergoing a major paradigm shift, embracing an innovation systems perspective while, at the same time, implementing large agricultural development initiatives based on the 'technology-push' model. This book, like the Innovation Africa Symposium (IAS) from which it is derived, focuses the spotlight on a subject that is attracting increasing attention from researchers and practitioners in Africa. The innovation systems approach shifts attention away from research and the supply of scientific knowledge and technologies to an interactive multi-stakeholder process of change in which technology dissemination and market development are only some elements of the system. The contributors to this book argue that creating opportunities in small-scale African agriculture suggests moving to a new notion of innovation, from pushing technologies to creating opportunities through institutional development. This implies that agricultural innovation and the articulation of its challenges and opportunities need to be framed as an integrated technical, organizational, institutional and policy issue.

The concept of innovation has become central in many research and development programmes in Africa. It has been endorsed by an array of international and national bodies, as well as non-governmental organizations (NGOs) and governments in many African countries. Since the IAS, there have been further international conferences on innovation and innovation systems. In December 2007, the Institute of Development Studies (IDS) at the University of Sussex, UK, brought together some 80 agricultural practitioners, researchers, farmer leaders and donor representatives to reflect on the theme of Innovation for Agricultural Research and Development (Scoones et al, 2008). In April 2008, the International Food Policy Research Institute (IFPRI) convened an international symposium on Advancing Agriculture through Knowledge and Innovation. The World Bank Institute also organized an international conference on Developing Agricultural and Agribusiness Innovation in

Africa in May 2008. We anticipate that many more such events will be held in the near future.

Many international agricultural research centres operating in Africa have also established programmes and projects on innovation. The concept of innovation and innovation systems appears prominently in many strategic documents and medium-term plans. The trend is similar in national agricultural research organizations that now use the agricultural innovation systems concept as their underlying framework (see Chapters 4, 20 and 21 in this volume). Other large regional initiatives such as the Comprehensive Africa Agricultural Development Programme (CAADP) and the Forum for Agricultural Research in Africa (FARA) and its sub-regional forums have embraced these concepts. Some African agricultural universities are now reforming their curricula to train their graduates on agricultural innovation systems. For example, Makerere University in Uganda has just launched an undergraduate programme on rural innovation. Many NGOs and NGO-facilitated multi-stakeholder platforms now focus on promoting local innovation processes that lead to farmer-led joint experimentation with researchers, development agents and other local actors (see Chapter 15 in this volume). Donor organizations such as the International Development Research Centre (IDRC), the UK Department for International Development (DFID), the Rockefeller Foundation and others are likewise establishing programmes on innovation and explicitly embracing the innovation systems concept. The World Bank has put the concept of innovation at the centre of its 2008 *World Development Report*, which focused on agriculture for development. The Bank has also facilitated consultations and studies that led to the publication of a potentially influential book on agricultural innovation systems (World Bank, 2006).

Despite this wide recognition and acceptance of the innovation systems concept, Africa is currently experiencing the return of the conventional 'diffusion of innovations' model (Rogers, 1962). Large initiatives, such as the Alliance for a Green Revolution in Africa, the Millennium Villages, the Bill and Melinda Gates Foundation and the Sasakawa–Global 2000 Programme, are reverting to the Green Revolution model. Their response to agricultural development challenges in Africa is to focus on investing much more on disseminating improved seed and fertilizers and improving the efficiency of rural markets (Adesina, 2007).

Given this backdrop, how can the concept of innovation and innovation systems, and the experiences and lessons learned in applying this concept in different contexts and in different organizations, have an impact on today's agricultural development challenges and opportunities in Africa? How can innovation professionals help to make the current agricultural development interventions more effective and equitable in enriching the livelihoods of small-scale farmers?

INNOVATION: A FAD
OR A FUNDAMENTAL CONCEPT?

With this increasing interest in innovation systems and the paradoxical mushrooming of the traditional technology-transfer models, the first question that emerges after reading the chapters in this book is whether innovation systems is just a new label for existing approaches or really a fundamentally new concept. Röling (see Chapter 2) cautions that the word 'innovation' is used with different meanings and can represent very different perspectives, leading to considerable confusion or a real 'battlefield of knowledge'. A radical critique may regard the term innovation as a catch-all concept, potentially including all social variables in whatever context, a concept that could mean more or less anything and that is therefore not analytically and practically useful. Hall (2007) pleads for the innovation concept to be taken seriously, not just as a fad.

An important contribution of this book is its focus on conceptual clarity and empirical application of the innovation systems concept in different contexts of African agriculture. Röling (see Chapter 2) has lived through the major approaches and perspectives and played a role in many of them during the past four decades, from the 1960s diffusion-of-innovations era to the current focus on agricultural innovation systems. He uses his autobiography to clarify the meanings of these concepts, their applications and implications for agricultural research and development in different African contexts. He distinguishes between innovation as a noun, innovation as a process and innovation as a system. Waters-Bayer and her colleagues (see Chapter 15) make a clear distinction between innovation (without an 's') and innovations (with an 's').

The authors of chapters in this book refer to innovation in the broad sense of activities and processes associated with generating, disseminating, adapting and using new technical, institutional and managerial knowledge that brings about technical, social and economic, and environmental change. The concept of 'innovation' refers to the search for, development, adaptation, imitation and putting into use of technologies, approaches and methodologies that are new to a specific context and that have social and economic significance (Hall, 2007). Innovation can concern new products, new technologies, new markets, new ways of doing things (institutions) or new policies.

A consensus seems to emerge from these contributions, shifting from the earlier focus on technologies and innovations (with an 's') to the wider concepts of innovation processes and innovation systems. The World Bank (2006) defines an innovation system as a system that comprises the organizations, enterprises and individuals that demand and supply knowledge and technologies, and the policies, rules and mechanisms that affect the way in which different agents interact to share, access, exchange and use knowledge. In this case, then, innovation can be seen as the emergent property of interaction among stakeholders in a system (Bawden and Packam, 1993).

The second major contribution of this book is the recognition and emphasis that innovation encompasses *both* technical *and* institutional innovation, and not only the one or the other. It is the nature of their combination that requires attention. To illustrate this, Röling (see Chapter 2) distinguishes between technological change at the farm level that leads to high productivity within existing windows of opportunity, and institutional change at higher system levels that stretches these windows of opportunity to achieve impacts at scale in order to benefit more people over a wider geographic area more quickly and more lastingly (IIRR, 2000). Technological innovation creates opportunity for institutional innovation. On the other hand, institutional innovation creates space for creativity by facilitating interactions between different stakeholders and their knowledge, and this can lead to technological innovation.

A third contribution of this book is the attempt to move beyond false dichotomies and unhelpful debates opposing different approaches and ideologies, for example, between endogenous or local innovation and induced or external innovation; between technical innovation or technology focus and institutional innovation or institutional change; between local or indigenous knowledge and scientific or modern knowledge; between technology push and demand drive; between technology and market pragmatists and innovation systems rhetoricians and idealists. This book presents examples of how these different dichotomies actually come together to form the system.

APPROACHES TO INNOVATION

A related concept widely used in this book is integrated agricultural research for development (IAR4D). This is an action research approach to investigating and facilitating the organization of groups of stakeholders (including researchers) to innovate more effectively in response to changing complex agricultural and natural resource management (NRM) contexts in order to achieve developmental outcomes. IAR4D is increasingly being used as synonymous to the agricultural innovation systems approach of many African organizations such as FARA and its sub-regional forums – for example, the Association for Strengthening Agricultural Research in Eastern and Central Africa (ASARECA) and the West and Central African Council for Agricultural Research and Development (CORAF), as well as national agricultural research and extension systems. The IAR4D framework provides an example of how actors within an innovation system can organize to achieve innovation.

Amanuel and his colleagues (see Chapter 3 in this volume) discuss the characteristics of different perspectives on innovation systems, with particular focus on comparing the agricultural innovation systems (AIS) and agricultural knowledge and information systems (AKIS) perspectives. Their analysis is echoed by the recent summary of the Farmer First Revisited workshop (Scoones et al, 2008) on the characteristics of changing approaches to agricultural research and development. Rather than seeing new approaches as opposing and vilifying the previous ones, the contributors to this book see them as additional, cumulative and complementary (see Table 25.1).

Table 25.1 *Changing approaches to agricultural research and development*

Characteristics	Diffusion of innovations / transfer of technology	Farming systems research (FSR)	Farmer participatory research (FPR)	Innovation systems (IS)
Era	Central since 1960s	Starting in the 1970s and 1980s	From 1990s	2000s
Mental model and activities	Supply technologies through pipeline	Learn farmers' constraints through surveys	Collaborate in research	Co-develop innovations involving multi-stakeholder processes and partnerships
Knowledge and disciplines	Single-discipline driven (breeding)	Multidisciplinary (agronomy plus agricultural economics)	Interdisciplinary (plus sociology and farmer experts)	Transdiciplinary, holistic systems perspective
Scope	Productivity increase	Efficiency gains (input–output relationships)	Farm-based livelihoods	Value chains Institutional change
Core elements	Technology packages	Modified packages to overcome constraints	Joint production of knowledge and technologies	Shared learning and change, politics of demand, social networks of innovators
Drivers	Supply-push from research	Diagnose farmers' constraints and needs	Demand–pull from farmers	Responsiveness to changing contexts, patterns of interaction
Innovators	Scientists	Scientists and extension	Farmers and scientists together	Multiple actors, innovation platforms

Characteristics	Diffusion of innovations / transfer of technology	Farming systems research (FSR)	Farmer participatory research (FPR)	Innovation systems (IS)
Role of farmers	Adopters or laggards	Sources of information	Experimenters	Partners, entrepreneurs, innovators exerting demands
Role of scientists	Innovators	Experts	Collaborators	Partners, one of many responding to demands
Key changes sought	Farmer behaviour	Removing farmers' constraints	Empowering farmers	Institutional change, innovation capacity
Intended outcomes	Technology adoption and uptake	Farming system fit	Co-evolved technologies with better fit to livelihood systems	Capacities to innovate, learn and change
Sustainability	Undefined	Important	Explicit	Championed, normative and multidimensional

Source: adapted from Scoones et al (2008)

However, although there appears to be a widening understanding of many principles of agricultural innovation systems, the approach has been incorporated in only a fragmentary way within mainstream research and development work. The notions of innovation systems have rarely been given priority in processes of reforming agricultural research and extension in Africa. Many opportunities for drawing lessons from experience and defining good practice continue to be missed.

LEARNING TO INNOVATE

Many contributors suggest that we are still in the early days of improving our understanding and practice of innovation as a process and as a system. Therefore, they regard learning as a critical component in enhancing innovation systems. Several chapters in this book present experiences in building and strengthening the innovation capacity of different stakeholders, including farmers and other resource users, extension workers, agricultural researchers and university lecturers, who can be called 'students of innovation' (Hagmann, 2002; Hall, 2007). These experiences often aimed to transform functions, structures and competences in organizations, and to strengthen the skills of current professionals and the ability of their institutions to work together to promote agricultural innovation. These capacity-building efforts leave behind the typical 'one-off' training workshops of the past and build learning alliances that convene stakeholders along the resources–consumption–policy continuum, with complementary skills and expertise. These alliances follow the principles of mutual learning, resource sharing and knowledge management in ways that facilitate institutional change.

It is important to recognize that learning to think and act in new ways and changing attitudes is a complex and slow process (see Chapter 20 in this volume). Much of this comes down to educational systems that set the parameters for professional and organizational behaviour. While some of the chapters describe efforts in rethinking agricultural education in universities, these efforts have been scattered and isolated. Reforming agricultural education for development is a major frontier for building innovation capacity. Numerous important lessons have already been learned from a range of recent experiments in higher education in Uganda, Kenya, South Africa and Benin, among others. These lessons need to be applied to help more universities design new curricula and to develop learning processes that are embedded in the concept of innovation systems. The challenge is also to sustain such initiatives beyond externally funded projects to become an integral part of agricultural research, development and education institutions.

Learning to innovate is the practice of building learning and reflectivity into development-oriented projects and organizations (Earl et al, 2001). Critical self-reflection is crucial because it can lead to more responsible and ethical innovation (see Chapter 14). At the heart of an innovation system is continuous structured learning and feedback, responding to this information through adjusting actions and tracking outcomes. A participatory monitoring and evaluation (PM&E) system is the means for generating and using this information. Njuki and her colleagues (see Chapter 24) describe efforts to build capacity for PM&E and participatory impact assessment.

TOWARDS SOCIAL LEARNING

Learning is an interactive and socially embedded process. In a number of chapters, it is emphasized that innovation emerges from interactions among stakeholders. It arises from insights that are gained through collaborative investigation, deliberation, relationship-building, communication and shared power. As Röling (2002) points out, it is necessary to move from individual 'multiple cognitions' to interrelated 'distributed cognition' and to an understanding of group processes to capture the essence of social learning.

Social learning refers to learning that takes place when divergent interests, norms, values and constructions of reality meet in an environment that is conducive to learning (Wals, 2007). It is the collective action and reflection that occurs among different individuals and groups as they work together to improve the management of human and environmental interrelations. Social learning is based on three key ideas. First, all relevant stakeholders should be involved in the innovation system. Typically, no single stakeholder has all the necessary information, legal competencies, funds and other resources to manage a natural resource to his or her satisfaction; therefore, the stakeholders need to collaborate. Second, practising agriculture and NRM requires some form of organization. To facilitate collaboration and coordinate their actions in a sustained way, the stakeholders need to enter into a long-term working relationship. Third, an innovation process is a learning process. It requires the development of new knowledge, attitudes, skills and behaviours to deal with differences constructively, to adapt to change and to cope with uncertainty. Social learning strives to provide opportunities for sharing experience and knowledge through active social networks, shared vision and goals, experimentation and reflection.

INNOVATION PLATFORMS

The focus in an innovation systems approach is on building partnerships and networks for innovation in which several different stakeholders work together towards a common goal, seeking to learn from one another and to change the norms and practices that comprise the institutional context. However, many chapters in this volume refer to only partial innovation systems as they still confine themselves to interactions between only two or three groups of stakeholders: farmers and farmer organizations, extensionists and extension organizations, and/or researchers and research organizations or universities.

In the Foreword to this book, Matlon remarks that there was little involvement of the private sector in the IAS and in the contributions to this book. Some contributors to the IAS have attempted to address the challenges of linking farmers to markets. Kaaria and her colleagues (see Chapter 11) discuss an approach to enabling rural innovation that empowers rural communities to identify and analyse market opportunities and develop agro-enterprises that benefit both men and women, and that particularly empower women to manage their resources better. Coppock et al (see Chapter 7) describe how, through collective action, pastoral women have made links to markets and improved their livelihoods. Similarly, Shiferay et al (2006) show how market institutional innovations improve the bargaining power of farmers. The IAS featured other examples from Ethiopia, Nigeria, Uganda and Zimbabwe using the concept of innovation systems for different value chains (for details, see www.innovationafrica. net). These contributions resonate with the 2008 *World Development Report*, which recognizes that the empowerment of producers, including women, through organization is vital in enabling them to get a fair deal from market opportunities (World Bank, 2007). However, there is a danger that farmer organizations are promoted as a new panacea to overcome market failures, and insufficient attention is paid to the role of the private sector.

Amanuel et al (see Chapter 3) note that the innovation systems concept is derived from industrialized countries and market-driven economies, where the private sector plays a critical role. However, in the context of smallholder farming systems in Africa, which have little access to capital and operate under very diverse conditions, the critical role of the private sector still remains to be seen. There are huge challenges in stimulating private-sector interest in small-scale farming, which seldom provides high or quick returns to investments. Spielman and Grebmer's (2004) analysis of public–private partnerships in agricultural research suggests that some of the challenges relate to differing incentives, cultures and interests. The private sector can engage in research that produces short-term results and products that appeal to paying consumers, while public research and development organizations are concerned with addressing the needs of small-scale farmers with poor market access. Most private-sector companies will prefer a contracting mode of partnership with farmers rather than developing a balanced partnership. Moreover, for all actors, engagement in multi-stakeholder partnerships involves transaction and opportunity costs for attending meetings, field visits and workshops, and private-sector actors often regard these costs as being too high (Sanginga et al, 2007). Learning how to build good links between small-scale farmers, public research and development organizations, and the private sector is still a key challenge for operationalizing the innovation systems concept in sub-Saharan Africa.

The concept of 'innovation platform' refers to a set of stakeholders bound together by their individual interests in a shared issue, objective, challenge or opportunity, dealing with which will improve livelihoods, enterprises and/or other interests (FARA, 2007). The leverage points for making a significant difference in an innovation platform lie mostly in the interaction between the different components or actors in the system, rather than in strengthening any one component on its own. A well-functioning and complete innovation platform would organize all relevant players, including farmers and their organizations, extensionists and their organizations, researchers and their organizations, higher learning institutions, civil society organizations and the private business sector in ways that facilitate the sharing of ideas, technology and learning.

INNOVATION PRINCIPLES AND PRAXIS

Sceptics doubt whether the agricultural innovation systems approach works outside the test environments and whether it can deliver more benefits to a large number of farmers more quickly than do conventional approaches (CGIAR, 2007). Hall et al (2004), among the key architects of the agricultural innovation systems concept, admit that it has not been widely applied to developing country issues and certainly not in African agricultural development. Spielman et al (see Chapter 5) observed that research theories about innovation have failed to fundamentally change the institutional and policy setting of public and private investment intended to promote innovation for development. They argue that the agricultural innovation systems approach has not yet matured to a point where it can inform policy in developing country agriculture of specific interventions needed to enhance the potential for innovation and to improve the distribution of gains from it.

The main challenges seem to be operational ones. This shortcoming lends credence to Omamo's (2003) conclusion that experts have failed to put Africa's agricultural problems on the policy and development agenda in more than abstract and generic fashions. We concur with Hall (2007) that further elaboration of the innovation systems concept is not the priority. In fact, that would be the easier part. The much more difficult and rather murkier part is to mobilize practitioners and governments to implement the concepts, methods and strategies. This will require focusing on innovative action research and social learning processes to be able to identify convincing 'how-to' answers. We need a simple narrative that makes these ideas accessible, along with user-friendly guidelines that help to put these ideas into practice.

However, a generic set of principles for facilitating innovation is still lacking, particularly in Africa. Thus, research programmes must address the need expressed by African agricultural research and development organizations for analytical tools, methodologies, information and policy instruments and a better understanding of what it takes (capacities and resources, but also contextual factors) to strengthen innovation capacity and institutional change. There are now some new research-for-development initiatives, such as the Research into Use programme of DFID, the Sub-Saharan Africa Challenge Programme of FARA, the innovations units of IFPRI and the International Livestock Research Institute (ILRI), the Innovation and Development Unit at the Agricultural Research Centre for International Development in France (CIRAD) and the Global Partnership programme Promoting Local Innovation in Ecologically Oriented Agriculture and Natural Resource Management (Prolinnova) under the umbrella of the Global Forum on Agricultural Research (GFAR), which could provide examples for organizing systemic innovation processes, among researchers, development practitioners, policy-makers, market-chain actors and agricultural communities, to have positive impacts on farmers' livelihoods.

In the Foreword to this book, Matlon points to some remaining and emerging challenges that we must overcome if the Innovation Africa concept is to take root. Issues of sustainability and scalability must be addressed with more rigour. Most chapters in this book refer to micro-level community interventions or to pioneering marginal work within agricultural research and extension and training organizations in specific contexts. In most of these experiences – as is the case with participatory research and development projects – success is often registered at a small scale where effective participation and interaction is possible. This poses significant challenges for the scaling up of innovation processes. Understanding the scaling up process, the outcomes and the conditions for the sustainability of such intensive social-learning processes is an important research challenge. This requires more systematic process documentation and outcome mapping to track changes in innovation capacity, including the ways in which innovation processes can be delayed or derailed altogether by other actors in the system. The concepts, methods and experiences of enabling innovation in the context of agricultural research and development, as discussed in the IAS and as presented in this book, will continue to evolve and need to be carefully analysed in order to draw lessons for improvement.

The innovation systems concept is often presented as a framework for analysis and planning (Hall et al, 2004), and for diagnosing and building innovation capacity (World Bank, 2006). Many authors emphasize the complexity of innovation systems and see innovation as an adaptive system, but fail to identify leverage points (i.e. places to intervene in the system) (Meadows, 1999). Presented this way, the concept does not appeal to development practitioners and policy-makers. Because the concept of agricultural innovation systems is relatively new, it is likely that most policy-makers have only a limited understanding of it. This often results in policy resistance, defensive routines and implementation failures that can defeat the entire process (Sterman, 2006). Policy resistance breeds cynicism about people's ability to change the world for the better. Defensive routines and implementation failures can hinder learning because powerful stakeholders tend to suppress dissent and seal themselves off from those with different views or possible disconfirming evidence.

With the current global food crisis and rocketing food prices, the urgent challenge – and opportunity – in front of us in Africa is to mobilize innovation systems researchers, practitioners and networks to exert effective influence on the large agricultural development and policy initiatives that aim to achieve a Green Revolution in Africa. Rather than positioning 'innovation professionals' as the protagonists of 'Green Revolution' and 'technology push', it might be possible to position the Innovation Africa concept as part of the solution in practical ways so that innovation systems perspectives can make a Green Revolution for Africa more effective, more equitable and more sustainable. Unless and until we do, our hopes and dreams for enriching farmers' livelihoods will remain unfulfilled.

REFERENCES

Adesina, A. (2007) 'Achieving the Green Revolution in Africa: The new development financing agenda', Keynote speech, 34th FAO Conference, 20 November, Rome, Italy

Bawden, R. J. and Packam, R. (1993) 'Systems praxis in the education of the agricultural systems practitioner', *Systems Practice*, vol 6, pp7–19

CGIAR Science Council (2007) 'Sub-Saharan Africa Challenge Program External Review', Science Council Secretariat, Rome, Italy

Earl, S., Carden, F. and Smutylo, T. (2001) *Outcome Mapping: Building Learning and Reflection into Development Programmes*, IDRC, Ottawa, Canada

FARA (Forum for Agricultural Research in Africa) (2007) *Sub-Saharan Africa Challenge Programme: Medium Term Plan 2008–2010*, FARA Secretariat, Accra, Ghana

Hagmann, J. (2002) '*Competence development in soft skills/personal mastery*', *Report on design of a learning programme at Makerere University*, Uganda, Rockefeller Foundation, Nairobi, Kenya

Hall, A. (2007) 'Challenges to strengthening agricultural innovation systems: Where do we go from here?', *Farmer First Revisited: Innovation for Agricultural Research and Development*, IDS, Brighton, UK

Hall, A. J., Yoganard, B., Sulaiman, R. V., Raina, R. S., Prasad, C. S., Naik, G. and Clark, N. G. (2004) *Innovations in Innovation: Reflections on Partnerships, Institutions and Learning*, ICRISAT, Pantacheru, India

IIRR (International Institute of Rural Reconstruction) (2000) *Going to Scale: Can We Bring More Benefits to More People More Quickly?* International Institute of Rural Reconstruction, Cavite, The Philippines

Meadows, D. (1999) *Leverage Points: Places to Intervene in a System*, The Sustainability Institute, Hartland, VT

Omamo, S. W. (2003) Policy Research on African Agriculture: Trends, Gaps, Challenges, Research Report 21, ISNAR, The Hague, The Netherlands

Rogers, E. M. (1962) *Diffusion of Innovations*, Free Press, New York, NY

Röling, N. (2002) 'Beyond the aggregation of individual preferences: Moving from multiple to distributed cognition in resource dilemmas', in C. Leeuwis and R. Pyburn (eds) *Wheelbarrows Full of Frogs: Social Learning in Rural Resource Management*, Koninklijke Van Gorcum, Assen, The Netherlands, pp25–47

Sanginga, P., Chitsike, C., Njuki, J., Kaaria, S. and Kanzikwera, R. (2007) 'Enhanced learning from multi-stakeholder partnerships: Lessons from the Enabling Rural Innovation in Africa', *Natural Resources Forum*, vol 31, no 4, pp273–285

Scoones, I., Thompson, J. and Chambers, R. (2008) Farmer First Revisited: Innovation for Agricultural Research and Development Workshop Summary, IDS, Brighton, www.future-agricultures.org/farmerfirst/files/Farmer_First_Revisited_Post_Workshop_Summary_Final.pdf, accessed 24 April 2008

Shiferay, B., Obare, G. and Muricho, G. (2006) 'Rural market imperfections and the role of institutions for collective action to improve markets for the poor', Innovation Africa Symposium, 20–23 November 2006, Kampala, Uganda

Spielman, D. J. and Grebmer, K. V. (2004) *Public–Private Partnerships in Agricultural Research: An Analysis of Challenges Facing Industry and the Consultative Group on International Agricultural Research*, International Food Policy Research Institute, Washington, DC

Sterman, J. D. (2006) 'Learning from evidence in a complex world', *American Journal of Public Health*, vol 96, pp505–514

Wals, A. E. J. (ed) (2007) *Social Learning Toward a Sustainable World: Principles, Perspectives and Praxis*, Wageningen Academic Publishers, Wageningen, The Netherlands

World Bank (2006) *Enhancing Agricultural Innovation: How to Go Beyond the Strengthening of Research Systems*, World Bank, Washington, DC

World Bank (2007) *World Development Report 2008: Agriculture for Development*, World Bank, Washington, DC

List of Contributors

Annet Abenakyo is with the International Centre for Tropical Agriculture (CIAT) in Kampala, Uganda.

Diana Akullo is with the National Agricultural Research Organization (NARO) in Kampala, Uganda.

Conny Almekinders is with the Wageningen University and Research Centre (WUR) in Wageningen, The Netherlands.

Winnie Alum is with the Bulindi Zonal Agriculture Research and Development Center of the National Agriculture Research Organization (NARO) in Bulindi, Uganda.

Amanuel Assefa is with Agri-Service Ethiopia, Addis Ababa, Ethiopia, and doctoral candidate with the University of KwaZulu-Natal, Pietermaritzburg, South Africa.

George Ayo is with the Lira District Local Government in Lira, Uganda.

Margaret Barwogeza is with the Bulindi Zonal Agriculture Research and Development Center of the National Agriculture Research Organization (NARO) in Bulindi, Uganda.

Pauline Birungi is with the Bulindi Zonal Agriculture Research and Development Center of the National Agriculture Research Organization (NARO) in Bulindi, Uganda.

Robert Booth is with the International Centre for development oriented Research in Agriculture (ICRA) in Wageningen, The Netherlands.

Juan Ceballos is with the International Centre for development oriented Research in Agriculture (ICRA) in Wageningen, The Netherlands.

Petronella Chaminuka is with the University of Limpopo in Sovenga, Limpopo Province, South Africa.

Colletah Chitsike is with the International Centre for development oriented Research in Agriculture (ICRA) and is based in Pretoria, South Africa.

D. Layne Coppock is with the Pastoral Risk Management (PARIMA) project under the Global Livestock Collaborative Research Support Program (GL-CRSP), Department of Environment and Society, Utah State University in Logan, USA.

Belew Damene is with Self-Help Development International in Addis Ababa, Ethiopia.

Kristin Davis is with the International Food Policy Research Institute (IFPRI) and is based in Addis Ababa, Ethiopia.

André de Jager is with the Wageningen University and Research Centre – LEI (Agricultural Economics Research Institute), The Hague, The Netherlands.

Robert Delve is with the Tropical Soil Biology and Fertility Institute (TSBF) of the International Center for Tropical Agriculture (CIAT) in Harare, Zimbabwe.

Solomon Desta is with the Pastoral Risk Management (PARIMA) project and is based at the International Livestock Research Institute (ILRI) in Nairobi, Kenya.

Michel Dulcire is with the Innovation and Development in Agriculture and the Agri-food Sector unit of the Agricultural Research Centre for International Development (CIRAD) in Montpellier, France.

Javier Ekboir, formerly with the International Food Policy Research Institute (IFPRI) working in San José, Costa Rica, is an independent consultant based in Mexico City, Mexico.

Andrew Farrow is with the International Centre for Tropical Agriculture (CIAT) and is based in Kampala, Uganda.

Guy Faure is with the Innovation and Development in Agriculture and the Agri-food Sector unit at the Agricultural Research Centre for International Development (CIRAD) in Montpellier, France.

Tesfahun Fenta is the Prolinnova-Ethiopia coordinator based at AgriService Ethiopia, Addis Ababa, Ethiopia.

Robert Fincham is retired Head of the Centre for Environment, Agriculture and Development (CEAD) at the University of KwaZulu-Natal in Pietermaritzburg, South Africa.

Louis Gachimibi is with the Kenya Agricultural Research Institute (KARI) in Nairobi, Kenya.

Gituii Njeru Gachini is with the Kenya Agricultural Research Institute (KARI) in Nairobi, Kenya.

Getachew Gebru is with the Pastoral Risk Management (PARIMA) project and is based at the International Livestock Research Institute (ILRI) in Addis Ababa, Ethiopia.

Peter Gildemacher is with the International Potato Centre (CIP) and is based in Nairobi, Kenya.

Henriette Gotoechan-Hodounou is with the Benin Agricultural Research Institute in Cotonou, Benin.

Jeanne Gradé works with the Christian Veterinary Mission (CVM)/World Concern, Karamoja, Uganda, and is a doctoral candidate at Ghent University, Belgium.

Jürgen Hagmann is with the Institute for People, Innovation and Change in Organizations (PICO) in Pretoria, South Africa.

Richard Hawkins is with the International Centre for development oriented Research in Agriculture (ICRA) in Wageningen , The Netherlands.

Willem Heemskerk is with the Royal Tropical Institute (KIT) in Amsterdam, The Netherlands.

Henri Hocdé is with the Collective Action, Policies and Markets unit at the Agricultural Research Centre for International Development (CIRAD) in Montpellier, France.

Susan Kaaria, formerly with the International Centre for Tropical Agriculture (CIAT) in Uganda, is now with Ford Foundation, Nairobi, Kenya.

Ignatius Kahiu is with Africare in Kampala, Uganda.

Rogers Kakuhenzire is with the Kachwekano Zonal Agricultural Research and Development Institute of the National Agriculture Research Organization (NARO) in Kachwekano, Uganda.

Geoffrey Kamau is with the Kenya Agricultural Research Institute (KARI) in Nairobi, Kenya.

Rick Kamugisha is with the Africa Highlands Initiative (AHI) in Kampala, Uganda.

Speciose Kantengwa, formerly with the Agricultural Technology Development and Transfer Project, is now with the Consortium for Improved Agricultural-based Livelihoods in Central Africa (CIALCA) in Kigali, Rwanda

Rogers Kanzikwera is with the Bulindi Zonal Agriculture Research and Development Center of the National Agriculture Research Organization (NARO) in Bulindi, Uganda.

George Karanja is with the Kenya Agricultural Research Institute (KARI) in Nairobi, Kenya.

Imelda Kashaija is with the Kachwekano Zonal Agricultural Research and Development Institute of the National Agriculture Research Organization (NARO) in Kachwekano, Uganda.

Peter Kinyae is with the Kenya Agricultural Research Institute (KARI) in Nairobi, Kenya.

Ninatubu Lema is with the Ministry of Agriculture and Food Security (MAFS) in Dar es Salaam, Tanzania.

Yohannes Lema is with the Ethiopian Institute for Agricultural Research (EIAR) in Addis Ababa, Ethiopia.

Kadenge Lewa is with the Mtwapa Regional Research Centre of the Kenya Agricultural Research Institute (KARI) in Mtwapa, Kenya.

Paul Maina is with the International Potato Centre (CIP) and is based in Nairobi, Kenya.

Adrienne Martin is with the Natural Resources Institute at the University of Greenwich in Greenwich, UK.

Peter Matlon is retired Managing Director of the Rockefeller Foundation, Nairobi, Kenya, currently living in Washington, DC, USA

Robert Mazur is with Iowa State University in Ames, USA.

Mike Morris is with the World Wide Fund for Nature-UK (WWF-UK) in Godalming, Surrey, UK.

Jeremias Mowo is with the African Highlands Initiative (AHI) in Kampala, Uganda.

Fred Muchena is with ETC-East Africa in Nairobi, Kenya.

Maxwell Mudhara is with the Farmer Support Group in the Centre for Environment, Agriculture and Development (CEAD) at the University of KwaZulu-Natal in Pietermaritzburg, South Africa.

Joseph Mudiope, formerly with Africare, is now with the Millennium Development Villages Project and is based in Kampala, Uganda.

Festus Murithi is with the Kenya Agricultural Research Institute (KARI) in Nairobi, Kenya.

Charles Musoke is with Africare in Kampala, Uganda.

Robert Muzira is with the International Centre for Tropical Agriculture (CIAT) in Kampala, Uganda.

Brighton M. Mvumi is with the University of Zimbabwe in Harare, Zimbabwe.

Thembi Ngcobo is with the Agricultural Research Council (ARC) in Pretoria, South Africa.

Hlamalani Ngwenya is with the Institute for People, Innovation and Change in Organizations (PICO) in Pretoria, South Africa.

Jemimah Njuki is with the International Centre for Tropical Agriculture (CIAT) and is based in Harare, Zimbabwe.

Michael Njunie is with the Mtwapa Regional Research Centre of the Kenya Agricultural Research Institute (KARI) in Mtwapa, Kenya.

Moses Nyongesa is with the Kenya Agricultural Research Institute (KARI) in Nairobi, Kenya.

Davies Onduru is with ETC-East Africa in Nairobi, Kenya.

Sheila Onzere is with Iowa State University in Ames, USA.

Chris Opondo is with the African Highlands Initiative (AHI) in Kampala, Uganda.

Oscar Ortiz is with the International Potato Centre (CIP) in Lima, Peru.

Silvia Andrea Pérez is a doctoral candidate in social sciences at Wageningen University and Research Centre (WUR) in Wageningen, The Netherlands

Johannes Ramaru is with the Provincial Department of Agriculture in Polokwane, Limpopo Province, South Africa.

William Riwa is with the Ministry of Agriculture and Food Security (MAFS) in Dar es Salaam, Tanzania.

Niels Röling is retired Chair of Communication and Innovation Studies, Wageningen University and Research Centre (WUR), Wageningen, The Netherlands.

Pascal C. Sanginga, formerly with the International Centre for Tropical Agriculture (CIAT) in Uganda, is now with the International Development Research Centre in Nairobi, Kenya.

David J. Spielman is with the International Food Policy Research Institute (IFPRI) and is based in Addis Ababa, Ethiopia.

Tanya E. Stathers is with the Natural Resources Institute in Chatham Maritime, UK.

John R. S. Tabuti is with Makerere University in Kampala, Uganda.

Shiferaw Tafesse is with Self-Help Development International in Addis Ababa, Ethiopia.

Amare Tegbaru, formerly with the Agricultural Technology Development and Transfer Project in Rwanda, is now with the International Institute of Tropical Agriculture (IITA) and is based in Borno State, Nigeria.

Moses Tenywa is with Makerere University in Kampala, Uganda.

Seyoum Tezera is with the Pastoral Risk Management (PARIMA) project and is based at the International Livestock Research Institute (ILRI) in Addis Ababa, Ethiopia.

Bernard Triomphe is with the Innovation and Development in Agriculture and the Agri-food Sector unit in the Agricultural Research Centre for International Development (CIRAD) in Montpellier, France.

Emily Twinamasiko is with the National Agricultural Research Organization (NARO) in Kampala, Uganda.

Christy Van Beek is with Alterra in Wageningen, The Netherlands.

Patrick Van Damme is with Ghent University in Ghent, Belgium.

Laurens van Veldhuizen is with the Prolinnova International Secretariat at ETC EcoCulture in Leusden, The Netherlands.

Aart-Jan Verschoor is with the Agricultural Research Council (ARC) in Pretoria, South Africa.

Arjen Wals is with the Wageningen University and Research Centre (WUR) in Wageningen, The Netherlands.

Ann Waters-Bayer is with the Prolinnova International Secretariat at ETC EcoCulture in Leusden, The Netherlands.

Bertus Wennink is with the Royal Tropical Institute (KIT) in Amsterdam, The Netherlands.

Chesha Wettasinha is with the Prolinnova International Secretariat at ETC EcoCulture in Leusden, The Netherlands.

Gebremedhin Woldegiorgis is with the Ethiopian Institute for Agricultural Research (EIAR) in Addis Ababa, Ethiopia.

Mariana Wongtschowski is with the Prolinnova International Secretariat at ETC EcoCulture in Leusden, The Netherlands.

Index